Brazilian Marine Biodiversity

Series Editor
Alexander Turra, Unesco Chair for Ocean Sustainability, Oceanographic Institute and Advanced Studies Institute, University of São Paulo, São Paulo, Brazil

The book series Brazilian Marine Biodiversity was designed to communicate to a broad and international readership the diversified marine and coastal habitats along the large Brazilian coast.

The diversity of marine habitats found in Brazil is astonishing and includes estuaries, coral reefs, rocky shores, sandy beaches, rhodolith beds, mangroves, salt marshes, deep-sea habitats, vegetated bottoms, and continental shelf. These habitats are addressed from an ecosystem perspective across the series, and characterized in terms of distribution and peculiarities along the Brazilian coast, records of relevant species, and information on the prevailing structuring ecological and oceanographic processes governing biodiversity.

This series is an initiative of the UNESCO Chair for Ocean Sustainability, hosted by the Oceanographic Institute and the Advanced Studies Institute of the University of São Paulo, within the Brazilian Network for Monitoring Coastal Benthic Habitats (ReBentos; rebentos.org), which is supported by the Brazilian National Council for Scientific and Technological Development (CNPq), the Research Program on Biodiversity Characterization, Conservation, Restoration and Sustainable Use of the São Paulo Research Foundation (BIOTA-FAPESP), the Coordination for the Improvement of Higher Education Personnel (CAPES) and the Brazilian Innovation Agency (FINEP). ReBentos is part of the Brazilian Network on Global Climate Change Research (Rede Clima) and the Science and Technology National Institute on Climate Changes (INCT Mudanças Climáticas) at the Ministry of Science, Technology, and Innovation (MCTI).

Antonia Cecilia Zacagnini Amaral
Helio Herminio Checon
Guilherme Nascimento Corte
Editors

Brazilian Sandy Beaches

Springer

Editors
Antonia Cecilia Zacagnini Amaral
Departamento de Biologia Animal
Instituto de Biologia Universidade Estadual de Campinas, IB/UNICAMP
Campinas, São Paulo, Brazil

Helio Herminio Checon
Departamento de Oceanografia Biológica
Instituto Oceanográfico da Universidade de São Paulo
São Paulo, SP, Brazil

Guilherme Nascimento Corte
College of Science and Mathematics
University of Virgin Islands
Saint Thomas, USVI, USA

ISSN 2520-1077 ISSN 2520-1085 (electronic)
Brazilian Marine Biodiversity
ISBN 978-3-031-30748-5 ISBN 978-3-031-30746-1 (eBook)
https://doi.org/10.1007/978-3-031-30746-1

This Springer imprint is published by the registered company Springer Nature Switzerland AG
The registered company address is: Gewerbestrasse 11, 6330 Cham, Switzerland

Paper in this product is recyclable.

Coordination

UNESCO Chair for Ocean Sustainability

Support

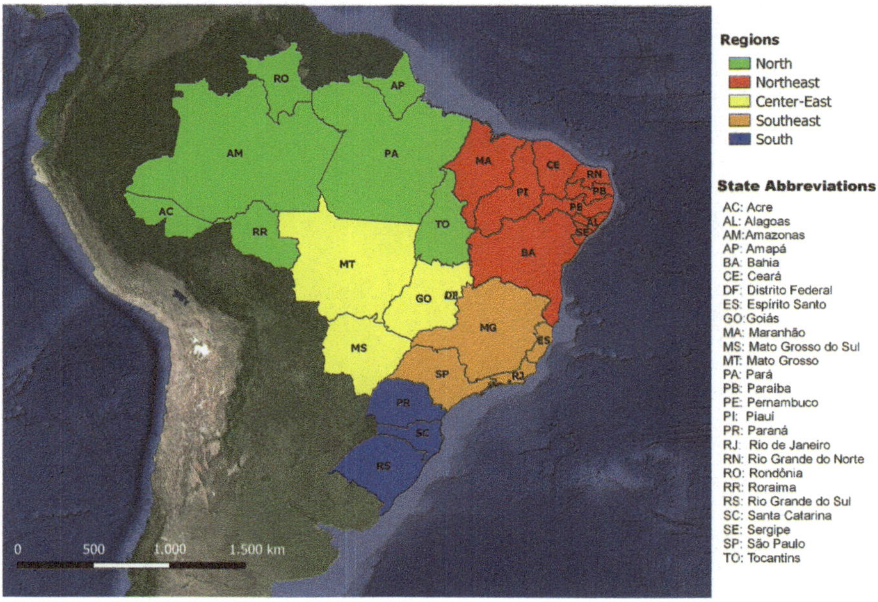

Geopolitical division of Brazil and Brazilian states. This map is intended to facilitate the international readership to identify the regions and states within the Brazilian territory cited along the chapters.

To my family (especially my grandchildren), friends, colleagues and students who have supported me during my career.

To my family for all the support and to my friends, the other family life gave me. You are all home to me. And to my love, Beatriz, for all that was and all that will be.

To my family, who have always been my support and strength. Especially to my father, Rineu Corte: I know you will always be with me in every wave of the ocean.

In memory of two great Brazilian oceanographers who contributed significantly to a better understanding of our beaches.

Paulo da Cunha Lana (20/04/1956–30/06/2022)
Full Professor – Center for Marine Studies, Federal University of Parana

Lauro Julio Calliari (14/08/1951–14/02/2023)
Full Professor – Federal University of Rio Grande

Preface

The more than 4000 beaches along the Brazilian coast are one of the country's main assets. They are distributed along more than 9000 km of coastline, from the equatorial Amazon coast (4°N) to the subtropical beaches of Rio Grande do Sul (34°S), and encompass every beach type, from wide tide-dominated flats in the North to wave-dominated reflective beaches in the Southeast and South. Brazilian sandy beaches harbor endemic and diverse biota and provide numerous goods and services that are essential to human populations such as shellfish harvesting and fishing, coastal infrastructure protection, sites for cultural manifestations, tourism, and recreation activities. Notwithstanding, they are under increasing pressure, trapped between the impacts of climate change and human activities in the terrestrial and marine environments.

Contrary to their great geographical extent and importance, knowledge about Brazilian beaches is still limited and insufficient to ensure their sustainable use. Thus far, there is no extensive work on their socio-ecological characteristics, an issue that prevents the necessary acknowledgment of their importance. This book comes to fill this void. Written by an interdisciplinary group of leading researchers from all coastal regions of Brazil, and having the contribution of international experts, it is the first-ever comprehensive work written about Brazil's sandy beaches addressing their physical, ecological, and social aspects.

The information synthesized within this book is organized into ten chapters, which follow a logical order but can also be read independently. Chapter 1 sets the background and shows the predominant physical characteristics of Brazilian beaches, classifying the Brazilian coast into distinct regions according to their geology, sediments, coastal processes, and beach types. The next chapters give an insight into the biodiversity recorded in Brazilian beaches, including algae (Chap. 2), meiofauna (Chap. 3), macrofauna (Chap. 4), and vertebrates (Chap. 5). An overview of the population biology and dynamics of the most studied species on Brazilian sandy shores is found in Chap. 6, while Chap. 7 summarizes the importance of biological interactions to the patterns and functioning of sandy beach ecosystems. Finally, the last three chapters discuss topics related to the social function of beaches as social-ecological systems such as the main services and goods provided to human populations (Chap. 8), the influence of natural and anthropic disturbance (Chap. 9), and the challenges to preserve these ecosystems (Chap. 10).

Our effort aims to reach university researchers and students, as well as coastal managers and policymakers. Nevertheless, the information included in this book is accessible to anyone who wants to know more about Brazilian coastal biodiversity and ecosystems. We trust that this book will provide insights for further investigations and represent a significant step towards the conservation of Brazilian sandy beaches, their biodiversity, and ecosystem services.

Campinas, SP, Brazil

Guilherme Nascimento Corte
Helio Herminio Checon
Antonia Cecilia Zacagnini Amaral

Acknowledgements

We are greatly indebted to all authors that contributed to this book. This work was only possible due to their expertise and wiliness to share the best knowledge on the Brazilian sandy beaches. We also thank Prof. Alexander Turra for the research initiative that led to the funding of the ReBentos, the network which allowed for the accomplishment of this book. We would also like to thank Beatriz Pereira Cunha – aka Bia – for editing all the pictures in this book. We acknowledge the support of Brazilian universities and research agencies, including federal (CNPq and CAPES), state (FAPs), and private agencies (Fundação Grupo Boticário). Finally, a big 'thank you' is also due to the students, technicians, and academic colleagues in Brazil and abroad investigating sandy shores. Their effort is the foundation wall that supports the understanding and preservation of beach ecosystems.

Contents

About the Editors

Antonia Cecilia Zacagnini Amaral is a Full Professor in the Biology Institute at the State University of Campinas, Brazil, which she joined in 1980. Her research focuses on marine benthic ecology and biodiversity, and she is a world expert in the taxonomy and ecology of annelids (polychaetes). Cecília actively participated in the development of Zoology in Brazil, and coordinated several multi-institutional projects such as the Assessment of the Sustainable Potential of Living Resources of the Brazilian Exclusive Economic Zone – REVIZEE/Score-Sul-Bentos; Biodiversity and Functioning of a Subtropical Coastal Ecosystem: Subsidies for Integrated Management – BIOTA/FAPESP; Biodiversity of Sandy Beaches (PROBIO); and Endangered Species of Brazilian Fauna – Aquatic Invertebrates (IBAMA, MMA, SISBIO/ICMBio). She currently coordinates the project Consolidation of Scientific Collections of Marine Invertebrates: Strategies for Biodiversity Conservation – BIOTA/FAPESP.

Helio Herminio Checon holds a Ph.D. in Ecology from the State University of Campinas and is currently a research collaborator at the State University of Campinas and an employee for the Secretariat of Green, Environment and Sustainable Development of the municipality of Campinas, where he works with the management of protected areas. During his career, he worked with different marine environments and benthic organisms, assessing how environmental variability affects benthic fauna. On sandy beaches, the best marine environment (sorry, mangroves), his work supports the management of coastal protected areas, especially for monitoring and evaluating environmental integrity.

Guilherme Nascimento Corte holds a PhD in Ecology from the State University of Campinas (Brazil, 2016) and is currently a Professor of Biology in the College of Science and Mathematics at the University of Virgin Islands, United States Virgin Islands, USA. As a marine ecologist, he has worked on the coasts of North and South America, the Caribbean, and Australia. Guilherme has studied several marine habitats such as mangroves, rocky shores, and open ocean; however, most of his research focuses on the biodiversity and functioning of sandy beach ecosystems. He has helped develop protocols to monitor these environments worldwide and can no longer differentiate between work and leisure when he is on a sandy beach.

Contributors

Rayane Romão Saad Abude Instituto de Biociências, Universidade Federal do Estado do Rio de Janeiro, Rio de Janeiro, RJ, Brazil

Antonia Cecilia Zacagnini Amaral Departamento de Biologia Animal, Instituto de Biologia, Universidade Estadual de Campinas, Campinas, SP, Brazil

Matheus Augusto Instituto de Biociências, Universidade Federal do Estado do Rio de Janeiro, Rio de Janeiro, RJ, Brazil

Daiane Aviz Faculdade de Oceanografia, Instituto de Geociências, Universidade Federal do Pará, Belem, PA, Brazil

Tatiana Medeiros Barbosa Cabrini Instituto de Biociências, Universidade Federal do Estado do Rio de Janeiro, Rio de Janeiro, RJ, Brazil

Carlos Henrique Soares Caetano Instituto de Biociências, Universidade Federal do Estado do Rio de Janeiro, Rio de Janeiro, RJ, Brazil

Ricardo Silva Cardoso Instituto de Biociências, Universidade Federal do Estado do Rio de Janeiro, Rio de Janeiro, RJ, Brazil

Helio Herminio Checon Departamento de Biologia Animal, Instituto de Biologia, Universidade Estadual de Campinas, Campinas, SP, Brazil

Marina Ribeiro Corrêa Instituto de Energia e Ambiente da Universidade de São Paulo, São Paulo, SP, Brazil

Guilherme Nascimento Corte College of Science and Mathematics, University of Virgin Islands, Saint Thomas, USVI, USA

Leonardo Lopes Costa Universidade Estadual do Norte Fluminense Darcy Ribeiro, Campos dos Goytacazes, RJ, Brazil

Beatriz Pereira Cunha Instituto de Biologia, Programa de Pós-graduação em Biologia Animal, Universidade Estadual de Campinas, Campinas, SP, Brazil

Antonio Henrique da Fontoura Klein Federal University of Santa Catarina, Campus Trindade, Florianópolis, SC, Brazil

Cristina de Almeida Rocha-Barreira Universidade Federal do Ceará, Instituto de Ciências do Mar, Fortaleza, CE, Brazil

Robson Henrique de Carvalho Instituto de Ciências Biológicas, Universidade Federal de Juiz de Fora, Juiz de Fora, MG, Brazil

Andrea de Oliveira da Rocha Franco Universidade Federal do Rio Grande, Instituto de Oceanografia, Rio Grande, RS, Brazil

Maikon Di Domenico Centro de Estudos do Mar, Universidade Federal do Paraná, Pontal do Sul, PR, Brazil

Yasmina Shah Esmaeili Departamento de Biologia Animal, Instituto de Biologia, Universidade Estadual de Campinas, Campinas, SP, Brazil

André Esteves Departamento de Zoologia, Universidade Federal de Pernambuco, Recife, PE, Brazil

Gustavo Fonseca Instituto do Mar, Universidade Federal de São Paulo, Santos, SP, Brazil

Fabiane Gallucci Instituto do Mar, Universidade Federal de São Paulo, Santos, SP, Brazil

André Garraffoni Departamento de Biologia Animal, Instituto de Biologia, Universidade Estadual de Campinas, Campinas, SP, Brazil

Leandra Regina Gonçalves Instituto do Mar da Universidade Federal de São Paulo, Santos, SP, Brazil

Paulo Horta Universidade Federal de Santa Catarina, Campus Trindade, Florianópolis, SC, Brazil

Phillipe Mota Machado Departamento de Biologia, Centro de Ciências Exatas, Naturais e da Saúde, Universidade Federal do Espírito Santo, Alegre, ES, Brazil

Nicole Malinconico Instituto Oceanográfico da Universidade de São Paulo, São Paulo, SP, Brazil

Tatiana Maria Departamento de Ecologia e Recursos Marinhos, Universidade Federal do Estado do Rio de Janeiro, Rio de Janeiro, RJ, Brazil

Sergio Netto Universidade do Sul de Santa Catarina, Tubarão, SC, Brazil

Clarisse Odebrecht Universidade Federal do Rio Grande, Instituto de Oceanografia, Rio Grande, RS, Brazil

Mayara Oliveira University of Queensland, Brisbane, QLD, Australia

Bruna Pagliani Instituto de Biodiversidade e Sustentabilidade – NUPEM, Universidade Federal do Rio de Janeiro, Macae, RJ, Brazil

Marcelo Petracco Faculdade de Oceanografia, Instituto de Geociências, Universidade Federal do Pará, Belem, PA, Brazil

Marcus Polette Universidade do Vale do Itajaí, Itajaí, SC, Brazil

Leonardo Rubi Rörig Universidade Federal de Santa Catarina, Campus Trindade, Florianópolis, SC, Brazil

José Souto Rosa-Filho Universidade Federal de Pernambuco, Centro de Tecnologia, Departamento de Oceanografia, Recife, PE, Brazil

Andrew D. Short School of Geosciences, University of Sydney, Camperdown, NSW, Australia

Abílio Soares-Gomes Departamento de Biologia Marinha, Universidade Federal Fluminense, Niterói, RJ, Brazil

Alexander Turra Instituto Oceanográfico da Universidade de São Paulo, São Paulo, SP, Brazil

Adriane Pereira Wandeness Departamento de Zoologia, Universidade Federal de Pernambuco, Recife, PE, Brazil

Luciana Yokoyama Xavier Instituto Oceanográfico da Universidade de São Paulo, São Paulo, SP, Brazil

Ilana Rosental Zalmon Universidade Estadual do Norte Fluminense Darcy Ribeiro, Campos dos Goytacazes, RJ, Brazil

Chapter 1
Physical Characteristics of Brazilian Sandy Beaches

Antonio Henrique da Fontoura Klein and Andrew D. Short

1.1 Introduction

The east coast of South America, including all of Brazil, is a classic trailing edge coast typified by numerous long meandering rivers, generally low gradient regressive coastal plains, an abundance of mature quartz sediment, and extensive beach-barrier systems, the antithesis of the rugged collision margin of the mountainous west coast. The Amazon, the world's largest river in terms of length, discharge, and its associated sediment supply, dominates the northern 1500 km of coast, maintaining a predominately mangrove-fringed mud-dominated shore, with scattered sandy beaches. South of the Amazon, however, sandy beaches increasingly dominate the shore with more than 4000 beaches comprising 2% (82,778 ha) of all coastal ecosystems in the country (Muehe 2003; Short and Klein 2016). The remaining coast is occupied by rocky shores, inlets, and, in sheltered locations, mangroves, as well as salt-marsh south of 27°S.

The entire coast extends over 38° of latitude, from 4°N to 34°S, and is bordered by the Atlantic Ocean. Coastal quartz-rich sediment has been derived from the Amazon River in the north and La Plata River in the south, discharging between Uruguay and Argentina, together with several moderate-sized rivers in between. This sediment has been deposited on the shelf at low sea levels and reworked onshore and longshore during the postglacial sea-level transgression, together with sediment supplied directly to the coast since the sea-level standstill.

The abundance of sand has allowed the formation of thousands of beaches, including some of the longest beach and barrier systems in the world, together with

---

A. H. F. Klein (✉)
Federal University of Santa Catarina (UFSC), Campus Universitário – Trindade, Florianópolis, SC, Brazil
e-mail: antonio.klein@ufsc.br

A. D. Short
School of Geosciences, University of Sydney, Camperdown, NSW, Australia

A. C. Z. Amaral et al. (eds.), *Brazilian Sandy Beaches*, Brazilian Marine Biodiversity, https://doi.org/10.1007/978-3-031-30746-1_1

extensive transgressive dune systems (Dillenburg and Hesp 2009). The considerable range in wave and tide energy maintains the full range of beach types and states along the open coast. In the north, the beaches are generally tide-dominated to tide-modified, while along the east coast, they range from tide-modified in the northeast to wave-dominated along with the central and southern sectors, with tide-dominated beaches predominating in sheltered bays and estuaries (Short and Klein 2016). The beaches span tropical to subtropical latitudes, and in the northeast, they are in places sheltered and modified by coral and beach rock reefs, which induce the formation of lower energy crenulate beaches in their lee. Also, in the north and east, the Barreiras Formation outcrops along sections of the coast, forming eroding cliffs that supply sediment directly to the beaches, while in the south, particularly between Cabo Frio (RJ) and Cabo de Santa Marta (SC), numerous igneous bedrock headlands result in many embayed and pocket beach systems.

This chapter is a summary of Brazilian beach systems (Short and Klein 2016), where the main highlight of Brazilian beaches are present.

1.2 Coastal Processes

Brazilian coastal processes are related to its latitudinal location and the associated tropical though temperate climate systems, particularly those that reside over the equatorial and South Atlantic (Rodrígues et al. 2016). In the north, the Intertropical Convergence Zone (ITCZ), which while residing over the equator, shifts seasonally and influences the region between 30°N and 30°S. Trade winds converge on the ITCZ, and during the southern hemisphere summer, when it moves to 14°N, it delivers southeast trades along the coast between the state of Amapá and as far south as the state of Santa Catarina, while during the southern hemisphere winter, it shifts to a maximum of 5–6°S, delivering northeast trades and heavy rainfall along the coast between the states of Amapá and Rio Grande do Norte.

Much of the coast is exposed to easterly trade winds and east through southerly seas and swell, which combined provide considerable energy to transport sediment and construct a wide range of beach, barrier, inlet, and deltaic forms. Wave energy ranges from low to moderate along the tide-dominated Amapá and Pará coasts, where considerable wave attenuation takes place across the shallow inshore, to moderate along the states of Maranhão, Ceará, and northern Rio Grande do Norte coast. The trade winds generate low to moderate seas, with waves usually between 1 and 2 m and period 7 and 13 s. South of the Cabo Frio municipality (RJ), the wave climate is dominated by subpolar lows that regularly cross the South Atlantic (40–60°S), generating moderate to occasionally high southerly swell. Waves can range from 1–4 m and periods from 10 to 16 s. Figure 1.1 illustrates the mean significant wave height (H_s) along the entire coast. The waves drive predominately northerly longshore sand transport, which has been calculated to reach one million m^3yr^{-1} in some southern locations (Motta and Toldo Jr 2013), together with some local and seasonal reversals in sediment transport.

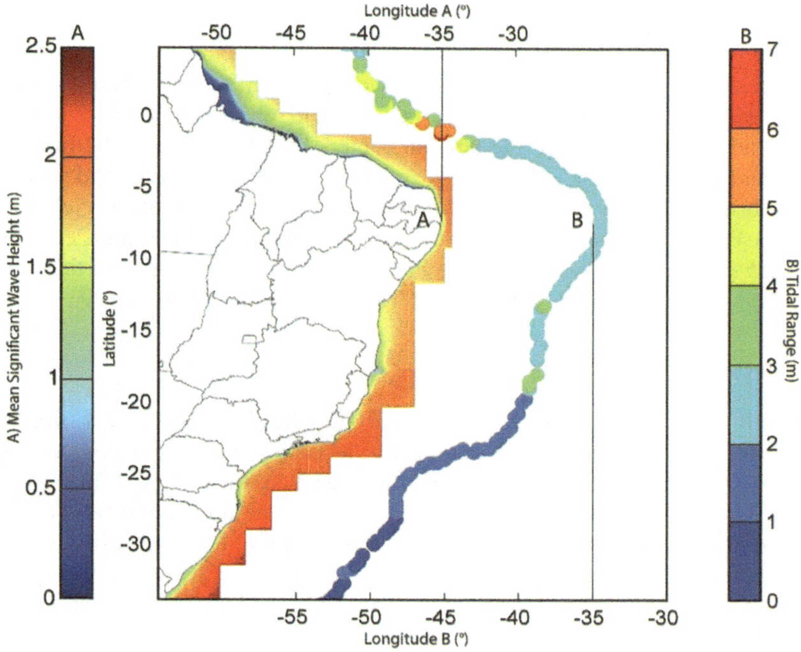

Fig. 1.1 Annual mean significant wave height (H_s) and (right) maximum tide range (m) along the Brazilian coast between French Guinea and Uruguay. (Source: Rodrígues et al. 2016)

Tides along the coast peak in the mouth of the Amazon, where they reach 11 m near Ilha do Maracá, the highest in Brazil. Tidal currents are significant along the north coast with the flood tides trending to the west reinforcing the easterly wind and wave-driven currents and the strong North Brazil current. The tide range decreases northward to 4 m along the northern Amapá coast. It also decreases to the east averaging 4–6 m along the Pará-Maranhão coast, then gradually decreasing to 3–4 m along Piauí and Ceará coast, and 2.5–3 m from Rio Grande do Norte down to Bahia. In the state of Espírito Santo, tides decrease to micro (0.8–1.9 m) and remain 1 m or less all down the southeast coast, with the smallest tides (0.5 m) along the coast of Rio Grande do Sul (Fig. 1.1). South from the state of Rio de Janeiro, storm surges up to 1.0 m high also contribute to coastal processes.

Winds are related to the regional pressure systems and are seasonally northeast and southeast along the Amapá-Pará coast, while southeast trades dominate the entire northeast coast in summer, extending as far south as northern Santa Catarina, with northerly winds dominating south of Santa Catarina Island. Winter brings periodic cold fronts, strong southerly winds, and frontal rain along the southeast coast usually reaching as far north as Sergipe. In all regions from Pará to the south, the winds move beach sand inland to develop in places extensive transgressive dune systems.

1.3 Beach Types and States

Beaches are accumulations of wave-deposited sediment (sand to boulders) at the shoreline usually in a tidal environment. As such, wave height and period, tide range, and sediment size are the key parameters in determining the nature of the beach, and based on these, all beaches can be classified by their type and state. This section briefly reviews the range of beach types and states that occur globally and along the Brazilian coast, based on the detailed description of Short and Klein (2016).

Beach type is based on the relative contribution of both waves and tides and is quantified using the relative tide range (RTR) (Masselink and Short 1993):

$$\mathrm{RTR} = \mathrm{TR} / H_{\mathrm{b}}$$

where TR = mean spring tide range (m) and H_{b} = breaker wave height (m). When waves are relatively high, tide low, and RTR < 3 beaches are *wave-dominated*. When between RTR 3 to ~10, they are *tide-modified*; and when waves are very low, tide relatively high, and RTR between ~10 to ~50, they become *tide-dominated* (Table 1.1 and Fig. 1.2; Short 2006). As RTR exceeds ~50, the tidal flats increasingly replace the beaches.

The Brazilian coast has tides ranging from 0.5 to 11 m and low through high waves (Klein and Short 2016). Therefore, it contains RTR's ranging from <1 to >50 and the full range of wave-dominated, tide-modified, and tide-dominated beach types (Klein and Short 2016). The RTR, however, does not consider the sand grain size and wave period. When using these two additional parameters, the three beach types can be further classified into 13 beach states using the dimensionless fall velocity (Ω) (Gourlay 1968), where

$$\Omega = H_{\mathrm{b}} / W_{\mathrm{s}} T$$

where W_{s} = sediment fall velocity (m s^{-1}) and T = wave period (s), the W_{s} acting as a surrogate for sand size.

Ω quantifies the relative contribution of wave height and period and sediment grain size [expressed as sediment fall velocity, see Gibbs et al. (1971) for a table to convert grain size to velocity] to beach morphodynamics.

In total, the three beach types that contain 13 beach states and together with two beaches dominated by rock/reef flats account for 15 different beach states ranging from the high energy wave-dominated multi-bar dissipative with 300–500 m wide surf zones, to barless reflective beaches; to with increasing tide range the tide-modified beaches with surf; to the very low energy tide-dominated beaches fronted by tidal flats (Klein and Short 2016; Short and Klein 2016). Table 1.1 lists the beach types and states, their abbreviations, and their general relationship to RTR, Ω, and H_{b} (Klein and Short 2016; Short and Klein 2016). Note that the actual relationship will vary with wave environments and need to be determined locally, while beaches

Table 1.1 List of the three beach types and 13 beach states, plus two states (14 and 15) dominated by reefs/rock), and some of their environmental characteristics

Beach type/No.	Beach state	Beach state	RTR	Ω	~H_b (m)
Wave-dominated (WD)			<3	1–6	
1	D	Dissipative	<1	>6	>2
2	LBT	Longshore Bar & Trough	<3	~5	<2
3	RBB	Rhythmic Bar & Beach	<3	~4	>1.5
4	TBR	Transverse Bar & Rip	<3	~3	~1.5
5	LTT	Low Tide Terrace	<3	~2	~1
6	R	Reflective	<3	~1	<1
Tide-modified (TM)			3~10	1–6	
7	R + LTT	R + Low Tide Terrace		~1	
8	R + LTR	R + Low Tide Rips		~3	
9	UD	Ultradissipative		>5	
Tide-dominated (TD)		Beach[a]	~10–50	<1	<<1
10	B + RSF	B + Ridged Sand Flats		<1	<0.5
11	B + SF	B + Sand Flats		<1	<0.3
12	B + TSF	B + Tidal Sand Flats		<1	<0.2
13	B + TMF	B + Tidal Mudflats		<1	<0.2
Rocks/reef[b]					
14	R + RF	R + Rock Flats	–	–	–
15	R + CF	R + Coral (reef) Flats	–	–	–

Source: Klein and Short (2016)

The relative tide range (RTR), dimensionless fall velocity (Ω), and breaker wave height (H_b) are all approximate and can vary between wave environments, while 14 and 15 states are independent of waves and tides. Also see Fig. 1.1

[a] Beach indicates a very low energy strip of high tide sand

[b]Rock and reef-fronted beaches form independently of RTR, Ω, and H_b

B high tide beach

fronted by rock flats or coral reefs are independent of waves and tides (Klein and Short 2016; Short and Klein 2016).

The relationship between the wave-dominated, tide-modified, and tide-dominated beach states and H_b, W_s, and RTR is also presented in Fig. 1.2 (Klein and Short 2016; Short and Klein 2016). Figure 1.2a–c plots the relation between beach state and H_b, sand size, and RTR, while Fig. 1.2d plots the impact of increasing Ω and RTR on beach type and state. It shows how wave-dominated beaches have the highest waves and lowest RTR with fine to medium sand that coarsens toward the reflective end (Klein and Short 2016; Short and Klein 2016). Tide-modified beaches have moderate waves, increasing RTR, and medium more poorly sorted sand, while tide-dominated beaches have low waves, high RTR, and the coarsest material, which is also very poorly sorted (Klein and Short 2016; Short and Klein 2016). The data in Table 1.1 are based on Australian beaches and may need modification in other coastal environments. Each of the beach types and states is briefly described below

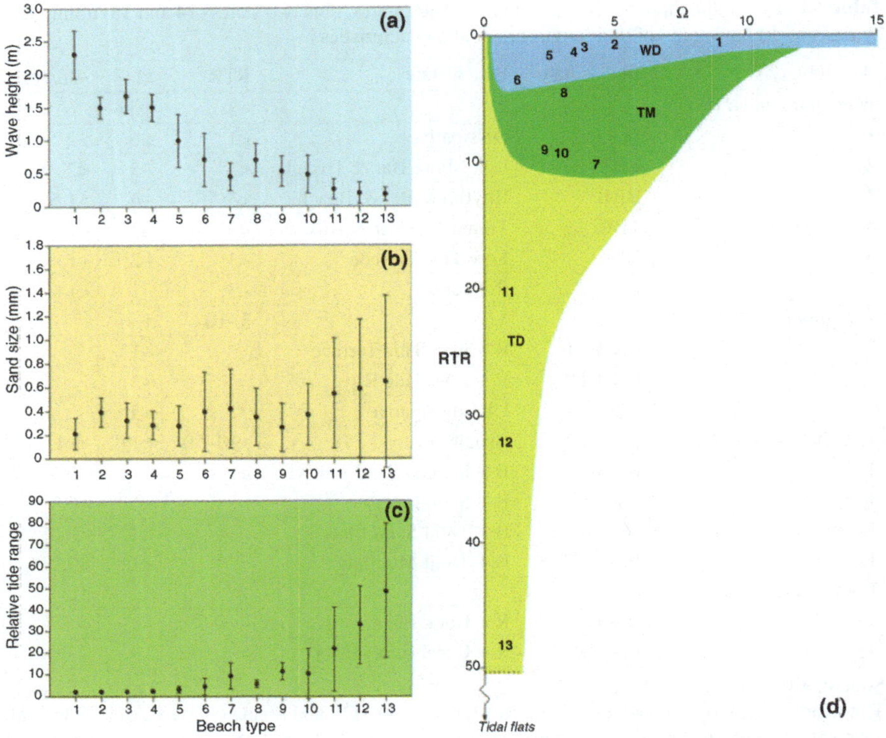

Fig. 1.2 The relationship (mean and standard deviation) between the wave-dominated (1–6), tide-modified (7–9), and tide-dominated (10–13) beach states and (**a**) breaker wave height (H_b); (**b**) sand size; (**c**) relative tide range (RTR). (Source: Short and Woodroffe 2009); and (**d**) relationship between the dimensionless fall velocity (Ω) and RTR in controlling beach type and state (WD wave-dominated, TM tide-modified, TD tide-dominated). (Source: Short and Jackson 2022). All data are based on Australian beaches and may need modification for other coastal environments. (Source: Klein and Short 2016). See Table 1.1 for the number (1–13) legend

and illustrated in Figs. 1.3 and 1.4. For a fuller description, see Short (1999), Short and Klein (2016), and Short and Jackson (2022).

As presented in the Klein and Short (2016) and Short and Klein (2016), using Ω, *wave-dominated beaches* can be classified into six beach states (1–6, Table 1.1; Fig. 1.2). When waves are relatively low, periods long, sand coarse, and $\Omega < 1$ the beaches are narrow and barless and are called *reflective*. When waves are moderate to high and $\Omega = 2$–5, the beaches become rip-dominated *intermediate* with usually one or two bars cut by rip channels and currents. When waves are high, sand is fine, and $\Omega > 6$, the beaches become wide and *dissipative* with often multiple (>2) shore-parallel sand bars. Figure 1.3 illustrates the six wave-dominated beach states. .

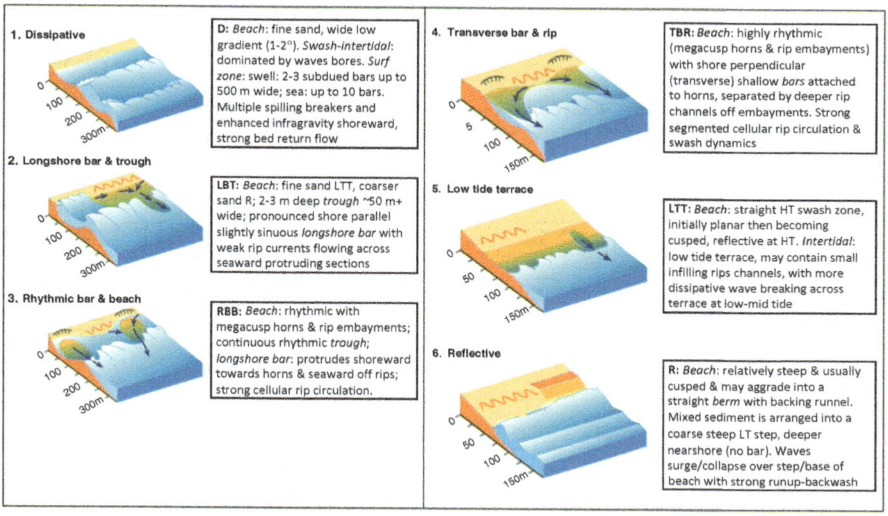

Fig. 1.3 Schematic sketch of the six (1–6) wave-dominated beach states and their general characteristics with states 2–5 encompassing the rip-dominated *intermediate* range. (Source: Modified from Short and Woodroffe 2009)

Tide-modified beaches have three beach states (7–9, Table 1.1; Fig. 1.2), each containing a usually wide intertidal zone grading into a low tide bar. The lower energy ($\Omega < 1$) *reflective plus low tide terrace* state consists of a reflective high tide beach plus a wide (~100 m+) low tide terrace. The intermediate state (*reflective plus low tide rips*, $\Omega = 2$–5) contains a reflective high tide beach and low tide bar cut by rip channels on its outer low tide sector, while the higher energy *ultradissipative* state ($\Omega > 6$) is composed of finer sand and features a very wide (>200 m) low gradient featureless concave inter to the subtidal beach (Fig. 1.4).

The *tide-dominated beaches* consist of four states each fronted by wide intertidal sand and/or mudflats (100s–1000s m wide). They range from a low energy high tide beach fronted by *ridged sand flats* under higher waves, through to very low energy *sand flats*, *tidal sand flats,* and finally, *tidal mudflats* (Fig. 1.4). For a full description of the beach types and states, see Short (1999, 2006).

In addition, two additional beach states can occur along the coast, which consist of high tide reflective sandy beaches fronted by intertidal *rocks flats* or beachrock reefs, and high tide beaches fringed by intertidal *coral reef flats* (Fig. 1.4). Both are common along with parts of the northeast Brazilian coast, particularly where beach rock and/or coral reefs fringe the shore.

Brazil has tides ranging from micro to mega, and waves from low to high, together with beach sand ranging from fine to coarse. As a result, all the above beach types and states are found along the Brazilian coast, as will be presented in the following sections.

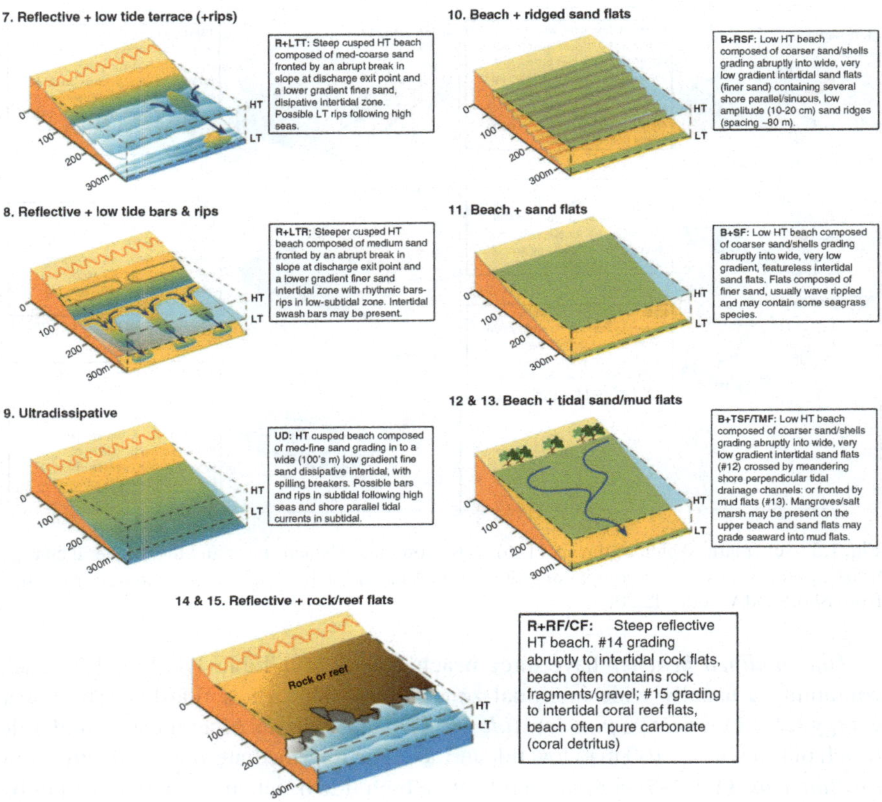

Fig. 1.4 Schematic sketch of the three (7–10) tide-modified (left), four tide-dominated (11–13), and two rock and reef flats (14 and 15) beach states (right) and their general characteristics. (Source: Modified from Short and Woodroffe 2009)

1.4 Brazilian Coastal Regions

The Brazilian coast can be divided into several regions based on geology, sediments, coastal processes, and beach types (Klein and Short 2016). The regions are illustrated in Fig. 1.5a, which largely follows the classification of Dominguez (2006, 2009). The following sections discuss the beach systems in each of the seven regions.

1.4.1 Region 1: Tide-Dominated Amazon Delta Coast

The Amazon coast is one of the longest (~3000 km) and the most complex and variable of Brazil's coastal regions. It extends either side of the world's greatest river, which flows into a gulf dominated by tides ranging up to 11 m (Klein and Short

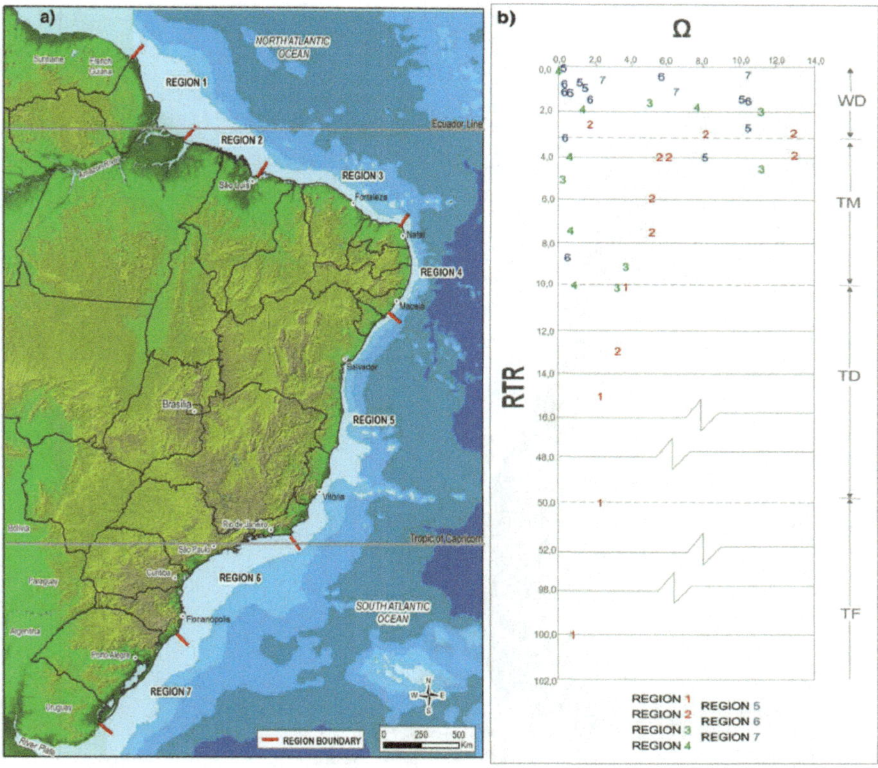

Fig. 1.5 (**a**) Brazil's seven coastal regions and their boundaries. (Source: Klein and Short 2016), and (**b**) plot of indicative regional (1–7) beach states using the Ω-RTR plot. The relative tide range (RTR) varies from 0.1 to 100 across the full-wave to tide-dominated beach types, and the dimensionless fall velocity (Ω) ranges from 0.1 to 13 across the full reflective to dissipative range. (Source: Short and Klein 2016). WD Wave-dominated, TM Tide-modified, TD Tide-dominated, TF Tidal flat

2016; Short and Klein 2016). Region 1 extends from the northern Amapá/French Guiana border at Cape Orange, down through the Amapá and Pará Amazon River mouth coast, including the Amazon's North and South channels and numerous distributaries and islands, to the northeast tip of Marajó Island (Figs. 1.5a and 1.6) (Klein and Short 2016; Short and Klein 2016). Tides range from 4 m in the north to 11 m at the mouth. Easterly trade winds generate moderate waves, which after they cross the wide shallow shelf and inter and subtidal zones, are usually very low to zero at the shore (Klein and Short 2016; Short and Klein 2016). While mud makes up 99% of the Amazon sediments, there is still sufficient sand (~6 M m^3 yr^{-1}, Dunne et al. 1998) to be deposited at the shore to form sand beaches and sand flats. As Fig. 1.5b indicates, the state of Amapá beaches (Region 1) are all tide-dominated with an RTR between 10 and 100 (Klein and Short 2016; Short and Klein 2016).

Fig. 1.6 Coastal region 1 – the tide-dominated Amazon delta and mud coast. (Source: Klein and Short 2016)

Santos et al. (2016) divided the state of *Amapá* coast into four sub-provinces each of which contains tide-dominated beaches. The 550-km-long northern Amapá coast, while dominated by mud, has low gradient tide-dominated sandy beaches (B + SF to B + TMF) occupying the intertidal along 20% of the coast (Fig. 1.7). The 500-km-long southern Amapá coast has tide-dominated sandy beaches fronted by wide tidal sand flats located on the eastern shores of the more exposed river mouth islands, while along the northern banks of the North Channel

Fig. 1.7 (**a**) High tide beach on state of Amapa's Parazinho island at the Amazon River mouth; and (**b**) wide exposed sand flat with mud hollows on Amapá's Farol Praia on Ilha Vitória. (Photos: AD Short; Source: Santos et al. 2016)

and some of the distributaries smaller pocket beaches occur usually hemmed in by mangroves and varzea forest and fronted by river-truncated mudflats (Klein and Short 2016; Short and Klein 2016). These beaches occupy only 9% and 1%, respectively, of this river and mud-dominated sub-province (Klein and Short 2016; Short and Klein 2016). The 1500-km-long *Pará* Amazon riverine-island sub-province is dominated by the mouth of the Amazon and includes the Amazon's South channel (Fig. 1.6) (Klein and Short 2016; Short and Klein 2016). About 50 tide-dominated beaches averaging 2 km in length and fronted by sand-mud flats, occupy about 7% of the shore, all located along more exposed eastern sections of the major river mouth islands (Klein and Short 2016; Short and Klein 2016). The final sub-province is the 200-km-long northern shore of Marajó Island, which faces north into the South Channel (Klein and Short 2016; Short and Klein 2016). It contains 39 tide-dominated sandy beaches fronted by both sand flats and ridged sand flats (B + SF and RSF) (Klein and Short 2016; Short and Klein 2016). The beaches average about 1 km in length and occupy about 20% of the coast, the remainder dominated by mangroves (Klein and Short 2016; Short and Klein 2016). All sub-provinces experience westerly longshore sediment transport. In total, about 260 tide-dominated (TD) beaches occupy about 420 km (~7%) of this predominantly river-dominated mud and mangrove-lined province (Klein and Short 2016; Short and Klein 2016).

1.4.2 Region 2: Tide-Dominated Beaches, Barriers, and Estuaries of Pará-Maranhão

Region 2 encompasses both sides of Marajó Bay and the open east coast of the state of *Pará* and western coast of the state of *Maranhão*. It commences at the mouth of Marajó Bay and extends 600 km east to São Marcos Bay, terminating at Baleia Island (Figs. 1.5a and 1.8) (Klein and Short 2016; Short and Klein 2016). The large tides that dominate the Amazon Gulf extend eastward along the east coast of Pará and Maranhão, decreasing from 10 m at the mouth of Marajó Bay (Baía de Marajó) to 3 m at Belém and on the eastern Maranhão coast (Klein and Short 2016; Short and Klein 2016). At the same time on the open coast, the easterly trade wind waves generate more energy at the shore particularly at high tide resulting in a combination of tide-dominated and tide-modified beaches (Pereira et al. 2016a).

This 1050-km-long region begins at Cape Maguari on the northeast tip of Marajó Island and includes the 285 beaches that lie on either side of the elongate 180-km-long Marajó Bay (Pereira et al. 2016a). The bay beaches are all tide-dominated and consist both of B + SF, and on some more exposed western shores, B + RSF. The "low" northern half of the bay is dominated more by tidal creeks and their associated ebb-tidal deltas and mangroves, while the southern "high" half of the bay is backed by the bluffs of the Barreiras Formation where the

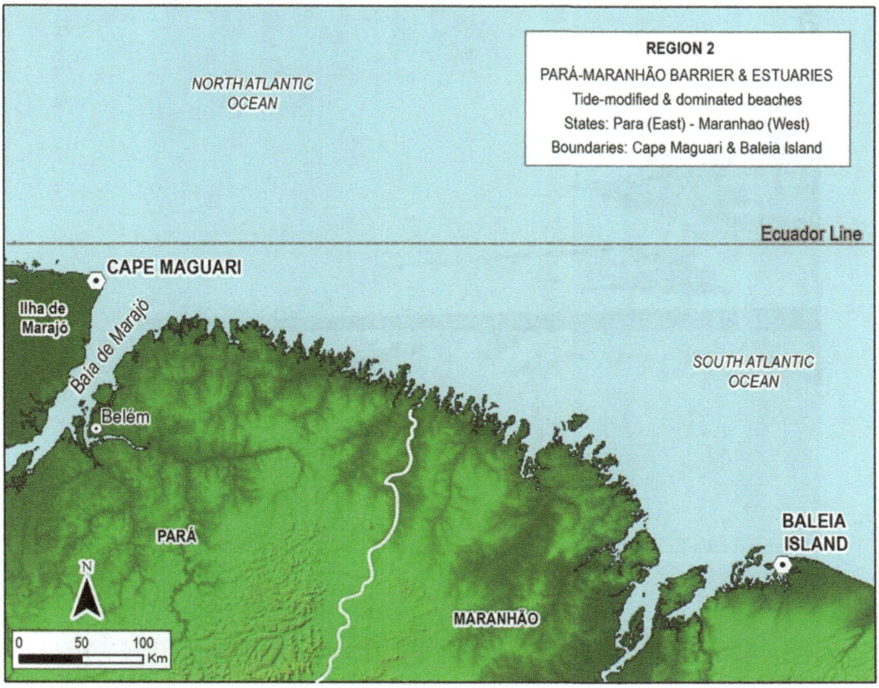

Fig. 1.8 Coastal region 2 – the highly indented tide-dominated Pará-Maranhão barriers and estuaries. (Source: Klein and Short 2016)

beaches are short and often backed and bordered by the bluffs. Mangroves increase southward into the bay and dominate the shore of the Tocantins River mouth at the base of the bay.

The mega to macro-tidal northern open coast consists of a series of 35 major promontories bordered by 5–10 km wide, up to 20-km-long funnel-shaped tide-dominated estuaries. Each of the promontories contains regressive barriers and "drumstick" barrier islands that recurve into the bay mouths. Sediment transport is predominantly to the west, though continually interrupted by the numerous estuaries. The easterly waves are modulated by the tides with moderate waves reaching the exposed beaches at high tide and little or no waves at low tide. The beaches range from tide-modified, including R + LTR (Fig. 1.9) on the most exposed tips of the promontories, to tide-dominated moving into the bay mouths and in the lee of the bay mouth shoals (Pereira et al. 2016b). Ninety beaches occupy the eastern Pará section and another 42 in the western Maranhão section. Figure 1.5b (region 2) locates these beaches in the tide-modified to the tide-dominated section of the Ω-RTR plot.

Fig. 1.9 (**a**) The wide tide-modified Atalaia beach shown here at low tide; and (**b**) view across Atalaia beach showing typical low tide surf conditions when rip currents prevail. (Photos: AD Short)

1.4.3 Region 3: Northeastern Tide-Modified Beach and Barrier-Dune Coast

Region 3 extends for 1300 km between Baleia Island at the mouth of São Marcos Bay and Cabo Calcanhar and includes the eastern open coast of the state of Maranhão, all of the states of Piauí and Ceará, and the northern (or equatorial) coast of the state of Rio Grande do Norte (Figs. 1.5a and 1.10). East of Baleia Island the coast changes dramatically from the highly irregular tide-dominated estuarine coast to the west, to a straighter and more continuous, more wave-dominated coast to its eastern boundary (Klein and Short 2016; Short and Klein 2016). The coast has meso-tides averaging about 3 m and easterly trade wind waves between 1 and 2 m. The trade winds also have a profound impact on the coast forming extensive transgressive dunes along much of the coast and together with the easterly waves drive substantial westerly longshore sand transport, including headland overpassing and bypassing, all estimated at 100,000s m^3 yr^{-1} (Pinheiro et al. 2016). Near the shore, the waves are modulated by outcrops of the Barreiras Formation, which forms headlands, cliffs, and scarps, and by shore-parallel beachrock reefs, together with numerous inlets (Klein and Short 2016; Short and Klein 2016). Combined, these structures

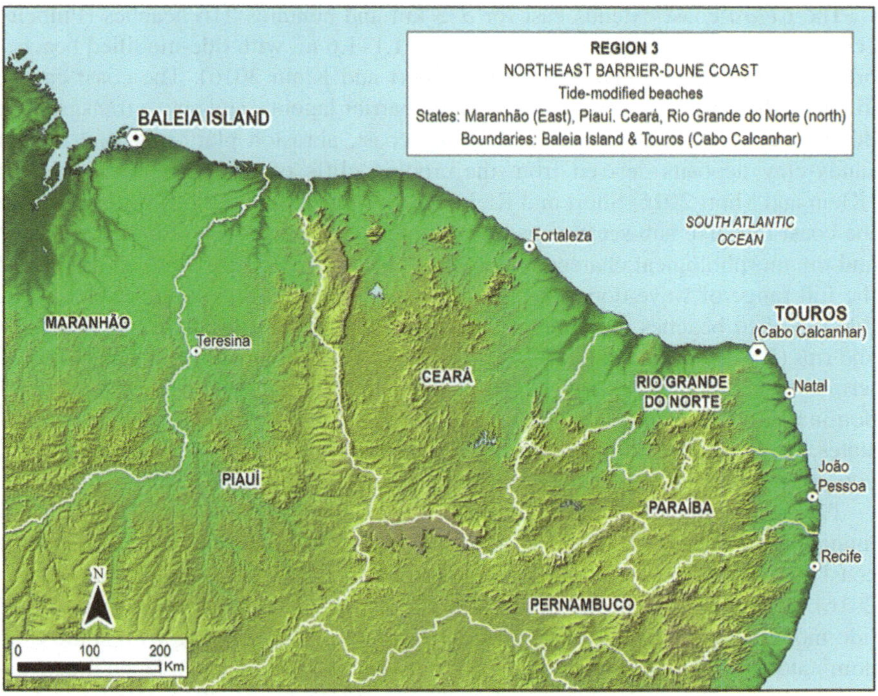

Fig. 1.10 Coastal region 3 – the northeast tide-modified barrier-dune coast. (Source: Klein and Short 2016)

produce a series of headland-bound embayed beaches, with wave height locally controlled by wave refraction and diffraction/attenuation around and over the reefs (Klein and Short 2016; Short and Klein 2016). Beaches range from wave-dominated in more exposed locations to tide-modified and tide-dominated in more sheltered locations (Fig. 1.5b, region 3) (Klein and Short 2016; Short and Klein 2016).

The 230 km eastern *Maranhão* coast has 26 generally long sandy beaches, the longest being 65-km-long (Pereira et al. 2016b). In the west, the more exposed beaches and those with fine sand are wave-dominated dissipative multi-bar beaches, all backed by massive active transgressive dunes, and all part of the large Lençóis Maranhenses National Park (Klein and Short 2016; Short and Klein 2016). Toward the east and the Parnaíba River mouth, wave height decreases and tide-modified and some tide-dominated beaches predominate, together with the only barrier islands in Brazil.

The 66-km-long *Piauí* coast consists of several larger headland-tied embayed beaches, together with some beach rock reefs. As wave energy increases from east to west within the embayments the full range of tide-modified reflective, intermediate, and dissipative beach states are found (Fig. 1.11a), with numerous well-developed low tide rips (Paula et al. 2016). Beachrock increases to the east with the shore-parallel rock usually backed by steep reflective beaches, which are only active at high tide. All the beaches are backed by massive active transgressive dunes, most of them without foredunes.

The *Ceará* coast extends east for 573 km and contains 216 beaches (Pinheiro et al. 2016). Tides are meso; waves average 1.1–1.6 m, with tide-modified beaches predominating (Klein and Short 2016; Short and Klein 2016). The coast can be divided into a western sector dominated by barrier lagoons and active transgressive dunes, and an eastern section with rocky coast, abrasion platforms, and tabular sandy-clay deposits derived from the eroding cliffs of the Barreiras Formation (Klein and Short 2016; Short and Klein 2016). Morais et al. (2006) further divided the coast into five sub-sectors, based on watershed boundaries, sediment delivery, and the morphological characteristics of the beaches. Pinheiro et al. (2016) found the full range of wave-dominated and tide-modified beaches along the coast. The tide-modified beaches predominate (48%) of which reflective with low tidal bars and rips (R + LTR) are the most common (26%), followed by reflective and low tide terrace (R + LTT) (21%) (Fig. 1.11b) with only 0.5% ultradissipative (UD). Wave-dominated beaches make up 14%, with LBT and TBR the most common beaches states, while reflective high tide beaches fronted by intertidal rocks flats (R + RF) comprise 38% of the beaches (Klein and Short 2016; Short and Klein 2016).

The northern coast of Rio Grande do Norte extends east for 244 km to Touros municipality (Fig. 1.10), where there is a dramatic change in the orientation of the coast as it swings south into the next region (Klein and Short 2016; Short and Klein 2016). Eighty percent of the northern coast consists of generally longer embayed tide-modified sandy beaches (predominately R + LTT), together with some tide-dominated beaches (Klein and Short 2016; Short and Klein 2016). Dune transgression increases westward within each embayment (Vital et al. 2016). Besides, there are outcrops of the Barreiras Formation (Fig. 1.11b) and beachrock reefs. Figure 1.5b shows the spread of Region 3 beach types between wave-dominated and tide-modified (Klein and Short 2016; Short and Klein 2016).

Fig. 1.11 (**a**) Low gradient dissipative Eólicas beach near Parnaiba River, which forms the border between the states of Maranhão and Piauí. (Photo: J E A Paula; Source: Paula et al. 2016); and (**b**) the cliffed Barreiras Formation and Nautico beach with low tide rips along beach – Fortaleza-City (CE). (Photo: L Pinheiro; Source: Pinheiro et al. 2016)

1.4.4 Region 4: Eastern Beachrock Coast

The eastern beachrock coast extends for 760 km between the municipality of Touros and the state of Alagoas and includes the eastern coast of the state of Rio Grande do Norte, and the coast of the states of Paraíba, Pernambuco, and Alagoas (Figs. 1.5a and 1.12) (Klein and Short 2016; Short and Klein 2016). The south then southeast trending coast faces into both the easterly trade winds and their waves, plus longer

Fig. 1.12 Coastal region 4 – the northeast wave-dominated beachrock coast. (Source: Klein and Short 2016)

swell arriving from the south, with waves averaging 1.5–2 m, while tides average 2.5 m (Klein and Short 2016; Short and Klein 2016). The coast consists of two sections: a northern 220-km-long more open coast between the municipalities of Touros and Baia da Traicão, and the 540 km central-southern sector, which is dominated by inshore beachrock and algal reefs (Klein and Short 2016; Short and Klein 2016).

The northern *Rio Grande do Norte* sector consists of a series of more than 20 headland-tied beaches, which occupy 61% of the coast (Vital et al. 2016), with wave energy increasing northward within each curving embayment and all backed by substantial dune transgression, dominated by longwalled parabolics. Beachrock, while present, is limited in extent and the beaches range from reflective to rip-dominated intermediate, with LTT and TBR being the most common beach states. The waves and wind drive northerly sand transport along the beaches with headland bypassing and/or overpassing occurring at most headlands. The Barreiras Formation occupies 39% of the coast and outcrops along the coast, forming numerous headland and cliffs and backing beaches (Fig. 1.13a).

The central-southern sector extends from just inside the state of Paraíba to the city of Coruripe in southern Alagoas. This section is dominated by extensive shore-parallel beachrock reefs (Fig. 1.13b, c), together with nearshore and inner shelf beachrock and algal reefs, that both attenuate the breaking waves as well as cause wave refraction and diffraction/attenuation which produces a more crenulate shoreline consisting of sandy bays and salients. The 145-km-long *Paraíba* coast is dominated by near-continuous exposed shore-parallel and submerged inner shelf reefs,

Fig. 1.13 (**a**) The Barreiras Formation at Baia Formosa (state of Rio Grande do Norte) is capped by dunes and fronted by a sheltered tide-dominated beach; (**b**) a prominent beachrock reef at Cumurupin (state of Rio Grande do Norte) shelters a low energy beach sheltered. (Photos: AD Short); (**c**) beachrock reef at Baía da Traíção (state of Paraíba) with attached salient and highly crenulate shoreline. (Photo: JML Dominguez; Dominguez et al. 2016a); and (**d**) Paiva Beach (state of Pernambuco) at low tide showing the well-developed rip current channels cutting the surf zone and separated by transverse bars. (Photo: RL Barcellos; Pereira et al. 2016)

particularly around the city of João Pessoa. The reefs lower wave energy and result in predominantly tide-modified R + LTT beaches which are largely backed by eroding cliffs of the Barreiras Formation which supplies sand directly to the beaches. Dominguez et al. (2016c) divide the coast into three compartments. The first extends from the border with Rio Grande do Norte for 55 km to the municipality of Lucena. This is a relatively straight coast largely free of reefs with active cliff erosion, the cliff blanketed by vegetated clifftop dunes. The second continues for 40 km to Cabo Branco and contains the Paraíba River mouth with sandy beaches to either side and some fringing and beachrock reefs, while the third extends for 50 km from Cabo Branco to the Pernambuco state border and consists of eroding cliffs with narrow beaches at their base sheltered by fringing reefs.

Pereira et al. (2016) divide the 200-km-long *Pernambuco* coast into three sections based on the level of beachrock control. The 70-km-long northern section has near-continuous beachrock reefs backed by lower energy tide-modified beaches (R + LTT); the 60-km-long central section, including Recife municipality, has moderate beachrock control with both protected R + LTT beaches and more exposed beaches with R + LTR (Fig. 1.13d) (Klein and Short 2016; Short and Klein 2016). Along the 70-km-long southern section the reefs are discontinuous and wave energy is higher at the shore resulting in predominantly rip-dominated R + LTR (Klein and Short 2016; Short and Klein 2016).

Dominguez et al. (2016a) divide the 250-km-long *Alagoas* coast into three sections. The 160 km northern-central sector is dominated by beachrock and algal reefs, lying between 1 and 4 km offshore (Klein and Short 2016; Short and Klein 2016). Like the Pernambuco coast, the reefs dominate the inshore wave climate, lowering waves at the shore to maintain tide-modified predominately R + LTT beaches and generating sandy bays and salients in the lee of the reefs (Klein and Short 2016; Short and Klein 2016). The exposed reefs essentially cease at Lagoa do Roterio beyond which is a 40-km-long section of higher energy wave and rip-dominated beaches, with long near-continuous beaches down to the municipality of Coruripe, where the next region begins (Klein and Short 2016; Short and Klein 2016). Region 4, while exposed to higher southerly waves, has considerable beachrock influence which lowers waves at the shore and maintains many lower energy reflective and some tide-modified beaches in amongst a generally higher energy wave-dominated coast (Fig. 1.5b, region 4).

1.4.5 Region 5: Eastern Wave-Dominated Deltaic Coast

Between the São Francisco delta and Cabo Frio lies the 2000-km-long "drip-feed" wave-dominated deltaic coast (Dominguez 2009), which includes the southern section of state of Alagoas and all of states of Sergipe, Bahia, Espírito Santo, and the northern coast of the state of Rio de Janeiro (Figs. 1.5a and 1.14). This coast is backed by a humid hinterland that feeds several rivers, including the large São Francisco, together with the Jequitinhonha, Doce, and Paraíba do Sul, which have deposited large protruding wave-dominated delta typified by the extensive regressive protruding beach and foredune ridge plains (Klein and Short 2016; Short and Klein 2016). These deltas "drip-feed" sediment to the coast. Between the deltas, however, the coast is sediment deficient and the coastal tableland (Barreiras Formation) and in places, Cretaceous and Precambrian rocks are exposed at the shore forming cliffs and rocky headlands with the Barreiras Formation actively

Fig. 1.14 Coastal region 5 – the eastern wave-dominated deltaic coast. (Source: Klein and Short 2016)

retreating in many places (Fig. 1.15a) (Klein and Short 2016; Short and Klein 2016). The coast is also dominated by a reef-covered shallow inner shelf which results in significant wave attenuation and refraction-diffraction over and around the reefs and generally low waves at the shore (Klein and Short 2016; Short and Klein 2016). While deepwater waves generated by the trades and southern swell average 1.5 m,

Fig. 1.15 (**a**) A long section of active sea-cliffs, carved into the Barreiras formation south of the city of Barra de São Miguel (state of Alagoas) (Dominguez et al. 2016b); (**b**) dissipative beach south of São Francisco river and north of the city of Aracaju (state of Sergipe) (Dominguez et al. 2016b); (**c**) example of the transverse bar and rip intermediate beach south of Conde (state of Bahia). (Photos: JML Dominguez; Dominguez et al. 2016c); and (**d**) beachrock reef at Jacuipe (state of Bahia). (Photo: AD Short)

they are significantly lowered in the lee of the many reefs. Tides range from meso in the north (2.5 m) to micro in the south (1.6 m), with beaches switching between wave-dominated on exposed sections to tide-modified along sheltered sections (Klein and Short 2016; Short and Klein 2016).

The wave-dominated São Francisco river delta dominates the southern 50 km of the state of *Alagoas* coast and the first 50 km of the state of Sergipe coast. The remaining 115 km of the state of *Sergipe* coast is dominated by three regressive barriers bordered by wide tidal inlets, with low gradient multi-bar dissipative beaches dominating the open shore (Fig. 1.15b).

The 1000-km-long state of *Bahia* coast contains 436 open coast beaches. As Dominguez et al. (2016b) stated "There are various coastal landforms along the Bahia coast, including cliffs and rivers, a wave-dominated delta, beach-ridge plains, coral reefs, and large bays." They go on to divide the coast into four sections. The 230-km-long Northern Littoral Coast extends down to city of Salvador. It trends north-northeast and contains generally narrow stable to slightly erosional foredune capped barriers and wave-dominated intermediate beaches (Fig. 1.15c), together with Precambrian rock outcrops and beachrock reefs particularly in the south (Fig. 1.15d). Its 90 beaches tend to be relatively short and bordered by reefs, rocks, and tidal inlets. The Mesozoic Rifts Coast commences at Itaparica Island and extends 230 km south to city of Itacaré and is characterized by exposures of eroding Mesozoic sedimentary rocks which result in a crenulate coastline fringed in places by coral reefs. The variation in both wave energy, sediment source (lithic to carbonate), and sediment size (0.1–0.9 mm) results in the 125 beaches ranging from reflective in the lee of reefs, to higher energy intermediate on exposed sections. The Jequitinhonha River Deltaic Coast occupies the next 240 km of the coast. This section has 70 generally longer beaches composed of fine sand and exposed to higher waves, which maintain higher energy dissipative and intermediate beaches (Figs. 1.15b, c), while near the river mouths coarser sand maintains more reflective beaches. The 300-km-long southern Sediment Starved Southern Coast is dominated by cliffed exposures of the Barreiras Formation (Dominguez et al. 2016b), which supplies most of the sediment to the adjoining beaches. Numerous coral reefs line the coast inducing wave attenuation and refraction and the formation of coastal salients, including the large Caravelas and Corumbau salients. Because of the lower wave energy, the 165 beaches are typically R to LTT.

The state of Espírito Santo has a 480-km-long coastline containing 327 beaches, which occupy 71% of the coast, of which 25% are embayed and pocket beaches, 22% located between eroding cliffs and abrasion terraces and 24% exposed sandy beaches (Albino et al. 2016). The coast contains two sedimentary plains totaling 180 km in length associated with the large Doce and Itabapoana river deltas. These contain exposed sandy beaches which, depending on grain size, are either higher energy intermediate (fine-medium sand) or dissipative (fine sand). South of the Doce River, the coast is more crenulated with outcrops of the Barreiras Formation, crystalline rocks, and Neogene sedimentary deposits forming the numerous smaller embayments and cliffed-abrasion sections. The beaches here range from intermediate to reflective.

The northern coast of the state of *Rio de Janeiro* contains two sub-provinces (Muehe and Lins-de-Barros 2016), the northern protruding Paraiba do Sul coastal plain extends 180 km south of the border. Its beaches, while well exposed, vary in state from intermediate to dissipative on the 90-km-long northern coast where sands are finer (0.4 mm) to LTT to R along the southern coast where sands are coarser (1.2 mm). The remaining 160 km of the coast down to municipality of Cabo Frio consists of two sectors, the Macaé-Búzios coastal plain, and outcrops of the rocky basement, which form major promontories at the municipalities of Macaé, Cape Búzios, and Cabo Frio, resulting in a series of smaller embayed beaches of variable energy and ranging from reflective to intermediate. Figure 1.5b indicates the generally high energy wave-dominated nature of the beaches in this region.

1.4.6 Region 6: Southeast Wave-Dominated Rocky-Embayed Coast

At Cabo Frio, there is a major change in coastal orientation, coastal processes, and beach systems. The east-facing often sheltered beaches to the north give way to the initially south-facing more exposed beaches to the south contained within the São

Paulo Bight. The moderate energy trade wind waves to the north are replaced by the persistent higher southerly swell. Also, the crystalline rocks of the Serra do Mar mountain range dominate much of the coast, controlling the coastal orientation, and result in considerable rocky shores and numerous embayed and pocket beaches. This 1700-km-long region includes the southern coast of the state of Rio de Janeiro, the coast of the states of São Paulo and Paraná, and the coast of the state of Santa Catarina down to Cabo Santa Marta (Figs. 1.5a and 1.16).

The southern coast of the state of *Rio de Janeiro* is dominated by the Serra do Mar mountains. The south-facing coast is heavily indented, containing the larger Guanabara, Sepetiba, and Ilha Grande bays and numerous rocky bays and embayments of varying lengths. In the rocky sections, there are 242 open coast beaches and 647 smaller beaches within the bays. The beaches range from small pockets of sand to several long barriers in the east and west, including the 42-km-long Marambaia beach. The open coast beaches are generally well exposed to the southerly waves and micro-tides are all wave-dominated. The main control on the beach state is sand particle size, which ranges from 0.15 to 1.0 mm (Muehe and

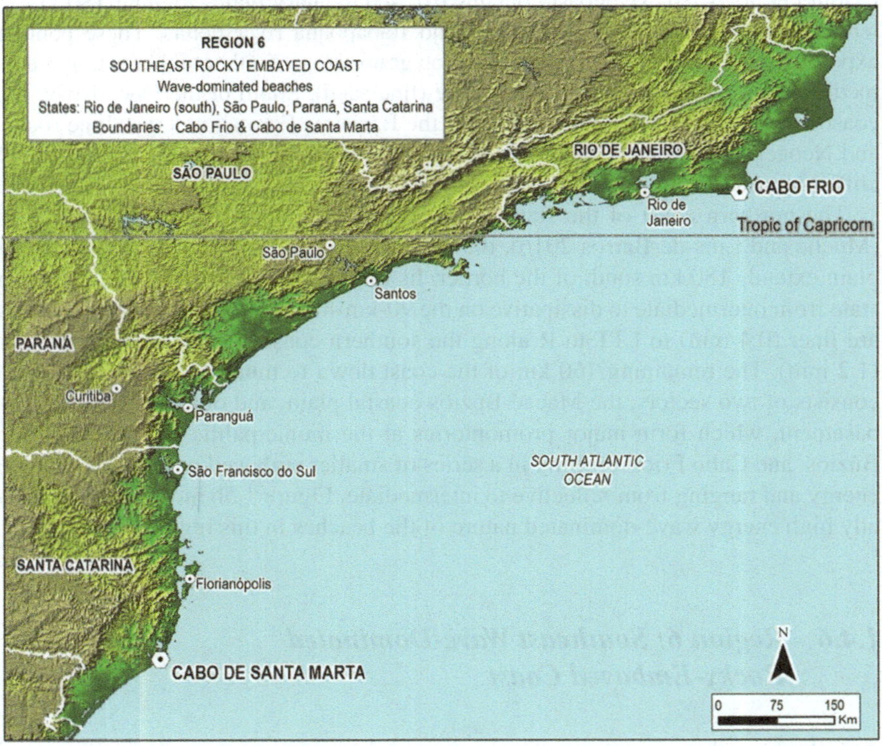

Fig. 1.16 Coastal region 1 – the southeast wave-dominated rocky-embayed coast. (Source: Klein and Short 2016)

Lins-de-Barros 2016). The coarser sand maintains steep reflective beaches, while the finer sand maintains intermediate beaches (Fig. 1.17a).

The state of *São Paulo* has 670 m of coastline containing 242 beaches. Mahiques et al. (2016) divide the coast into six sectors, all of which are dominated by the Serra do Mar crystalline massif. The highly indented Northern Scarps Coast contains a series of small bays and islands, with the 280 km of mainland shore containing 155 small wave-dominated beaches, which, depending on exposure and sand size, range from reflective to dissipative. In contrast, the 50-km-long Bertioga coastal plain contains just nine beaches, dominated by four higher energy intermediate to dissipative beaches. The 90-km-long Santos sector is more heavily indented, with variable orientation, and contains 35 embayed beaches which range from reflective to intermediate depending on exposure. South of Santos (Fig. 1.16) is 260 km of southeast-facing coast containing two long coastal plains (Itanhaém-Peruíbe and Cananéia-Iguape) separated by rocky-embayed sections. The long beaches are intermediate to dissipative, while the intermediate type tend to dominate the embayed beaches.

Fig. 1.17 (**a**) Barra da Tijuca beach (state of Rio de Janeiro) with transverse bars and well-developed rip channels. (Photo: Flavia Lins-de-Barros; Source: Muehe and Lins-de-Barros 2016); (**b**) beach and mud flats at Tijucas (Santa Catarina state). (Source: Klein et al. 2016); (**c**) embayed beach with transverse bars and rips at Galheta (state of Santa Catarina). (Source: Klein et al. 2016); and (**d**) reflective beach and cusps at Estalerio (state of Santa Catarina). (Photos: AD Short; Source: Klein et al. 2016)

The state of *Paraná* coast is just 150 km long and dominated by a wide regressive coastal plain, with the Serra do Mar mountains located up to 50 km inland. The coast has just 32 beaches separated by inlets and some rocky headlands. The exposed beaches dominate the coast with beach type depending on grain size ranging from predominantly fine sand dissipative to intermediate. The sandy beaches extend into the large tidal inlets where they become more tide-modified to tide-dominated (Angulo et al. 2016) (Fig. 1.5b).

The Serra do Mar mountains return to the coast along much of the state of *Santa Catarina* coast down to Cabo Santa Marta (Fig. 1.16), the southern boundary of this region. This 576 km section contains 240 beaches, which occupy 60% of the coast, the remainder being rocky. Klein et al. (2016) divide the Santa Catarina coast into five sub-provinces. The first from the border to Farol Cabeçudas tends to have longer exposed intermediate to dissipative beaches. This is followed by a heavily indented rock-dominated section that extends down to the entrance to North Bay. This sub-province contains numerous smaller embayed reflective beaches with variable orientation as well as the tide-dominated Tijucas located in a very sheltered mud-rich embayment (Fig. 1.17b). The rock-dominated Santa Catarina Island forms the next section that has a mix of exposed intermediate and sheltered reflective beaches (Fig. 1.17c, d). The backing South Bay (fourth sub-province) contains the southernmost mangroves in South America and 392 generally short tide-modified to tide-dominated beaches. The final section down to Cabo Santa Marta remains rock-dominated with 37 generally well-exposed embayed intermediate beaches occupying the 87 km of east-facing coast. Figure 1.5b (Region 6) illustrates the dominance of higher energy wave-dominated beaches, together with some tide-modified beaches in sheltered locations.

1.4.7 Region 7: Wave-Dominated Beaches and Barrier Coast of Rio Grande do Sul

The southernmost region extends from Santa Catarina's Cabo Santa Marta down to Chuí at the border between state of Rio Grande do Sul and Uruguay a distance of 735 km (Figs. 1.5a and 1.18). It contains 12 long beaches all composed of fine sand and facing east-southeast into the prevailing southerly swell, which drives substantial northerly longshore sand transport (Motta and Toldo Jr 2013). The beaches are all higher energy dissipative (Fig. 1.19a), with bar numbers ranging from two to four depending on grain size. Rips can however dominate the inner bar (Fig. 1.19b; Calliari and Toldo Jr 2016). The beaches are part of the longest and largest barrier systems in South America and are all backed by episodes of massive dune transgression which trends south under the prevailing northerly winds. The wave-dominated beach system is essentially fully dissipative along its entire length, the only variation being the sand size, which ranged from 0.125 to 1.4 mm and the consequent number of bars. The coarser sand (0.35–1.4 mm) maintains two shore-parallel bars,

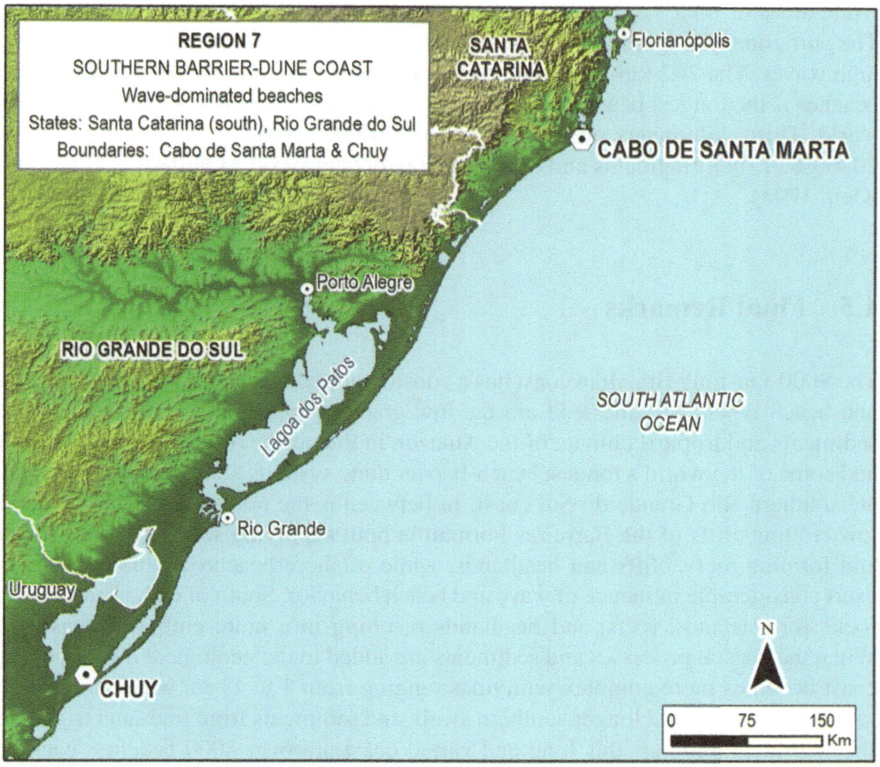

Fig. 1.18 Coastal region 1 – the southern wave-dominated barrier coast. (Source: Klein and Short 2016)

Fig. 1.19 (**a**) Dissipative beach with three bars (1, 2, 3) at Camacho (SC) typical of the long beaches of region 5; and (**b**) high energy intermediate beach at Cabo Santa Marta. (Photos: AD Short)

while areas of finer sand (0.125–0.2 mm) result in three to occasionally four bars. The surf zone averages 300–400 m in width increasing to up to a kilometer during high waves. The 242-km-long stretch of beach between Hermenegildo and Cassino beaches is the longest beach in South America and one of the longest beaches in the world. Thirty kilometers north of Hermenegildo beach the beach is composed of 20–60% of shell fragments and coarse sand (Concheiros do Albardão) (Calliari and Klein 1993).

1.5 Final Remarks

The 9000-km-long Brazilian coast has a considerable range in coastal environments and beach types. At either end are the low gradient coasts dominated by the fine sediments and tropical climate of the Amazon in the north (Amapá and Pará states) and some of the world's longest beach-barrier dune systems along the more temperate southern Rio Grande do Sul coast. In between is the northeast coast where the low eroding cliffs of the Barreiras Formation both supplying sediment to the coast and forming reefs, cliffs and headlands, while offshore beachrock and coral reefs exert considerable influence of wave and beach behavior. South of Cabo Frio granite rocks form islands, reefs, and headlands resulting in a more embayed coastline. When the coastal processes and sediments are added to the geological influence, the coast becomes more complex with tides ranging from 1 to 11 m, waves from very low seas to higher and longer southern swell, and sediments from fine sand to coarse shelly grit. In amongst this long and varied coast are over 4000 beaches, each of which reflects the interaction of the climate, geology, waves, tides and sediment, and more recently human impact.

The beaches can be classified by the tide range and wave height into wave-dominated, tide-modified, and tide-dominated which contain 13 separate beach states, with two additional states covering beaches fronted by rock or (coral) reefs flats. All three beach types and 15 beach states are found along the Brazilian coast. However, regional variation in waves and tides allows the coast to be divided into seven distinct regions each of which reflects the combination of these processes acting on the regional geology and sediment availability.

Brazil has a magnificent coast dominated by its many beach systems, a coast which attracts both Brazilian and international visitors. Since the 1970s, this attraction has seen the rapid expansion of most coastal towns and communities, proliferation of long ribbon strips of "second homes," numerous resorts and infrastructure, much extending onto the beach and foredune. The often hazardous location of many of these developments has led to the construction of often protective works, such as rocky groynes and seawalls, while many discharge sewer onto the beach, taller structures overshadow the beach, and congestion, noise, and pollution reins during the summer month. Humans are now having a massive impact on Brazil's beaches, most of it adverse. In addition, visitors are attracted to beaches that are inherently dangerous, with rip currents common and strong tidal flows at the many inlets, all leading to a large increase in public risk, injuries, and drowning.

As Short and Klein (2016) concluded "The future of Brazil's coasts and beaches requires greater scientific knowledge of the coast and its behavior; better coastal management that is regulated and enforced; and an expansion of beach safety resources to ensure Brazilians can enjoy the beach in greater safety." Given the length of the coast and with a wide variety of coastal environments and beach settings, this will require more detailed regional and local studies so that management plans can be adapted to the wide range of coastal and environments.

Acknowledgments We thank the reviewers Alexander Turra, Helio Checon, Cecflia Amaral, and Guilherme Corte for their many helpful comments, which substantially improved the chapter.

References

Albino J, Neto NC, Oliveria TCA (2016) The beaches of Espírito Santo. In: Short AD, Klein AHF (eds) Brazilian beach systems. Springer, Dordrecht

Angulo RJ, Borozone CA, Noenberg MA et al (2016) The state of Paraná beaches. In: Short AD, Klein AHF (eds) Brazilian beach systems. Springer, Dordrecht

Calliari JJ, Klein AHF (1993) Características morfodinâmicas e sedimentológicas das praias oceânicas entre Rio Grande e Chui -RS. Pesqui em Geocienc 20(1):48–56

Calliari LJ, Toldo EE Jr (2016) Ocean beaches of Rio Grande do Sul. In: Short AD, Klein AHF (eds) Brazilian beach systems. Springer, Dordrecht

Dillenburg SR, Hesp PA (2009) Geology and geomorphology of Holocene coastal barriers of Brazil. Springer, Berlin

Dominguez JML (2006) The coastal zone of Brazil: an overview. J Coast Res SI39:16–20

Dominguez JML (2009) The coastal zone of Brazil. In: Dillenburg SR, Hesp PA (eds) Geology and geomorphology of Holocene coastal barriers of Brazil. Springer, Berlin

Dominguez JML, Bittencourt ACSP, Nascimento AS et al (2016a) The sandy beaches of the states of Sergipe and Alagoas. In: Short AD, Klein AHF (eds) Brazilian beach systems. Springer, Dordrecht

Dominguez JML, Bittercourt ACS, Nascimento L et al (2016b) Beaches in the state of Bahia- the importance of geological setting. In: Short AD, Klein AHF (eds) Brazilian beach systems. Springer, Dordrecht

Dominguez JML, Neves AS, Bittencourt ACSP (2016c) Sandy beaches of the states of Paraíba: the importance of geological heritage. In: Short AD, Klein AHF (eds) Brazilian beach systems. Springer, Dordrecht

Dunne T, Mertes LAK, Meade RH et al (1998) Exchanges of sediment between the flood plain and channel of the Amazon River in Brazil. Geol Soc Am Bull 110:450–467

Gibbs RJ, Matthews MD, Link DA (1971) The relationship between sphere size and settling velocity. J Sediment Petrol 41:7–18

Gourlay MR (1968) Beach and dune erosion tests. Delft Hydraulics Laboratory, Report No.M935/M936

Klein AHF, Short AD (2016) Brazilian beach systems. In: Short AD, Klein AHF (eds) Brazilian beach systems. Springer, Dordrecht

Klein AHF, Short AD, Bonetti J (2016) Santa Caterina beach systems. In: Short AD, Klein AHF (eds) Brazilian beach systems. Springer, Dordrecht

Mahiques MM, Siegle E, Alcántara-Carrió J et al (2016) The beaches of the state of São Paulo. In: Short AD, Klein AHF (eds) Brazilian beach systems. Springer, Dordrecht, pp 397–418
Masselink G, Short AD (1993) The effect of tide range on beach morphodynamics and morphology: a conceptual model. J Coast Res 9:785–800
Morais JO, Pinheiro LS, Medeiros C et al (2006) The gabions for the protection of Caponga Beach, Ceará, Brazil: hazards and management. J Coastal Res SI 39:848–851
Motta LM, Toldo EE Jr (2013) Regional sediment budget in wave-dominated dissipative-intermediate beaches. J Mar Res 73:49–69
Muehe D (2003) Beach morphodynamics research in Brazil: evolution and applicability. J Coast Res SI 35:32–42
Muehe D, Lins-de-Barros F (2016) The beaches of Rio de Janerio. In: Short AD, Klein AHF (eds) Brazilian beach systems. Coastal Research Library, Springer, Dordrecht, pp 363–369
Paula JEA, Morais JO, Baptista EMC, Gomes ML, Pinheriro LS (2016) Piauí beach systems. In: Short AD, Klein AHF (eds) Brazilian beach systems. Springer, Dordrecht, pp 153–174
Pereira PS, Araújo TCM, Manso VA (2016) Tropical sandy beaches of Pernambuco state. In: Short AD, Klein AHF (eds) Brazilian beach systems. Coastal Research Library, Springer, Dordrecht, pp 251–279
Pereira LCC, Vila-Concejo A, Short AD (2016a) Coastal morphodynamic processes on the macro-tidal beaches of Pará state under tidally modulated wave conditions. In: Short AD, Klein AHF (eds) Brazilian beach systems. Springer, Dordrecht, pp 95–124
Pereira LCC, Nascimento WT, Silva IR et al (2016b) Maranhão beach systems, including the human impact on São Luís beaches. In: Short AD, Klein AHF (eds) Brazilian beach systems. Springer, Dordrecht, pp 125–152
Pinheiro LS, Morais JO, Maia LP (2016) The beaches of Ceará. In: Short AD, Klein AHF (eds) Brazilian beach systems. Springer, Dordrecht, pp 175–199
Rodrígues MG, Nicolodi JL, Gutiérrez OQ et al (2016) Brazilian coastal processes: wind, wave climate and sea level. In: Short AD, Klein AHF (eds) Brazilian beach systems. Springer, Dordrecht, pp 37–66
Santos VF, Short AD, Mendes AC (2016) Beaches of the Amazon coast: Amapá and West Pará. In: Short AD, Klein AHF (eds) Brazilian beach systems. Coastal Research Library, Springer, Dordrecht, pp 67–93
Short AD (1999) Wave-dominated beaches. In: Short AD (ed) Beach and shoreface morphodynamics. Wiley, Chichester, pp 173–203
Short AD (2006) Australian beach systems – nature and distribution. J Coast Res 22:11–27
Short AD, Jackson DWT (2022) Beach morphodynamics. In: Shroder JJF (ed) Treatise on geomorphology, vol 8. Elsevier, Academic Press, pp 199–229. https://doi.org/10.1016/B978-0-12-818234-5.00052-3
Short AD, Klein AHF (2016) Brazilian beach systems: review and overview. In: Short AD, Klein AHF (eds) Brazilian beach systems. Springer, Dordrecht, pp 573–608
Short AD, Woodroffe CD (2009) The coast of Australia. Cambridge University Press, Melbourne
Vital H, Silveira IM, Tabosa WF et al (2016) The beaches of Rio Grande do Norte. In: Short AD, Klein AHF (eds) Brazilian beach systems. Springer, Dordrecht, pp 201–229

Chapter 2
Primary Producers

Clarisse Odebrecht, Andrea de Oliveira da Rocha Franco, Paulo Horta, and Leonardo Rubi Rörig

2.1 Introduction

The great extension and physiographic diversity of the Brazilian coast (Chap. 1) favor the presence and predominance of different groups of primary producers according to the local characteristics. On sandy beaches, microscopic unicellular microalgae provide the organic matter and energy fueling the diversification of food webs, but the absence of visible vegetation leads to the wrong assumption of marine deserts.

Under optimal conditions, such as adequate availability of nutrients and sunlight, microalgae present a high growth rate due to their reproduction predominantly by vegetative division, up to more than two divisions per day or about ten generations per week. They are present in the water column (phytoplankton) and on the sediment (microphytobenthos), the latter including psammic species (i.e., living among sand grains) and epipsammic (i.e., living attached to grains of sand). Microphytobenthos can accumulate on the surface of the sediment at high densities forming a photosynthetic biofilm that performs important ecological functions including participating in the flow and cycling of carbon and nutrients, underpinning water marine food webs, and stabilizing sediments. In the surf zone of sandy beaches, due to the turbulent wave action that suspends material from the bottom, the microalgae generally include benthic species, that is, organisms present in the plankton whose main habitat is benthic (tychoplankton). Primary producers may be composed of neritic and oceanic species, and this varied set of organisms provides

C. Odebrecht (✉) · A. de Oliveira da Rocha Franco
Universidade Federal do Rio Grande-FURG, Instituto de Oceanografia, Rio Grande, RS, Brazil
e-mail: doclar@furg.br

P. Horta · L. R. Rörig
Universidade Federal de Santa Catarina-UFSC, Laboratório de Ficologia, Campus Trindade, Florianópolis, SC, Brazil

A. C. Z. Amaral et al. (eds.), *Brazilian Sandy Beaches*, Brazilian Marine Biodiversity, https://doi.org/10.1007/978-3-031-30746-1_2

an important indicator of environmental conditions, including water quality, the influence of ocean currents in the coastal region, the presence of invasive species, and ecosystem changes due to anthropogenic impacts and the global climate change.

In addition, several macroalgae can be found forming deposits on sandy beaches along the Brazilian coast. Macroalgae grow especially on rocky shores and rocky bottoms of shallow and lighted areas of the continental shelf. Eventually, they are torn and dragged and transported to the coast, forming wracks that cause ecological disruption and stress to the human population, but they also provide an input of organic matter that may shape the structure and functioning of sandy-beach ecosystems (Hyndes et al. 2022).

In this chapter, we review what is known about primary producers at sandy beaches in Brazil, as well as the impacts of coastal eutrophication in these environments, and knowledge gaps. First, we synthesize the main results to find some patterns regarding the spatial and temporal variability of biomass, species composition of microalgae along the Brazilian coast, and evaluate how they are influenced by environmental factors. Next, we discuss the presence and consequences of potential harmful microalgae, as well as the quite usual phenomenon of macroalgae beach stranding. Finally, we address the consequences of eutrophication on the beach flora and fauna through a case study.

2.2 Microalgae Studies in Brazilian Sandy Beaches

The microalgae of sandy beaches have been studied in all geographical regions of the Brazilian coast. The earliest records refer to diatom species by Karl Zimmermann in a series of studies developed between 1913 and 1919 and published in Brotéria, Revista Luso-Brazileira: Botanica. The importance of this early contribution to the taxonomy and distribution of diatoms in Brazil was highlighted in some studies such as Gomes et al. (2012); however, the names of the sampled beaches were not registered. Checklists of marine and estuarine diatoms and species ecological data were provided in the 1990s by Moreira Filho and collaborators, including sandy beaches from the northern to the southern regions of Brazil (Moreira Filho et al. 1990, 1993, 1996, 1999). Species lists of all microalgae groups and, more recently, their fine taxonomy based on electron microscopy are available for some beaches. Since the 1970s, the taxonomic focus has changed to the functional role of microalgae as the base of sandy-beach food webs, although their importance has been recognized very early (Gliesch 1925).

In this chapter, we review and analyze scientific articles, thesis and dissertations, proceedings, and scientific news dealing with planktonic, tychoplanktonic, and psammic microalgae of Brazilian sandy beaches (Table 2.1), except taxonomic studies of a single species, which were not included in this chapter. The microalgae biodiversity has been studied on 93 beaches of different types (embayments, exposed, dissipative to reflective, variable river influence), subjected to contrasting tidal regimes from macrotidal (Pará, N Brazil, 7 beaches), to

Table 2.1 Microalgae studies in Brazilian beaches, including the beach name, location, sampling year and frequency, parameters, presence of patches/deposits at the surf zone or sand and respective references

Beach name and location	Sampling year and frequency	Analysis	Patches	Reference
North Brazil				
Princesa, PA, 00°34.5′S	2008, 2009. Wet, dry, nict	PhyPl QN, QL (mesh 64 µm), Chl, PH, CH	No	Matos et al. (2012, 2013)
Atalaia, PA, 00°37′11″S	2008, 2009. Wet, dry, nict	PhyPl QN, QL (mesh 64 µm), Chl, PH, CH	No	Costa et al. (2013)
Cajuúna, PA, 00°38′52.08″S	2006	SED DIAT: semi-QN	No	Ribeiro et al. (2007)
Garrote, PA, 00°43′45.6″S	2006	SED DIAT: semi-QN	No	Ribeiro et al. (2007)
Ajuruteua, PA, 00°49′9.4″S	1994, 1995, 2004, 2005, 2008, 2009. Wet, dry, nict, monthly	PhyPl QN, QL (mesh 64 µm), Chl, PH, CH	No	Costa et al. (2011), Santana et al. (2005), Melo et al. (2005), and Matos et al. (2016)
Itupanema, PA, 01°31′03″S	2006	SED DIAT: semi-QN	No	Ribeiro et al. (2007, 2008)
Canela, PA, 01°46′37.2″S	2004, 2005. Wet, dry, nict	PhyPl QN, QL (mesh 64 µm), Chl, PH	No	Sousa et al. (2008, 2009)
Northeast Brazil				
Futuro, CE, 03°44′S	1990, 1995–1998, 2011–2012, 2014	PhyPl QN, QL, Chl, PH, CH	DIAT	Tahim et al. (1990), Pereira da Costa et al. (1997, 1998), and Franco et al. (2016, 2018)
Carne de Vaca, PE, 07°34′40–53″S	1998. Biweekly	PhyPl QN, QL (mesh 65 µm)	No	Campelo et al. (2001/2002)
Casa Caiada-Rio Doce, PE, 07°57′26″S–07°59′55″S	1998, 1999. Wet, dry	PhyPl QN, QL (mesh 65 µm), PH, CH	No	Pereira et al. (2005)
Brasília Teimosa, Boa Viagem, Piedade, PE, 08°04′S–08°11′S	2005, 2006. Wet, dry	PhyPl QN, QL (mesh 45 µm), Chl, PH, CH	No	Ferreira et al. (2010, 2013)
Tamandaré, PE, 08°45′S	1961, 1963, 1989–1990, 1998–1999. Monthly	PhyPl QL, (mesh 50 µm), Chl, PrPr, AssNr, PH	CY	Satô et al. (1963/1964), Moura and Passavante (1993, 1994/1995), and Rosevel da Silva et al. (2005)

(continued)

Table 2.1 (continued)

Beach name and location	Sampling year and frequency	Analysis	Patches	Reference
Barra Propriá to Vaza-Barris to Barra Piauí Fundo Real, SE, 10°50′–11°28′S	1994	Water filtered (mesh 50 μm), PhyPl QN, QL, PH	DIAT	Franco (1998)
Adjacent to Rio Real, BA, 11°29′S	2013, 2014, 2015. Wet, dry	PhyPl QN, QL (mesh 30 μm), Chl, PH, CH	No	Alves (2017)
Sargi, Ponta da Tulha, Milionários, Cururupe, BA, 14°30′–15°05′S	2004–2005. Biweekly	PhyPl QN, QL (mesh 20 μm), Chl, PH, CH	DIAT	Tedesco (2006) and Tedesco et al. (2017)
East Brazil				
[a]Beaches, RJ, 22°59′–23°00′S	2007	250 L. DINO QL. Phycotoxins HPLC	No	Menezes et al. (2007)
[b]Beaches, SP, 23°20′–25°S	2005–2006. Monthly	PhyPl semi-QN, QL (mesh 20 μm), T, S	No	Villac and Noronha (2008) and Villac et al. (2008)
Itanhaém, SP, 24°11′S	1978	PhyPl QN, QL	DIAT	Zavala-Camin and Yamanaka (1980)
[c]Beaches, SP, 23°45′–23°55′S	1980	PhyPl QN, Chl	DIAT	Tommasi and Navas-Pereira (1983)
South Brazil				
Pontal do Sul, PR, 25°34′16″S	1988–1989. Weekly	PhyPl QN, QL (mesh 50 μm), PH	No	Rezende and Brandini (1997)
Balneário Village, PR, 25°36′S	2008	SED PhyBe, QL, QN, Chl, PH	FL	Braga (2008)
Navegantes, SC, 26°52′S	1996–1997, 2006–2007	PhyPl QN, QL, Chl, PH	DIAT	Rörig et al. (1997) and Macedo (2007)
Navegantes, SC, 26°52′S Balneário Rincão, SC, 28°50′S	2009, 2013	*As. glacialis, An. Australis* strains ECOPH	DIAT	Macedo (1999) and Rörig et al. (2017a)
Camboriú, SC, 27°S	2011–2013	DIAT QL (mesh 20 μm), semi-QN, SED, DIAT strains. Lipids, TOC, ECOPH	DIAT	Rörig et al. (2017b)
Ponta das Canas, SC, 27°23′S	1991–1992	PhyPl QN, QL (mesh 25 μm), PH	No	Cardoso (1993, 1994, 1998)
Sol, Gi, Iró, Mar Grosso, SC, 28°25′–28°29′S	2006	SED DIAT in/out deposits. Chl, PH, CH	DIAT	Netto and Meneghel (2014) and Dutra and Garcia (2016)
[d]Beaches, SC and RS	1995, 1996, 2001, 2002	SED PhyBe DIAT		Garcia et al. (2012) and Garcia (2016)

(continued)

Table 2.1 (continued)

Beach name and location	Sampling year and frequency	Analysis	Patches	Reference
Azul, RS, 29°29′S	1990–1991. Monthly, annual cycle	SED, PhyBe, QN, QL, PH, CH	DIAT, FL, CY	Garcia-Baptista (1993, 1995)
Jardim Beira-Mar, RS, 29°43′S	1987–1988. Monthly	SED, PhyBe DIAT	No	Garcia-Baptista and Baptista (1992)
Atlântida to Rainha do Mar, RS, 29°44′–29°52′S	1971. Monthly	Visual, patches. PhyPl, DIAT QL	DIAT	Aguiar and Corte-Real (1973)
Tramandaí, RS, 29°59′S	1976–1977, 2011	PhyPl QL (mesh 48/64 μm), PH	No	Rosa (1982), Kremer and Rosa (1983), and Jardim and Cardoso (2013, 2015)
Capão da Canoa, Cidreira, Pinhal, RS, 29°46′–30°15′S	2008	Visual, PhyPl QN, QL (mesh 64 μm)	DINO	Cardoso (2012)
Mar Grosso, Tavares, Mostardas, RS 30°38′–32°03′S	2003, 2004	Chl, PrPr, PH, CH	Yes	Niencheski et al. (2007)
Cassino to Chuí, RS, 32°12′–33°45′S	1992	PhyPl QN, QL	DIAT	Rörig (1997)
Cassino, RS, 32°12′22″S	1992–2017. Every 2 days, weekly, monthly, long term	PhyPl QN, QL, Chl, PrPr, PH, CH, ECOPH	DIAT	Gianuca (1983), Reynaldi (2000), Rörig and Garcia (2003), Rörig et al. (2006), Maria et al. (2016), Odebrecht et al. (1995, 2003, 2010, 2014), Garcia and Odebrecht (2009), Piedras and Odebrecht (2012), and Abreu et al. (2017)
Hermenegildo, RS, 33°40′–33°45′S	1978, 1979, 1980, 1981	PhyPl QN, QL (mesh 48 μm)		Machado (1979) and Rosa and Buselato (1981)

Abbreviations: *Nict* nictemeral, *PhyPl* phytoplankton, *PhyBe* phytobenthos, *QN* quantitative, *QL* qualitative, *SED* sediment, *DIAT* diatoms, *DINO* dinoflagellates, *FL* flagellates, *CY* cyanobacteria, *Chl* chlorophyll a, *PrPr* primary production, *AssNr* assimilation number, *PH* physical factors, *CH* chemical factors, *ECOPH* ecophysiology. States in Brazil: *PA* Pará, *CE* Ceará, *PE* Pernambuco, *SE* Sergipe, *BA* Bahia, *RJ* Rio de Janeiro, *SP* São Paulo, *PR* Paraná, *SC* Santa Catarina, *RS* Rio Grande do Sul

[a]RJ: Arpoador, Ipanema, Leblon, São Conrado, Joatinga Amores, Pepê, Barra da Tijuca

[b]SP: Picinguaba, Itaguá, Enseada, Lagoinha-Eng Velho, Tabatinga, Indaiá, Porto Grande, Barequeçaba, Juquei-Faustino, Sino, Boraceia, Enseada-Indaiá, Enseada, Estr Pernambuco, Gonzaga, Gonzaguinha, Ocian, Parque Balneário, Guaraú, Juréia, Pontal

[c]SP: Milionários, Itararé, José Menino, Ponta da Praia, Enseada, Pernambuco, Bertioga, Indaiá, S. Lourenço, Boracéia

[d]SC: Penha, São Miguel, Navegantes, Perequê, Araçá, Bombas, Quatro Ilhas, Mariscal, Canto Grande, Zimbros, Guarda do Embaú, Gamboa, Siriú, Garopaba, Itapirubá, Iró, Mar Grosso; RS: Tramandaí, Cassino

mesotidal (Piauí, NE Brazil, to Pernambuco, NE Brazil, 8 beaches) and microtidal (Alagoas, NE Brazil, to Rio Grande do Sul, S Brazil, 78 beaches), and diverse trophic conditions from oligotrophic to highly eutrophic sandy beaches (Table 2.1). A great heterogeneity of sampling strategies and frequency is evident among these studies and varies according to the objectives. In most studies, one to three sites were sampled at a single or few beaches, but it is noteworthy that some studies included as much as 20 beaches in the states of São Paulo (Villac and Noronha 2008; Villac et al. 2008), Santa Catarina, and Rio Grande do Sul (Garcia 2016). The length of the beaches varies greatly, from a few tens of meters to extensive beaches with hundreds of kilometers at the extreme southern coast (Rörig 1997; Rörig and Garcia 2003; Rörig et al. 2006; Niencheski et al. 2007; Odebrecht et al. 2010, 2014).

The various studies targeting surf diatoms involved sampling different sectors of the beach ecosystem, including the water column (plankton) in the inner surf zone and behind the surf zone (nearshore), the bottom sediment behind the surf zone and the sediment from intertidal areas, from the swash zone to near the foredunes. Sampling devices used for qualitative and quantitative analysis of phytoplankton were, respectively, horizontal tows with plankton nets (25, 48, or 64 μm mesh-size), and bottles, buckets, or flasks. Rörig and Garcia (2003) collected vertically integrated (1 m) water samples with a hose. Sediment samples were taken with PVC tubes or scraping the sand surface. The main fixatives used were formaldehyde (0.4–4%) and Lugol's solution, but also Transeu (Rosa and Buselato 1981; Alves 2017). Cell density was estimated mostly using sedimentation chambers and the Utermöhl method, and more rarely Sedgewick Rafter chambers. Semi-quantitative analyses of diatoms were performed on the chambers, permanent or semi-permanent mounts.

The sampling frequency comprised several time scales ranging from high frequency (e.g., every few hours for the determination of nictemeral variation or twice a day considering the ebb/flood cycle) to seasonal (e.g., wet and dry seasons) or annual variations (Table 2.1). One ongoing long-term study was set in 1994 at Cassino Beach, southern Brazil (Odebrecht et al. 2010, 2014; Abreu et al. 2017), contributing to the Brazilian Long-Term Ecological Research (BR-LTER) program.

The aim of the studies varied from the identification of a group of species and floristic lists to ecological approaches aiming to explain spatial and temporal distribution patterns of microalgae based on physical and chemical variables. Density and biomass data are more common, while primary production estimates are scarce and only available on the Northeast (Moura and Passavante 1993) and South Brazilian coast (Reynaldi 2000; Niencheski et al. 2007). Studies regarding the eco-physiology of microalgae and trophic interactions are more recent (Piedras and Odebrecht 2012; Rörig et al. 2017a, b).

2.3 Microalgae Biomass and Distribution

The biomass yield of the microalgae growing in the surf zone and sediments of sandy beaches depends on several factors, such as irradiance, salinity, the proximity to nutrient sources (mainly nitrogenous elements, silicate, and phosphate, but also on microelements such as iron), and the grazing intensity. However, the occurrence and growth of microalgae species are mainly determined by physical factors, such as morpho- and hydrodynamics (e.g., grain size, wave regime, currents, and tides), meteorological patterns (e.g., wind direction and intensity), the degree of beach exposure to wave action, and the orientation of the shoreline. High wave energy conditions and extensive surf zones with flat slope and fine sediments are located at one end of the beach morphodynamic gradient, the so-called dissipative state. At the other extreme are beaches with steeper slope, coarser sediment size, lower exposure, and narrow surf zone, typical of reflective states. In between these extremes, intermediate and/or alternating conditions are common (Chap. 1). Dissipative and intermediate sandy beaches with high wave energy and a high degree of exposure to wave action tend to favor the occurrence of diatom accumulations (Fig. 2.1a–c), characterizing highly productive ecosystems (McLachlan and Defeo 2018).

The phenomenon of diatom accumulations, reaching up to 10^9 cells L^{-1}, is observed in the surf zone of dissipative and intermediate sandy beaches around the world (Odebrecht et al. 2014). In Brazil, the accumulation of diatoms forming patches in the surf zone was first mentioned by Buckup (1967) in South Brazil. However, brownish to greenish patches are formed in several intermediate to dissipative sandy beaches from the Northeast to the extreme South, due to the accumulation of a few phylogenetically unrelated diatom species in the surf zone (Odebrecht et al. 2014). On the other hand, reflective beaches lack resident surf-zone microalgae populations.

Microphytobenthos is more important in sheltered beaches subjected to a large tidal range where the taxonomic composition results from an interaction of factors, but typically, muddy sediments are dominated by diatoms while more diverse populations, dominated by euglenoids and cyanobacteria, prevail in sandy sediments (Jesus et al. 2009)

2.3.1 Inorganic Nutrients, Chlorophyll a, and Primary Production

Microalgae biomass, as chlorophyll *a* (Chl) concentration, was measured in the surf zone water or sediments of 22 sandy beaches from the tropics to the warm temperate regions, while dissolved inorganic nutrients (DIN) was assessed in ten beaches and primary production in two beaches only (Table 2.1). Four beaches were located on the macrotidal beaches of the state of Pará, four on the mesotidal coast of the state of Ceará (1) and of the state of Pernambuco (3), and 14 beaches on the microtidal

Fig. 2.1 Surf-zone diatom accumulations: *Asterionellopsis guyunusae* at (**a**) Torres Beach. (Photo: Stela Valenti Raupp) and (**b**) Cassino Beach. (Photo: Ella Soares Pereira), state of Rio Grande do Sul. (**c**) *Anaulus australis* at Futuro Beach, state of Ceará. (Photo: Andrea Oliveira Rocha Franco). Note that water color is associated with distinct diatom species. (**d**) Bioluminescence at night of the dinoflagellate *Noctiluca scintillans* at Tramandaí Beach, state of Rio Grande do Sul. (Photo: Eduardo Beleske)

coastal region of Brazil extending from the state of Bahia (NE Brazil) to Rio Grande do Sul (S Brazil). Despite the great heterogeneity, some spatial and temporal patterns were recognized. In the North region, a strong association was observed between higher phytoplankton biomass (average Chl maxima 9–60 μg/L) and the rainfall season in the four dissipative beaches studied. The rainy period was associated with high values of dissolved inorganic nutrients (DIN), particularly of silicate, which presented the highest average maxima (248–332 μM), compared to values of nitrite (average min-max 0.1–2.1 μM), nitrate (1.1–16.2 μM), and phosphate (<0.1–1.2 μM) (Costa et al. 2013; Matos et al. 2012, 2016).

In the northeastern region, at the mesotidal Futuro Beach, state of Ceará, Chl was also higher during the rainy season (mean: 150.5 ± 201 μg/L; median: 68.8 μg/L) and was associated with diatom accumulations (Fig. 2.1c) frequently observed in the surf zone (81% of samples from the rainfall season) (Franco et al. 2018). However, the diatom accumulations began 2 months before the rainfall season and were triggered by changes in wave period from sea waves to swell. The phytoplankton biomass in the absence of diatom accumulations (57% of all samples) ranged between 1.4 and 12.9 μg/L (Franco et al. 2018). An opposite pattern was observed at the beaches of Brasília Teimosa, Boa Viagem, and Piedade, located in the metropolitan region of Recife, capital of the state of Pernambuco, where the highest Chl values (12–33.5 μg/L) occurred in the dry season when compared to the wet period (Chl 1.55–9 μg/L; Ferreira et al. 2010, 2013). These beaches presented lower silicate (4–62 μM) values than the beaches at the state of Pará although nitrogen (nitrite + nitrate 0.10–22.1 μM) and phosphate (0.03–1 μM) values were comparable (Ferreira et al. 2010, 2013). The oligotrophic Brazil Current in conjunction with coastal processes probably influenced the seasonal pattern of phytoplankton biomass in the surf zone of these beaches, which present sandstone reef lines and a narrow surf zone (Ferreira et al. 2010). Southward at Tamandaré (PE), Chl (0.14–2.3 μg/L) and DIN were low at two sampling periods, characteristic of oligotrophic conditions (Moura and Passavante 1994/1995). Primary production and the assimilation number (mean 5.8 mg C/mg Chl/h) tended to be inversely associated with precipitation (Moura and Passavante 1993), as observed for Chl at the beaches near Recife.

The influence of oligotrophic water was also evident southward at the beach adjacent to Real River (state of Bahia, NE Brazil), where extremely low Chl (0.08–0.24 μg/L) and DIN values (average silicate 19.6 μM, nitrate 2.6 μM, ammonium 1.5 μM, phosphate 0.7 μM) were recorded in both the rainy and dry seasons (Alves 2017). However, even lower DIN (average silicate 7.4 μM; nitrate 0.2 μM, nitrite 0.05 μM, ammonium 0.18 μM, phosphate 0.18 μM) was associated with higher Chl values (1.5–106.5 μg/L, average 16.3 μg/L) at the exposed Cururupe Beach, southward of Ilhéus, state of Bahia state (Tedesco et al. 2017). There, diatom accumulations (28% of the samples) were registered from April to August, concomitantly with southeastern and eastern winds associated with the inflow of frontal systems during autumn-winter. Diatom patches sampled daily by the local community along the Ilhéus coast at the beaches of Sargi, Ponta da Tulha, and Milionários verified the extent of this phenomenon (Tedesco et al. 2017). It is likely that, when available, DIN is quickly absorbed by the microalgae leading to the apparent low nutrient concentration. The authors suggested that the strong southern and southeastern winds suspend sediments and nutrients from the bottom favoring the higher microalgae biomass.

In the southeastern region of Brazil, at the state of São Paulo, the beaches of Bertioga, Indaiá, São Lourenço, and Boracéia presented extremely high Chl values (217–9195 μg/L) also associated with diatom patches in the years 1979 and 1980 (Tommasi and Navas-Pereira 1983). In contrast, at the nearby Araçá Beach, São Sebastião Channel, low values of both Chl (1–6 μg/L) and DIN (silicate 4–6 μM, phosphate 0.2–0.7 μM, nitrate nitrite 0.5–1.2 μM, ammonium 0.6–3.0 μM) were registered in a recent study (Ciotti et al. 2018). The authors pointed out the importance of short-term variability (daily scale), driven by changes in wind speed and direction. The contrasting results between the two studies concerning phytoplankton biomass indicate the need of combining short-term coastal observations and near-real-time regional hydrodynamic models in future studies to obtain a more complete picture of such environments (Ciotti et al. 2018).

Diatom patches with extremely high biomass (Chl up to 10,230 μg/L) were registered in the state of Santa Catarina (S Brazil), at Navegantes Beach in the year 1995 (Rörig et al. 1997), and at the beaches of Sol, Gi, Giró, and Mar Grosso at the municipality of Laguna (Netto and Meneghel 2014). In the latter, diatom accumulations were common in 2006 (70% out of 365 days) leading to biomass ten times higher in the beach deposits (average Chl 6.9 mg/cm^3, maxima of 1 g/cm^3) than outside. Southward, based on daily observations over 2 years, diatom patches were observed in 18.5% of the days at Cassino Beach, state of Rio Grande do Sul (Rörig and Garcia 2003; Rörig et al. 2006). Chl values in the integrated water column, when patches were present (4.2–352 μg/L, average 103.1 μg/L), were two orders of magnitude higher than without patches, while surface patch samples presented maxima (2220 μg/L) in the same order of those observed in the beaches of the state of São Paulo (Tommasi and Navas-Pereira 1983) and Navegantes beach (Rörig et al. 1997). Based on monthly samplings between 1992 and 2007, the frequency of biomass accumulations in the surf zone of Cassino Beach (Fig. 2.1b) varied between 10% and 40% (1992–1998) to none (1999, 2003, 2005–2007), according to environmental conditions (Odebrecht et al. 2010). A major impact on this ecosystem was due to large-scale mud deposition events in the surf zone after strong rainfall and storm events, changing DIN and microalgae biomass. Consequently, DIN concentrations varied intra- and inter-annually (ammonium 0.07–47 μM, mean 3.5 μM; nitrate + nitrite 0.09–19.4 μM, mean 2.1 μM; phosphate <0.001–10.6 μM, mean 0.9 μM; silicate 0.8–159 μM, mean 25 μM). Chl values and variability showed no clear relationship with nutrient variables in the surf zone of Cassino Beach (Abreu et al. 2017). Instead, mud deposition and storm wave periodicity associated with southerly frontal systems in southern Brazil were recognized as important factors driving biomass accumulations and variability (Rörig and Garcia 2003; Rörig et al. 2006; Odebrecht et al. 2010, 2014). In general, the low atomic N:P (mean 10.1) and N:Si (mean 0.6) ratios indicate that nitrogen is the least available DIN in this area. The input of groundwater nitrogen was associated with high Chl values (6–470 μg/L, median 109 μg/L) along an extension of 240 km northward to the Patos Lagoon outflow (Niencheski et al. 2007). Based on the fluxes of DIN in both the groundwater and surf zone, the authors estimated a primary production rate of 3 kg C m^{-2} y^{-1} for this region, equivalent to the production of salt marshes and mangroves. The authors pointed out that only about half of the production is "new production" (supported by advected nutrients from terrestrial, deep water. and atmosphere sources), as much of the DIN in groundwater discharge is from recycled seawater. Reynaldi (2000) measured the primary production of the particulate and dissolved fractions at Cassino Beach using the ^{14}C method after incubating patch water sampled in the morning and the afternoon. The transformation of the daily data resulted in a primary production rate of 7.7 kg C m^{-2} y^{-1}, that is, in the same magnitude estimated by Garcia and Gianuca (1997) and Niencheski et al. (2007). However, the estimate of Niencheski et al. (2007) is probably closer to the real value, as they considered the groundwater flux over a wide area (240 km).

2.3.2 Biodiversity

Microalgae are a heterogeneous group of autotrophic microscopic organisms that belong to distinct groups as cyanobacteria, diatoms, dinoflagellates, haptophytes, cryptophytes, raphidophytes, silicoflagellates, chlorophytes, and euglenoids. In common, these evolutionarily distant organisms grow due to the photosynthesis performed by pigments, in particular, chlorophyll *a* (Chl), in the presence of light (energy), CO_2, and nutrients. Even so, many species (especially dinoflagellates, haptophytes, and euglenoids) present mixotrophic habits, that is, in addition to photosynthesis, they feed on other organisms, with evident advantages under unfavorable environmental conditions.

Diatoms, which are mainly photoautotrophic, present the highest contribution (number of species, frequency, dominance, and abundance) in Brazilian sandy beaches. In terms of the number of species compared to other groups, diatoms ranged from 62% on the coast of the state of São Paulo (Villac, et al. 2008) to 90–99% at the state of Pará (Matos et al. 2012; Costa et al. 2011, 2013; state of Bahia, Alves 2017). Dinoflagellates were the second most important group, while silicoflagellates, cryptophytes, haptophytes, euglenoids, and chlorophytes were less expressive. An unexpected increase in the contribution of dinoflagellates from the beaches in the North and Northeast (1–12%, Costa et al. 2011, 2013; Matos et al. 2012, 2016; Sousa et al. 2008; Alves 2017; Ferreira et al. 2013; Rosevel da Silva et al. 2005) to the East and South Brazilian regions (23–34%, Rosa 1982; Kremer and Rosa 1983; Rörig et al. 2006; Villac et al. 2008; Odebrecht et al. 2010) was observed. This is likely related to the gradual decrease of the tidal range (see Chap. 1) and speed of tidal currents (Knauss 2005), which may decrease water turbulence and favor flagellates (Margalef 1978). Cyanobacteria contributed to around 20% of taxa on the beaches of Carne de Vaca (state of Pernambuco) and Azul (state of Rio Grande do Sul). In the former, chlorophytes also attained a high number, possibly due to their location near mangroves and calcareous macroalgae deposits (Campelo et al. 2001/2002). At Azul Beach, sediment was sampled in a transect from the backshore to the swash, influencing the distinct microalgae composition (Garcia-Baptista 1995).

In most studies, species identification relied on morphological characteristics based on optical microscopy, but electron microscopy analysis of the ultrastructure may also be necessary. For example, the identification of the worldwide distributed *Skeletonema* requires ultrastructure analysis; at least four species were observed on Brazilian sandy beaches (*S. costatum*, *S. pseudocostatum*, *S. tropicum*, and cf. *S. grethae*, Bergesch et al. 2009). Cryptic species, however, are only reliably identified using molecular biology tools, as shown for the common surf zone diatom *Asterionellopsis glacialis* (previously *Asterionella japonica*, *Asterionella glacialis*), which represents a species complex (Kaczmarska et al. 2014). Up to now, two species of *Asterionellopsis* have been reported in Brazilian surf zones, *A. tropicalis* and *A. guyunusae*, respectively, on the Northeast and South coast (Franco et al. 2016). The identification of the potentially toxic dinoflagellate *Alexandrium*, which was registered at beaches of the state of Rio de Janeiro, also requires genetic analysis (Menezes et al. 2007, 2018). These examples illustrate the need for basic biodiversity studies not only on sandy beaches but in the Brazilian marine environment in general.

Diatom accumulations visible to the naked eye in the surf zone are named by the local population as "verdetes" (greenish) or chocolate water (dark brownish), the color determined by the pigments of the main species, *Anaulus australis* or *Asterionellopsis* spp. (Fig. 2.1a–c). Patches of the former on beaches of the Northeast, East, and South Brazil are greenish/orange-yellowish. The latter form dark brownish patches from the North to the South Brazilian coast. Other surf diatoms like *Aulacodiscus* spp. are found in high densities in the beaches of the Northeast only (Franco 1998; Franco et al. 2018). Surf diatoms benefit food webs and diverse trophic interactions (Netto and Meneghel 2014).

Diatom accumulations in the surf zone are identified as local interferences in the succession patterns of regional neritic phytoplankton, increasing biomass while decreasing diversity (Rörig et al. 2006). Most of the time, several neritic, oceanic, and tychoplanktonic diatoms are frequent and/or abundant in the surf zone of Brazilian beaches, the most frequently cited are *Actinoptychus senarius*, *Asterionellopsis glacialis* complex, *Bacteriastrum* spp., *Bellerochea malleus*, *Campylosira cymbelliformis*, *Cerataulina pelagica*, *Chaetoceros affinis*, *C. atlanticus*, *C. brevis*, *C. compressus*, *C. curvisetus*, *C. decipiens*, *C. didymus*, *C. lorenzianus*, *C. peruvianus*, *Coscinodiscus* spp., *Cylindrotheca closterium*, *Cymatosira belgica*, *C. lorenziana*, *Dactyliosolen fragilissimus*, *Ditylum brightwellii*, *Guinardia delicatula*, *G. flaccida*, *G. striata*, *Gyrosigma* sp., *Haslea wawrikae*, *Helicotheca tamesis*, *Hemiaulus membranaceus*, *Leptocylindrus danicus*, *Melosira moniliformis*, *Navicula* spp., *Neocalyptrella robusta*, *Nitzschia* spp., *Odontella aurita*, *Trieres sinensis*, *T. mobiliensis*, *T. regia*, *Paralia sulcata*, *Pleurosigma* spp., *Proboscia alata*, *Pseudo-nitzschia* spp., *P. pungens*, *Rhizosolenia imbricata*, *R. hebetata*, *R. pungens*, *R. setigera*, *R. styliformis*, *Thalassionema nitzschioides*, *T. frauenfeldii*, *Thalassiosira* spp., *Skeletonema costatum* complex, and *Triceratium* spp.

Among cyanobacteria, *Oscillatoria princeps*, *Trichodesmium erythraeum*, *Komvophoron* sp., and *Synechococcus elongatus* are commonly cited, while main dinoflagellates are Gymnodiniales *Amphidinium*, *Gymnodinium*, *Gyrodinium*, *Alexandrium fraterculus*, *Tripos furca*, *T. fusus*, *T. hircus*, *T. kofoidii*, *Dinophysis caudata*, *Protoperidinium* spp., *Scrippsiella* sp., *Peridinium quinquecorne* Abé, *Prorocentrum* spp., *Pyrocystis lunula*, *Pyrophacus horologicum*, and *P. steinii*. The silicoflagellate *Dictyocha fibula* also presents a wide distribution. On certain occasions, brackish and freshwater diatoms (*Bacillaria paxillifera*, *Chaetoceros subtilis*, *Aulacoseira granulata, and Cyclotella meneghiniana*), cyanobacteria (*Microcystis aeruginosa* and *M. wesenbergii*), euglenophytes (*Euglena acus*), chlorophytes (*Scenedesmus*), and Zygnematales (*Mougeotia*, *Eudorina*) are associated with rainy periods.

2.4 Harmful Algae

Potentially toxic cyanobacteria, dinoflagellates, and diatoms were reported at sandy beaches in Brazil. Accumulation of *Trichodesmium erythraeum*, a genus of harmful marine cyanobacteria, was registered in the 1960s at Tamandaré Beach, state of Pernambuco, and associated with human respiratory diseases (Satô et al. 1963/1964); this is a recurrent phenomenon at some beaches in the Northeast coast. Analyzing *Trichodesmium* blooms from Brazilian offshore oligotrophic waters, Detoni et al. (2016) detected neurotoxins (saxitoxins), which can be released from the cells when the colonies are transported and deposited on sandy beaches becoming a risk to the plankton, benthos, and fish communities in the surf zone. Toxic freshwater cyanobacteria may also eventually reach beaches adjacent to river discharges, as *Microcystis* species in the Northeast (Ferreira et al. 2010) and South Brazilian coast (Matthiensen et al. 1999).

Dinoflagellates comprise the main group of harmful algae, and blooms associated with these organisms are frequently reported in newspapers as red tides, sometimes associated with environmental impacts. In Southern Brazil, at Hermenegildo Beach, state of Rio Grande do Sul, in April 1978 (Machado 1979; Rosa and Buselato 1981), suspected harmful dinoflagellates (identified as *Gyrodinium aureolum*) were detected in water sampled days after human respiratory problems begun and the death of animals (mollusks, crustaceans, fishes, and mammals) on the southern Brazilian coast was publicized. However, another possible cause of this event was the transport of chemicals toward the coast from a shipwreck, leading to controversies, which were not elucidated. Nevertheless, the presence of harmful dinoflagellates *Alexandrium* and *Dinophysis* was confirmed on several beaches in Brazil. Species of *Alexandrium* are associated with paralytic shellfish poisoning due to the production of neurotoxins, and its presence in coastal waters represents a risk mainly due to the ingestion of contaminated seafood (Menezes et al. 2007, 2018). Diarrhetic shellfish poisoning is another seafood contamination associated with the presence of *Dinophysis acuminata* complex, which is widely distributed in coastal waters, including sandy beaches (Villac et al. 2008; Odebrecht et al. 2010). The heterotrophic *Noctiluca scintillans* is considered non-toxic, but its food vacuoles may contain toxigenic microalgae, acting as a vector of phycotoxins to higher trophic levels (Escalera et al. 2007). The accumulation of *Noctiluca* in the surf zone can occasionally be observed by the naked eye either during the daytime due to their red patches (Cardoso 2012) or at night due to their bioluminescence (Fig. 2.1d). Recently, the high concentration (up to 10^6 cells L^{-1}) of *Margalefidinium* at São Sebastião, state of São Paulo, was considered unusual and caught the attention of researchers mainly for its harmful potential as a fish killer (USP 2019). Other toxic pelagic dinoflagellates, growing in offshore waters, can be transported to the beaches as a response to oceanographic processes and meteorological and climatic shifts. The attachment of epiphytic and epibenthic species to different plastic litter items indicates an alternate route for toxin transfer (Tibiriçá et al. 2019), which may reach the beaches. Besides the availability of nutrients and light, the growth of dinoflagellates depends on seeding mechanisms and the advection to the surf zone according to oceanographic currents and weather conditions, with wind calm periods favoring their success.

Among marine diatoms, species of *Pseudo-nitzschia* produce domoic acid, associated with amnesic shellfish poisoning; several potentially toxic species thrive in Brazilian coastal waters and beaches (Villac et al. 2008; Hagström et al. 2011; Tibiriçá et al. 2015). Identification of *Pseudo-nitzschia* species requires electron microscopy analysis of their ultrastructure, highlighting again the need for the qualification of human resources to face the demands of future microalgae harmful events in Brazil. It is also important to implement harmful algae monitoring programs including species identification and toxin analysis along the Brazilian coast.

2.5 Macroalgae

Arriving on Brazilian tropical beaches early in the morning, it is usual to observe abundant wrack deposits in the supralittoral, intertidal, or drifting in the surf zone. Most of this stranding organic allochthonous material is composed of macroalgae, marine multicellular autotrophic organisms represented by green (Chlorophyta), brown (Phaeophyceae), and red algae (Rhodophyta). However, recent events, known as green and golden tides, called attention to these organisms and process worldwide (Smetacek and Zingone 2013) as well in the Brazilian warm temperate coast (Martins et al. 2016), mainly due their intrinsic economic and ecological relevance (Arroyo and Bonsdorff 2016). Positive economic aspects are related, for example, to the beach cleaning that can contemplate an eventual use of the algal biomass as fertilizers (Vila Nova et al. 2014). Nevertheless, this should be weighted considering all possible impacts and their magnitude on local fish and benthic assemblages, as well as on sandy beach and circumjacent environments (see below, the case of Camboriú Beach, Santa Catarina state).

The source of these allochthonous resources can be adjacent or even a distant environment such as biogenic reefs, rocky shores, seagrass meadows, and mangroves. Macroalgae can drift a few meters of travel for thousands of kilometers, for several days or weeks (Batista et al. 2018). Storms or processes, such as bioerosion, can detach the organisms or their fragments, which can reach the shoreline. The abundance and species composition of allochthonous material may vary according to the donor-habitat productivity, the site exposure rate, beach slope, wave height, type of substratum, swash environment, and buoyancy of the drifting species. These organisms floating on the surface, drifting on the bottom, or even buried on the beaches produce an important niche for an underestimated diversity of fishes (Gomes et al. 2018), invertebrates (Ruiz-Delgado et al. 2014), and macro- and microalgae (Martins et al. 2016) and represent an important component in the carbon dynamics and the beach food web.

The presence of wrack material shaped the density of invertebrate community structure in warm temperate sites (Ruiz-Delgado et al. 2014). In the same region, a high abundance of macroalgae was directly related to the species richness, density, biomass, and diversity of fish assemblages (Gaelzer and Zalmon 2003). A similar result was observed in Brazilian tropical sandy beaches, where the high abundance of drift algae had strong effects on the fish composition, and increased density, biomass, and species richness. The high abundance of allochthonous algae observed during the autumn and winter on beaches of southeastern Brazil represents an important provision of shelter and food that could increase fish survival during settlement (Andrades et al. 2014).

Recent initiatives have investigated the use of the algal biomass as fertilizers (Sousa et al. 2022) or in animal feeding (Tejada-Tejada et al. 2021). In the Brazilian Northeast coast, for example, field and laboratory experiments on macroalgae degradation reveal that this material is an important source of nitrate, nitrite, and phosphate, which can favor phytoplankton growth, reinforcing their ecological importance (Sassi et al. 1988).

Despite the importance of macroalgae deposits, few studies characterized their composition. On the southern coast, in the southern limit of the Brazilian warm temperate province, the filamentous red algae *Aglaothamnion uruguayense* was the dominant species among the other 27 taxa that reached 8.35 tons (dry weight) in an estimated area of 52,770 m^2 (Martins et al. 2016). On the other hand, in the tropical region, the richness and abundance of drifting macroalgae are frequently higher. In seven beaches around the latitude of 13°S, 123 infrageneric taxa were identified, with 52–66 taxa per evaluated beach (Santos et al. 2013). These species totalized an abundance that varied between 2269.7 and 5765.0 $g.m^{-2}$ of wet weight (Santos et al. 2013). On the Amazonian coast, the stranding of huge *Sargassum natans* and *S. fluitans* produced the accumulation of 98,000 $g.m^{-2}$ of wet weight (Fig. 2.2), a similar value as observed in golden tide events in the Caribbean (Sissini et al. 2017). This represents the consequence of a growing event related to the formation of a great Atlantic *Sargassum* belt (GASB), observed in satellite imagery since 2011. In 2018, >20 million metric tons of *Sargassum* extended from West Africa to the Gulf of Mexico. This unusual phenomenon might be a result of shifts in upwelling events off West Africa during boreal winter and changes in Amazon River discharge nutrient concentration (Wang et al. 2019).

Brazilian beaches host important and diversified assemblages of drifting macroalgae. The management of this biomass should be considered with caution since they have ecological and evolutionary importance (Suarez-Castillo et al. 2013), with an unequivocal role in the biogeochemical carbon balance. The economic exploitation of these natural resources (Kirkman and Kendrick 1997) should also be considered with caution once its complete suppression can cause negative impacts with unknown consequences on associated communities, and ecological and evolutionary processes acting on sandy beaches and beyond. The increased frequency of macroalgae population blooms, resulting in golden and green tide events, despite all socio-economic conflicts, should be contemplated as an opportunity to innovate in ocean management, with potential mitigation of socio-environmental crises related to climate change mitigation/adaptation.

Fig. 2.2 Different aspects of wrack macroalgae, dominated by *Sargassum natans* and *S. fluitans*, that produced socio-economic impacts on the Brazilian northern coast. (**a**) Wrack of *Sargassum* on a pristine beach. (**b**) Sampling for biomass evaluation. (**c**) Wrack of *Sargassum* on an urbanized beach. (**d**) Cleaning process evidencing losses of sand and associated biodiversity. (Photos: Prof. José Eduardo Martinelli Filho (UFPA))

2.6 Beach Eutrophication: The Case of Camboriú Beach

Sandy beaches are intensely used by human civilizations for living, recreation, and extraction of natural resources. Therefore, sandy beaches are severely impacted and suffer environmental degradation, including eutrophication caused by domestic, industrial, aquaculture, or agricultural effluents. These wastes can reach the beaches through runoff (i.e., when stormwater carries contaminants) (He and He 2008); from nearby rivers and estuaries, from the sea, and also from pore water, when contaminated continental waters flow from the sub-surface to the beach environments (Niencheski et al. 2007). The tendency for eutrophication in sandy beaches is related to the intensity and quality of these flows and morphodynamic features, where sheltered beaches are more susceptible due to the more limited pollutant dispersion (Schlacher et al. 2008). Beach anthropogenic driven eutrophication (ADE) symptoms may include fertilization with organic matter and inorganic nutrients; biomass accumulation of algae and other organisms; high frequency of fecal contamination; changes in benthic, planktonic, and nektonic community structure; death of bottom-dwelling fish and invertebrates; and the occurrence of hypoxia or anoxia.

It is important to distinguish natural processes of biomass accumulation, such as the case of surf diatoms and regular wrack deposits, from cases where nuisance of invasive or uncommon organisms suddenly appear and become persistent. Worldwide, one of the most iconic cases of ADE on beaches is represented by the green tides of the macroalgae *Ulva* that occurred in Qingdao Beach, China, in 2008 (Smetacek and Zingone 2013) and those affecting the Brittany coast in France, in 2009 (Diaz et al. 2013). In these cases, the biomass does not necessarily originate from the beach, but from nearshore. Plucked and accumulated on the beach, biomass degradation increases the availability of organic matter and nutrients and the formation of anoxic zones, which even aggravate the tendency of eutrophication (Dugan et al. 2011).

In Brazil, a case has drawn attention due the intensity of the nuisance phenomena: Camboriú Beach (Balneário Camboriú, state of Santa Catarina) – a highly urbanized semi-protected beach influenced by river discharges (Fig. 2.3). Since 2002, Camboriú Beach has been undergoing dredging and jetty construction works, which altered the sedimentation pattern, resulting in increased water turbidity. According to Pezzuto et al. (2006), the sediment mobilization caused the elimination of dense populations of the filter-feeding bivalves *Tagelus plebeius* and *Anomalocardia brasiliana*. During 2003, mass mortalities of the dominant suspension feeder bivalve *Tivela mactroides* were recorded along the beach, probably as a result of suffocation by the fine sediment migration from the sediment disposal area to deeper parts of the bay. Since December 2003, wracks of non-native bryozoans *Membraniporopsis tubigera* and *Arbocuspis bellula* and epibenthic diatoms *Amphitetras antediluviana* and *Biddulphia biddulphiana* started to occur, sometimes reaching the order of hundreds of tons a day (Rörig et al. 2017b). More recently, in the years 2018 and 2019, large green algae wracks (*Bryopsis* sp.) were also recorded, interspersed with bryozoan and diatom events (C. Resgalla Jr., personal communication). The dominance of organisms varies seasonally, with bryozoans being more abundant in warmer and diatoms in colder months. Bryozoans grow on substrates near the beach and are torn off by hydrodynamics, being dragged by currents and tides to the beach face. The two chain-forming diatom species grow on bryozoans without any trophic relationships among these organisms, and the origin of the bryozoans and diatoms is not yet known (Rörig et al. 2017b; Ottonelli 2014). The absence of such phenomena on nearby beaches calls attention, and the hypothesis to explain this pattern is the unique history of sequential ecological changes at Camboriú Beach in conjunction with the sufficient pollutant load to maintain the growth of the diatoms and bryozoans, culminating in their dominance. Although the municipality provides a sewage treatment system, it is not efficient enough, and the beach receives untreated wastes from upstream cities and rural areas. The dense urban occupation (see Fig. 2.3a) and the consequent land sealing favor runoff and pore water flow, which provide additional organic matter and nutrients to the beach system.

Camboriú Beach is a typical example of how complex the degradation and eutrophication of a beach ecosystem can be. It is clear that a high level of urbanization, if unavoidable, must necessarily be accompanied by an efficient waste and sewage

Fig. 2.3 (**a**, **b**) Wracks of epibenthic diatoms and non-native bryozoans along the touristic Camboriú Beach, Santa Catarina. (**c**, **d**) Daily beach cleaning activity. (Photos: Fernando Luiz Diehl)

management system, both at the local level and at the river basins that influence the region. The greater the land use and occupation of the beach region, the more difficult and expensive the management and restoration of ecosystem health become.

2.7 Final Remarks

Primary producers on sandy beaches, whether from resident assemblages or nearshore environments, play a critical ecological and economic role in the coastal zone. The relatively narrow extent of beach ecosystems, and their position at the interface between the ocean and the continent, makes them a place of intense ecological processes. With resident primary producers, as in the case of surf diatoms and microphytobenthos assemblages, high primary productivity sustains local food chains and organic matter export. However, when high biomasses of micro- and macroalgae from other systems are transported and accumulated in the beach environment, a rapid ecological collapse can take place, with repercussions on adjacent ecosystems.

Intense urban occupation, low pollution control, and the increase in coastal works in Brazil, in conjunction with climate changes and ocean acidification, are significantly altering the natural processes on sandy beaches. The fragility of these ecosystems draws attention to the need for monitoring and integrated management practices aimed at maintaining natural processes. The negative consequences are also economic as they accelerate biodiversity losses, affecting fishing, and tourism imposing increased public resources to mitigate the problems.

As hot spots for future studies involving primary producers on sandy beaches in Brazil, the following topics are suggested: (i) basic biodiversity studies including genetic analysis, key species, and indicators; (ii) the understanding of the photoacclimation processes, primary production, and life cycles of surf diatoms; (iii) the knowledge of the flows of nutrients and organic matter in different regions and temporal scales; (iv) the evaluation of stranded biomass management reducing local impact and fostering regional green economy; (v) ecological conservation and restoration practices; (vi) macrodiagnostic and monitoring programs of pollution/contamination/toxins aiming at the evaluation of ecological and sanitary risks; (vii) integrated studies that allow understanding of the origin, fate, and impacts of the carbon/oxygen balance of beach communities and large biomasses of floating macroalgae, anomalous phytoplankton blooms, and invasive species.

Acknowledgments The authors would like to thank the Brazilian development funding agencies CNPQ, CAPES, FAPERGS, FAPESC, and Instituto Boticário for the support received in research projects and scholarships, enabling studies on the ecosystem of sandy beaches in Brazil.

References

Abreu PC, Marangoni J, Odebrecht C (2017) So close, so far: differences in long-term chlorophyll *a* variability in three nearby estuarine-coastal stations. Mar Biol Res 13:9–21

Aguiar LW, Corte-Real M (1973) Sobre uma floração de *Asterionella japonica* Cleve (1978) na costa do Rio Grande do Sul. Iheringia Ser Bot 17:18–27

Alves LD (2017) Variabilidade espaço-temporal da comunidade fitoplanctônica em um complexo estuarino de região tropical, no Nordeste do Brasil. Undergraduate Thesis Universidade Federal da Bahia (UFBA), 55 p

Andrades R, Gomes MP, Pereira-Filho GH et al (2014) The influence of allochthonous macroalgae on the fish communities of tropical sandy beaches. Estuar Coast Shelf Sci 144:75–81

Arroyo NL, Bonsdorff E (2016) The role of drifting algae for marine biodiversity. In: Ólafsson E (ed) Marine macrophytes as foundation species. CRC Press, Boca Raton, pp 100–129

Batista MB, Anderson AB, Sanches PF et al (2018) Kelps' long-distance dispersal: role of ecological/oceanographic processes and implications to marine forest conservation. Diversity 10(1):11

Bergesch M, Garcia M, Odebrecht C (2009) Diversity and morphology of *Skeletonema* species in southern Brazil, Southwestern Atlantic Ocean. J Phycol 45:1348–1352

Braga VZ (2008) Composição e variabilidade espaço-temporal do microfitobentos na praia arenosa do Village (Pontal do Sul, PR, Brasil). Undergraduate Thesis Universidade Federal do Paraná (UFPR), 60 p

Buckup L (1967) Quando o mar se torna pardacento: iodo ou plâncton marinho? Correio do Povo, Suplemento Rural:8–11

Campelo MJA, Koening ML, Passavante JZO (2001/2002) Microalgas da Praia de Carne de Vaca – Goiana, Pernambuco, Brasil. Bol Lab Hidrobiol 14(15):1–17

Cardoso LS (1993) Dinoflagelados da Ilha do Arvoredo e da Praia de Ponta das Canas, Santa Catarina, Brasil (setembro de 1991 a fevereiro de 1992): considerações taxonômicas e ecológicas. Dissertation, Universidade Federal do Rio Grande do Sul (UFRGS), 210 p

Cardoso LS (1994) Dinoflagelados da Ilha do Arvoredo e da Praia de Ponta das Canas, Santa Catarina, Brasil (setembro de 1991 a fevereiro de 1992): condições hidrológicas e lista de espécies. Iheringia Ser Bot 45:143–155

Cardoso LS (1998) Dinoflagelados da Ilha do Arvoredo e da Praia de Ponta das Canas, Santa Catarina, Brasil. Biociencias 6:3–54

Cardoso LS (2012) Bloom of *Noctiluca scintillans* (Macartney) Kofoid & Swezy (Dinophyceae) in southern Brazil. Braz J Oceanogr 60:267–271

Ciotti AM, Ferreira A, Giannini MFC (2018) Seasonal and event-driven changes in the phytoplankton communities in the Araçá Bay and adjacent waters. Ocean Coast Manag 164:14–31

Costa VB, Sousa EB, Pinheiro SCC et al (2011) Effects of a high energy coastal environment on the structure and dynamics of phytoplankton communities (Brazilian Amazon littoral). J Coast Res SI 64:354–358

Costa RM, Matos JB, Pinto KST et al (2013) Phytoplankton of a dynamic Amazon sandy beach. Proceedings 12th international coastal symposium (Plymouth, England). J Coast Res SI65:1751–1756

Detoni AMS, Costa LDF, Pacheco LA et al (2016) Toxic *Trichodesmium* bloom occurrence in the southwestern South Atlantic Ocean. Toxicon 110:51–55

Diaz M, Darnhofer I, Darrot C et al (2013) Green tides in Brittany: what can we learn about niche–regime interactions? Environ Innov Soc Trans 8:62–75

Dugan JE, Hubbard DM, Page HM et al (2011) Marine macrophyte wrack inputs and dissolved nutrients in beach sands. Estuar Coasts 34:839–850

Dutra DB, Garcia M (2016) Diatomáceas fitoplanctônicas da Praia do Gi, Laguna, Santa Catarina, Brasil. Acta Biol Catarin 3(2):102–120

Escalera L, Pazos Y, Moroño Á et al (2007) *Noctiluca scintillans* may act as a vector of toxigenic microalgae. Harmful Algae 6(3):317–320

Ferreira LC, Cunha MGGS, Koening ML et al (2010) Variação temporal do fitoplâncton em três praias urbanas do litoral do estado de Pernambuco, Nordeste do Brasil. Acta Bot Bras 24(1):214–224

Ferreira LC, Cunha MGGS, Firmo ALB et al (2013) Fitoplâncton como ferramenta de gestão ambiental na praia de Brasília Teimosa, Pernambuco, Nordeste do Brasil. Trop Oceanogr 41(1–2):120–131

Franco CRP (1998) Notas sobre florescimentos de diatomáceas na costa de Sergipe, Brasil: ocorrências de *Asterionellopsis glacialis* (Castracane) Round, *Aulacodiscus crux* Ehrenberg e *Odontella* spp. Publ Avul Centro Acad Liv Biol 2:43–47

Franco AOR, They NH, Canani LGC et al (2016) *Asterionellopsis tropicalis* (Bacillariophyceae): a new tropical species found in diatom accumulations. J Phycol 52:888–895

Franco AOR, Soares MO, Moreira MOP (2018) Diatom accumulations on a tropical meso-tidal beach: environmental drivers on phytoplankton biomass. Estuar Coast Shelf Sci 207:414–421

Gaelzer LR, Zalmon IR (2003) The influence of wave gradient on the ichthyofauna of southeastern Brazil: focusing the community structure in surf-zone. J Coast Res 35:456–462

Garcia M (2016) Taxonomy, morphology and distribution of Cymatosiraceae (Bacillariophyceae) in the littorals of Santa Catarina and Rio Grande do Sul. Biota Neotrop 16(2):e20150139

Garcia VMT, Gianuca NM (1997) The beach and surf-zone. In: Seeliger U, Odebrecht C, Castello JP (eds) Subtropical convergence environments: the coast and sea in the Southwestern Atlantic. Springer, Berlin, pp 166–170

Garcia M, Odebrecht C (2009) Morphology and ecology of *Thalassiosira* Cleve (Bacillariophyta) species rarely recorded in Brazilian coastal waters. Braz J Biol 69(4):631–637

Garcia M, Poncet GMF, Rodrigues LBV (2012) Morfologia e distribuição de *Caloneis*, *Lyrella*, *Moreneis*, *Petroneis* e *Pinnularia* em praias arenosas de Santa Catarina, Sul do Brasil. Insula Rev Bot 41:5–22

Garcia-Baptista M (1993) Psammic algae from Praia Azul, Brazil, vol 94. Bibliotheca Phycologica, J. Cramer, 167 p

Garcia-Baptista M (1995) The distribution of psammic algae on a marine beach at Praia Azul, Brazil. In: Marino D, Montresor M (eds) Proceedings of the 13th international diatom symposium. Maratea, Italy, pp 183–205

Garcia-Baptista M, Baptista LR (1992) Algas psâmicas de Jardim Beira-Mar, Capão da Canoa, Rio Grande do Sul. Rev Bras Biol 52(2):325–342

Gianuca NM (1983) A preliminary account of the ecology of sandy beaches in southern Brazil. In: McLachlan A, Erasmus T (eds) Sandy beaches as ecosystems. Junk, The Hague, pp 413–419

Gliesch R (1925) O plâncton de Torres. Egatea 10(4):294–301

Gomes DF, Caldas O, Silva EM et al (2012) Father Zimmermann (1871–1950): the first Brazilian diatomist. Diatom Res 27(3):177–188

Gomes MP, de Albuquerque CQ, Andrades R et al (2018) Influence of detached macroalgae on fish size and condition of nearshore habitats. Estuar Coast Shelf Sci 211:227–237

Hagström JA, Granéli E, Moreira MOP et al (2011) Domoic acid production and elemental composition of two *Pseudo-nitzschia multiseries* strains, from the NW and SW Atlantic Ocean, growing in phosphorus- or nitrogen-limited chemostat cultures. J Plankton Res 33(2):297–308

He LL, He ZL (2008) Water quality prediction of marine recreational beaches receiving watershed baseflow and stormwater runoff in southern California, USA. Water Res 42:2563–2573

Hyndes GA, Berdan EL, Duarte C et al (2022) The role of inputs of marine wrack and carrion in sandy-beach ecosystems: a global review. Biol Rev 97:2127. https://doi.org/10.1111/brv.12886

Jardim PFG, Cardoso LS (2013) New distribution records of Dinophyta in Brazilian waters. Checklist 9(3):631–639

Jardim PFG, Cardoso LS (2015) Contribuição taxonômica e ecológica para o gênero *Protoperidinium* Bergh (Dinophyta, Peridiniales) em águas sul-brasileiras (Tramandaí, RS, Brasil). Iheringia Ser Bot 70:229–244

Jesus B, Brotas V, Ribeiro L et al (2009) Adaptations of microphytobenthos assemblages to sediment type and tidal position. Cont Shelf Res 29(13):1624–1634

Kaczmarska I, Mather L, Luddington IA et al (2014) Cryptic diversity in a cosmopolitan diatom known as *Asterionellopsis glacialis* (Fragilariaceae): implications for ecology, biogeography, and taxonomy. Am J Bot 101(2):267–286

Kirkman H, Kendrick GA (1997) Ecological significance and commercial harvesting of drifting and beach-cast macroalgae and seagrasses in Australia: a review. J Appl Phycol 9:311–326

Knauss JA (2005) Introduction to physical oceanography. Waveland Press, Illinois, 309 p

Kremer LM, Rosa ZM (1983) Dinoflagelados do microplâncton de Tramandaí, Rio Grande do Sul, Brasil. Iheringia Ser Bot 30:3–35

Macedo CX (1999) Implantação de cultivos unialgais das diatomáceas *Asterionellopsis glacialis* e *Anaulus australis* e suas características de crescimento em diferentes salinidades. Undergraduate Thesis Universidade do Vale do Itajaí (UNIVALI), 40 p

Macedo CX (2007) Dinâmica temporal de microalgas de zona de arrebentação na Praia de Navegantes, SC. Disseratation, Universidade Federal do Paraná (UFPR), 93 p

Machado PA (1979) Dinoflagellate bloom on the Brazilian South Atlantic coast. In: Taylor DL, Seeliger HH (eds) Toxic dinoflagellate blooms. Elsevier, New York, pp 29–32

Margalef R (1978) Life-forms of phytoplankton as survival alternatives in an unstable environment. Oceanol Acta 1:493–509

Maria LS, Franco AOR, Odebrecht C et al (2016) Carbohydrates produced in batch cultures of the surf zone diatom *Asterionellopsis glacialis* sensu lato: influence in vertical migration of the microalga and in bacterial abundance. J Exp Mar Biol Ecol 474:126–132

Martins MS, Massocato TF, Horta PA et al (2016) First record of red macroalgae bloom in Southern Atlantic Brazil. Algae 31(1):33–39

Matos JB, Silva NIS, Pereira LCC et al (2012) Caracterização quali-quantitativa do fitoplâncton da zona de arrebentação de uma praia amazônica. Acta Bot Bras 26(4):979–990

Matos JB, Cardoso EHN, Pereira LCC et al (2013) Diatomáceas cêntricas da zona de arrebentação de uma ilha amazônica. Trop Oceanogr 41(1–2):54–66

Matos JB, Oliveira SMO, Pereira LCC et al (2016) Structure and temporal variation of the phytoplankton of a macrotidal beach from the Amazon coastal zone. An Acad Bras Cienc 88(3):1325–1339

Matthiensen A, Yunes JS, Codd GA (1999) Ocorrência, distribuição e toxicidade de cianobactérias no estuário da Lagoa dos Patos, RS. Rev Bras Biol 59(3):361–376

McLachlan A, Defeo O (2018) The ecology of sandy shores, 3rd edn. Elsevier, Amsterdam, 572 p

Melo NFAC, Paiva RS, Silva MMT (2005) Variação diurna da densidade planctônica na região intertidal da praia de Ajuruteua (Bragança, Pará). Bol Mus Para Emílio Goeldi Ser Cienc Nat 1(3):189–200

Menezes M, Branco S, Proença LAO et al (2007) Bloom of *Alexandrium minutum* Halim on Rio de Janeiro coast: occurrence and toxicity. Harmful Algae News 34:7–9

Menezes MM, Branco S, Miotto MC et al (2018) The genus *Alexandrium* (Dinophyceae, Dinophyta) in Brazilian coastal waters. Front Mar Sci 15:421

Moreira Filho H, Valente-Moreira IM, Mosimann RS et al (1990) Avaliação florística e ecológica das diatomáceas (Chrysophyta-Bacillariophyceae) marinhas e estuarinas nos Estados do Paraná, Santa Catarina e Rio Grande do Sul. Estudos de Biol 25:5–48

Moreira Filho H, Valente-Moreira IM, Cunha JÁ et al (1993) Estudos preliminares sobre a avaliação taxonômica e ecológica das diatomáceas (Chrysophyta-Bacillariophyceae) marinhas e estuarinas nos Estados de São Paulo e Rio de Janeiro. Estudos de Biol 22:5–52

Moreira Filho H, Eskinazi-Leça E, Valente-Moreira IM et al (1996) Avaliação taxonômica e ecológica das diatomáceas (Chrysophyta- Bacillariophyceae) marinhas e estuarinas nos Estados do Espírito Santo, Bahia, Sergipe e Alagoas. Biol Bras 6(1, 2):87–110

Moreira Filho H, Eskinazi-Leça E, Valente-Moreira IM (1999) Avaliação taxonômica e ecológica das diatomáceas (Chrysophyta-Bacillariophyceae) marinhas e estuarinas nos estados de Pernambuco, Paraiba, Rio Grande do Norte, Ceará, Piauí, Maranhão, Pará e Amapá, Brasil. Trab Oceanogr Univ Fed PE 27(1):55–90

Moura T, Passavante JZO (1993) Taxa de assimilação do fitoplâncton da Baía de Tamandaré, Rio Formoso, PE, Brasil. Boletim Técnico Científico CEPENE 1(1):7–23
Moura RT, Passavante JZO (1994/1995) Biomassa fitoplanctônicada Baía de Tamandaré, Rio Formoso, PE, Brasil. Trab Oceanogr Univ Fed PE 23:1–15
Netto SA, Meneghel A (2014) Pulse of marine subsidies: the role of surf diatom *Asterionellopsis glacialis* accumulations in structuring the meiofauna of sandy beaches. Mar Biodivers 44(3):445–457
Niencheski LFH, Windom HL, Moore WS et al (2007) Submarine groundwater discharge of nutrients to the ocean along a coastal lagoon barrier, Southern Brazil. Mar Chem 106:546–561
Odebrecht C, Segatto AZ, Freitas CA (1995) Surf-zone chlorophyll a variability at Cassino Beach, Southern Brazil. Estuar Coast Shelf Sci 41:81–90
Odebrecht C, Abreu PC, Fujita C et al (2003) The impact of mud deposition on the long-term variability of the surf-zone diatom *Asterionellopsis glacialis* (Castacane) Round at Cassino Beach, Brazil. J Coast Res SI35:486–491
Odebrecht C, Bergesch M, Rörig LR et al (2010) Phytoplankton interannual variability at Cassino Beach, Southern Brazil (1992–2007), with emphasis on the surf zone diatom *Asterionellopsis glacialis*. Estuar Coasts 33:70–583
Odebrecht C, Du Preez DR, Abreu PC et al (2014) Surf zone diatoms: a review of the drivers, patterns and role in sandy beaches food chains. Estuar Coast Shelf Sci 150:24–35
Ottonelli M (2014) Florações de diatomáceas epibênticas associadas briozoários na Enseada de Camboriú – SC: características químicas da biomassa, ecofisiologia das microalgas envolvidas e hipóteses sobre as causas. Dissertation, Universidade Federal de Santa Catarina (UFSC), 85 p
Pereira da Costa FA, Moreira MOP, Sales MF (1997) Monitoramento dos florescimentos de fitoplâncton na Praia do Futuro, Fortaleza, CE: primeiro registro do gênero *Anaulus* Ehrenberg como responsável pelo fenômeno na costa Nordeste. 48° Congresso Nacional de Botânica. Sociedade Botânica do Brasil, Crato, Brazil, p 22
Pereira da Costa FA, Moreira MOP, Sales MF (1998) Florescimentos de diatomáceas no litoral leste de Fortaleza, Ceará (Praia do Futuro). 5° Encontro Brasileiro de Ecotoxicologia, 1° Colóquio Brasileiro de Algas Nocivas. Universidade do Vale do Itajaí-UNIVALI, Itajaí, Brazil, p 67
Pereira LCC, Jiménez JA, Koening ML et al (2005) Effect of coastline properties and wastewater on plankton composition and distribution in a stressed environment on the north coast of Olinda, PE (Brazil). Braz Arch Biol Technol 48(6):1013–1026
Pezzuto PR, Resgalla C Jr, Abreu JGN, Menezes JT (2006) Environmental impacts of the nourishment of Balneário Camboriú Beach, SC, Brazil. J Coast Res SI39:863–868
Piedras FR, Odebrecht C (2012) The response of surf-zone phytoplankton to nutrient enrichment (Cassino Beach, Brazil). J Exp Mar Biol Ecol 432-433:156–161
Reynaldi S (2000) Efeito da diatomácea de zona de arrebentação *Asterionellopsis glacialis* (Castracane) Round, sobre o crescimento bacteriano na praia do Cassino, RS, Brasil. Dissertation, Universidade Federal dio Rio Grande (FURG), 70 p
Rezende KRV, Brandini FP (1997) Variação sazonal do fitoplâncton na zona de arrebentação da Praia de Pontal do Sul (Paranaguá, Brasil). Nerítica 11:49–62
Ribeiro FCP, Senna CSF, Costa BO (2007) Análise de diatomáceas e relação água doce/água salgada na planície costeira de Soure, PA. In: Anais do XI Congresso da Associação Brasileira de Estudos do Quaternário, ABEQUA
Ribeiro FCP, Senna CSF, Torgan LC (2008) Diatomáceas em sedimentos superficiais na planície de maré da praia de Itupanema, estado do Pará, Amazônia. Rodriguésia 59(2):309–324
Rörig LR (1997) Acumulações da diatomácea *Asterionellopsis glacialis* e sua relação com variáveis ambientais na Praia do Cassino, Rio Grande, RS. Dissertation, Universidade Federal do Rio Grande (FURG), 126 p
Rörig LR, Garcia VMT (2003) Accumulations of the surf-zone diatom *Asterionellopsis glacialis* (Castracane) Round in Cassino Beach, Southern Brazil, and its relationship with environmental factors. J Coast Res 19:167–177

Rörig LR, Resgalla C Jr, Pezzuto PR et al (1997) Análise ecológica de um processo de acumulação da diatomácea *Anaulus* sp. na zona de arrebentação da Praia de Navegantes (Santa Catarina, Brasil). In: Absalão R, Esteves AM (eds) Ecologia de Praias Arenosas do litoral brasileiro. Oecologia Brasiliensis 3. Federal University of Rio de Janeiro, pp 29–43

Rörig LR, Almeida TCM, Garcia VMT (2006) Structure and succession of the surf-zone phytoplankton in Cassino Beach, Southern Brazil. J Coast Res SI39:1246–1250

Rörig RR, Honorato M, Itokazu AG et al (2017a) Ecophysiological and biochemical variation of the surf zone diatom *Asterionellopsis glacialis* sensu lato from Santa Catarina, Southern Brazil. Braz J Oceanogr 65:702–715

Rörig LR, Ottonelli M, Itokazu AG et al (2017b) Blooms of bryozoans and epibenthic diatoms in an urbanized sandy beach (Balneário Camboriú – SC – Brazil): dynamics, possible causes and biomass characterization. Braz J Oceanogr 65:685–701

Rosa ZM (1982) Diatomáceas marinhas e estuarinas de Tramandaí, Rio Grande do Sul, Brasil. Iheringia Ser Bot 29:49–145

Rosa ZM, Buselato TC (1981) Sobre a ocorrência de floração de *Gyrodinium aureolum* Hulburt (Dinophyceae) no litoral sul do estado do Rio Grande do Sul, Brasil. Iheringia Ser Bot 28:169–179

Rosevel da Silva M, Silva-Cunha MGG, Feitosa FAN et al (2005) Estrutura da comunidade fitoplanctônica na Baía de Tamandaré (PE, Nordeste do Brasil). Trop Oceanogr 33(2):163–181

Ruiz-Delgado C, Vieira JV, Veloso VG et al (2014) The role of wrack deposits for supralittoral arthropods: an example using Atlantic sandy beaches of Brazil and Spain. Estuar Coast Shelf Sci 136:61–71

Santana DS, Paiva RS, Melo NFAC (2005) Diatomáceas cêntricas da região entre marés da praia de Ajuruteua (Bragança, Pará). Bol Mus Para Emílio Goeldi Ser Cienc Nat 1(3):109–116

Santos GN, Nascimento OS, Pedreira FA et al (2013) Análise quali-quantitativa das algas arribadas do norte do estado da Bahia, Brasil. Acta Bot Bras 38:13–24

Sassi R, Kutner MBB, Moura GF (1988) Studies on the decomposition of drift seaweed from the northeast Brazilian coastal reefs. Hydrobiologia 157(2):187–192

Satô S, Paranaguá MN, Eskinazi E (1963/1964) On the mechanism of red tide of *Trichodesmium* in Recife, Northeastern Brazil, with some considerations of the relation to the human disease, "Tamandaré fever". Trab Oceanogr Univ Fed PE 5(6):7–49

Schlacher TA, Schoeman DS, Dugan J et al (2008) Sandy beach ecosystems: key features, sampling issues, management challenges and climate change impacts. Mar Ecol 29(1):70–90

Sissini MN, Barreto MBBB, Széchy MTM et al (2017) The floating *Sargassum* (Phaeophyceae) of the South Atlantic Ocean–likely scenarios. Phycologia 56:321–328

Smetacek V, Zingone A (2013) Green and golden seaweed tides on the rise. Nature 504(7478):84

Sousa EB, Costa VB, Pereira LCC et al (2008) Microfitoplâncton de águas costeiras amazônicas: Ilha Canela (Bragança, PA, Brasil). Acta Bot Bras 22(3):626–636

Sousa EB, Costa VB, Pereira LCC et al (2009) Variação temporal do fitoplâncton e dos parâmetros hidrológicos da zona de arrebentação da Ilha Canela (Bragança, Pará, Brasil). Acta Bot Bras 23(4):1084–1095

Sousa F, Martins M, Sousa B et al (2022) The potential of beach wrack as plant biostimulant to mitigate metal toxicity: mineral composition, antioxidant properties and effects against Cu-induced stress. J Appl Phycol 34(1):667–678

Suarez-Castillo AN, Riosmena-Rodriguez R, Rojo Amaya M et al (2013) Bosques de algas pardas en el golfo de California: *Sargassum*, unhabitat esencial. Biodiversitas 108:12–16

Tahim EF, Verde NGL, Eskinazi-Leça E (1990) Florescimento de algas planctônicas na Praia do Futuro (Fortaleza, CE). Anais do IV Encontro Brasileiro de Plâncton. Editora Universitária da UFPE, Recife, pp 117–121

Tedesco EC (2006) Estudo das diatomáceas da zona de arrebentação da Praia do Cururupe, Ilhéus, Ba. Dissertation, Universidade Estadual de Santa Cruz (UESC), 95 p

Tedesco EC, Susini SMMS, Pompeu M et al (2017) Low-latitude accumulation of the surf-zone diatoms *Anaulus australis* Drebes & Schulz and *Asterionellopsis glacialis* (Castracane) Round species complex in the eastern coast of Brazil. Braz J Oceanogr 65(2):324–331

Tejada-Tejada P, Rodriguez-Rodriguez Y, de Francisco LER et al (2021) Lead, chromium, nickel, copper and zinc levels in *Sargassum* species reached the coasts of Dominican Republic during 2019: a preliminary evaluation for the use of algal biomass as fertilizer and animal feeding. Tecnol y Cienc del Agua 12(3):124–163

Tibiriçá CEJA, Fernandes LF, Mafra LL Jr (2015) Seasonal and spatial patterns of toxigenic species of *Dinophysis* and *Pseudo-nitzschia* in a subtropical Brazilian estuary. Braz J Oceanogr 63:17–32

Tibiriçá CEJA, Leite IP, Batista TVV et al (2019) *Ostreopsis* cf. *ovata* bloom in Currais, Brazil: phylogeny, toxin profile and contamination of mussels and marine plastic litter. Toxins 11:446

Tommasi LR, Navas-Pereira D (1983) Nota sobre a ocorrência de florescimentos de diatomáceas, na Baía de Santos (SP) e adjacências (Estado de São Paulo, Brasil). Cien Cult 35(4):507–512

USP (2019) Jornal da USP. https://jornal.usp.br/ciencias/ciencias-biologicas/mare-vermelha-traz-microalga-incomum-ao-canal-de-sao-sebastiao/. Accessed 29 Sept 2019

Vila Nova LLM, Costa MMS, Costa JG, Amorim ECS, Guedes EAC (2014) Utilização de "algas arribadas" como alternativa para adubação orgânica em cultivo de Moringa oleifera Lam. Revista Ouricuri 4(3):68–81

Villac MC, Noronha VAP (2008) The surf-zone phytoplankton of the state of São Paulo, Brazil. I. Trends in space-time distribution with emphasis on *Asterionellopsis glacialis* and *Anaulus australis* (Bacillariophyceae). Nova Hedwigia Beih 133:115–129

Villac MC, Cabral-Noronha VAP, Pinto TO (2008) The phytoplankton biodiversity of the coast of the state of São Paulo, Brazil. Biota Neotrop 8(3):151–173

Wang M, Hu C, Barnes BB et al (2019) The great Atlantic *Sargassum* belt. Science 365(6448):83–87

Zavala-Camin LA, Yamanaka N (1980) Notas sobre um caso de mortalidade de peixes, ocorrida em Itanhaém, São Paulo, Brasil. Bol Inst Oceanogr 29(2):377

Chapter 3
Meiofauna Biodiversity

Tatiana Maria, André Esteves, André Garraffoni, Fabiane Gallucci, Adriane Pereira Wandeness, Beatriz Pereira Cunha, Gustavo Fonseca, Sergio Netto, and Maikon Di Domenico

3.1 Introduction

Small animals belonging to the meiofauna, defined as organisms smaller than 0.5 mm and larger than 0.044 mm (Giere 2009), are at the threshold of the optical resolution for routine identification (Curini-Galletti et al. 2012). Due to the inherent difficulties in working with microscopic organisms, our knowledge of meiofauna biodiversity is unfortunately still scarce. This group includes 25 metazoan phyla. Among them, few are known to be exclusively meiofaunal (i.e., Loricifera,

T. Maria
Departamento de Ecologia e Recursos Marinhos, Universidade Federal do Estado do Rio de Janeiro, Rio de Janeiro, RJ, Brazil

A. Esteves · A. P. Wandeness
Departamento de Zoologia, Universidade Federal de Pernambuco, Centro de Biociências, Recife, PE, Brazil

A. Garraffoni
Departamento de Biologia Animal, Instituto de Biologia, Universidade Estadual de Campinas, Campinas, SP, Brazil

F. Gallucci · G. Fonseca
Instituto do Mar, Universidade Federal de São Paulo, Santos, SP, Brazil

B. P. Cunha
Instituto de Biologia, Programa de Pós-graduação em Biologia Animal, Universidade Estadual de Campinas, Campinas, SP, Brazil

S. Netto
Laboratório de Ciências Marinhas, Universidade do Sul de Santa Catarina, Tubarão, SC, Brazil

M. Di Domenico (✉)
Centro de Estudos do Mar, Universidade Federal do Paraná, Pontal do Paraná, Pontal do Sul, PR, Brazil
e-mail: didomenico@ufpr.br

A. C. Z. Amaral et al. (eds.), *Brazilian Sandy Beaches*, Brazilian Marine Biodiversity, https://doi.org/10.1007/978-3-031-30746-1_3

Tardigrada, Kinorhyncha, and Gastrotricha), while many are considered temporary meiofauna because they belong to this group only during early life stages (e.g., Annelida, Arthropoda, and Mollusca) (Giere 2009). Meiofauna organisms play a key role in the functioning of sandy beach ecosystems by being involved in the remineralization processes and in the energy transfer for higher trophic levels and across ecosystems (McLachlan and Defeo 2018; Schratzberger and Ingels 2017). Meiofauna species have also been used to monitor ecosystem health (e.g., Gallucci et al. 2015), as they are known to respond in the short and long term to anthropogenic stress (Fonseca and Gallucci 2016).

Ecological meiofaunal studies on sandy beaches started at the end of the last century and increased over the last decade (Table 3.1). Despite the vast extension of the Brazilian coastline, which includes 17 states and more than 4000 beaches (Chap. 1), meiofaunal investigations were performed only on nine states (Espírito Santo, Pará, Paraná, Pernambuco, Rio de Janeiro, Rio Grande do Sul, Santa Catarina, and São Paulo) (Fig. 3.1). Most of these investigations (19/42) are restricted to the Southeast (Espírito Santo, Rio de Janeiro, and São Paulo) and South regions (12 studies in Santa Catarina and Rio Grande do Sul). Meiofaunal investigations in the Northeast and Northern regions account for only 10 and 4 of the 42 studies. The small number of studies performed on the Northern region is associated with the recent exploitation of this geographic region by meiobenthologists, which initiated their studies at the beginning of this century (Fig. 3.1).

This chapter presents the biodiversity of the most studied meiofaunal groups along the Brazilian coastline and provides general information about their occurrence and distribution patterns. A compendium of papers published in indexed scientific journals, considering articles dealing with community studies and published until early 2022, is provided (see Table 3.1).

3.2 Spatial Patterns

The occurrence and distribution of dominant meiofaunal groups in sandy beaches, such as nematodes and copepods, are correlated to beach morphodynamics but also depend on biological interactions and environmental changes related to processes of urbanization and tourism (Medeiros 1987; Moellmann and Corbisier 2003; Rodríguez et al. 2003; Gheskiere et al. 2006; Kotwicki et al. 2005; Moreno et al. 2006). Meiofaunal organisms usually surpass the macrofauna density and occur on all sandy beach morphodynamic types; however, several studies indicate that interstitial fauna is more representative in reflective sandy beaches (while also occurring in dissipative ones) than are macrofauna and bacteria (Moore and Bett 1989), and this is one of the reasons why this group, not considering species alone but the community structure, is claimed as the best bioindicator for sandy beach impacts in this reflective beaches.

The environmental optimum for the occurrence of interstitial fauna, in terms of diversity and abundance, is a balance between physical and chemical features of the

Table 3.1 Overview of meiofauna studies after 1970 in sandy beaches, with an indication of the focus of the study and the number of beaches included

ID	Author	Focus	Fauna	N° beaches	State	Main finding
1	Silva et al. (1991)	DT	M	1	RJ	Granulometry and temperature determine the spatial distribution
2	Bezerra et al. (1996)	DT/T		1	PE	Higher meiofauna densities during the dry season and at 5–10 cm layer
3	Bezerra et al. (1997)	DT	M	1	PE	Granulometry just influences the distribution of harpacticoids copepodes and ostracodes
4	Wandeness et al. (1997)	EI	M	1	RJ	Meiofauna densities and composition varies according to the source of pollution
5	Esteves et al. (1998)	DT/T	M	1	RJ	Upper layer (0–15 cm) showed higher meiofauna density and temperature and salinity are responsible for the temporal distribution
6	Victor-Castro et al. (1999)	T	Ta	1	PE	Sedimentologic processes directly influence the distribution of *Batillipes pennaki*
7	Curvelo and Corbisier (2000)	BI	M	1	SP	Higher meiofauna density and biomass may be explained by the habitat complexity created by the algae
8	da Rocha et al. (2002)	T	Ta	3	PE	Specimens of *Batillipes pennaki* showed higher density at the upper mesolittoral, while they were almost absent in the infralittoral
9	Corgosinho et al. (2003)	DT	M	2	PR	High nematode densities in the sublittoral of exposed sandy beaches are explained by their high vertical migratory capacity or high resistance to turbidity
10	Moellmann and Corbisier (2003)	P	M	2	SP	Nematodes migrate to the deeper layers of the sediment in intense trampling areas
11	Oliveira and Soares-Gomes (2003)	P	M	1	RJ	No relationship between sewage disposal and meiofauna
12	Souza-Santos et al. (2003)	T	M	1	PE	Environmental factors (e.g., grain size, Chl-*a*, phaeopigments, sediment skewness) are the main factors structuring the meiofauna community seasonally
13	Todaro and Rocha (2004)	DT	G	23	SP	A diversified gastrotrich community occurred in beaches characterized by medium or fine sand with little detritus and clear water

(continued)

Table 3.1 (continued)

ID	Author	Focus	Fauna	N° beaches	State	Main finding
14	Pinto and Santos (2006)	DT/T	M	1	PE	Meiofauna spatial distribution is structured by environmental factors (e.g., Chl-a, RPD layer and grain size sorting. High densities occurred in the wet period of the year
15	Albuquerque et al. (2007)	DT/T	M	1	RJ	Highest meiofauna densities in the retention zone at 0–10 cm layer. Seasonality is taxa dependent
16	Netto et al. (2007)	BI	M/N	1	SC	Correlation between pigments (chla and the ratio chla/feo a) and meiofauna richness and abundance
17	Di Domenico et al. (2009)	DT	P	6	SC, PR	High annelids diversity on exposed beaches than any other morphodynamic states
18	Maria et al. (2008)	DT	N	3	RJ	Highest richness in very coarse sandy beach
19	Gomes and Rosa Filho (2009)	DT/T	M	1	PA	Highest meiofauna densities were found in the mid-intertidal zone. No seasonality
20	Verçosa et al. (2009)	DT/T	Ta	1	PE	*Stygarctus Bradypus* distribution is neither influenced by the size of the sediment grains nor by the presence of interstitial waters and the specimens preferably colonizing the upper mid-littoral of the studied beach
21	Rosa Filho et al. (2011)	DT	M	3	PA	Highest density and diversity in the mid-tide level; the existence of zonation
22	Maria et al. (2013a)	DT	N	2	RJ	Highest density in the upper tide level; gravel is the most important variable that explains the nematode community
23	Maria et al. (2013b)	BI	N	1	SP	Annelids did not affect the colonization of azoic sediments by nematodes
24	Di Domenico et al. (2013)	DT	A	7	SC, SP, RJ	Species of Protodrilidae showed preferences for sheltered and exposed beaches
25	Di Domenico et al. (2014a)	DT	A	7	SC, PR	*Saccocirrus pussicus* distribution correlated with the presence of a steep slope, large waves, and coarse sand
26	Di Domenico et al. (2014b)	DT	A	11	SC, PR, SP, RJ	*Saccocirrus pussicus* prefers reflective beaches, and *Pharyngocirrus gabriellae* sheltered intertidal and shallow areas
27	Netto and Meneghel (2014)	BI	M	3	PR	Surf zone diatoms are an important food source for the meiofaunal organism of high-energy sandy beaches

(continued)

Table 3.1 (continued)

ID	Author	Focus	Fauna	N° beaches	State	Main finding
28	Venekey et al. (2014a)	DT	N	1	PE	Lowest density in dry months; no influence of the tidal cycle in the nematode density
29	Venekey et al. (2014b)	DT	N	1	PE	Seasonal and spatial variation of nematode communities in a tropical sandy beach
30	Martins et al. (2015)	DT	M/N	2	SC	Lowest densities of meiofauna and nematodes were recorded for the upper stratum (0–10 cm) in reflective sandy beaches
31	Garraffoni et al. (2016)	DT	G	28	SP	Gastrotrichs composition and beta diversity were correlated with tidal zone, location, and average grain size, and the highest turnover of species was expected on islands
32	Felix et al. (2016)	EI	M	2	SC	Beaches with lower environmental quality possess higher oligochaete densities
33	Gusmão et al. (2016)	EI	M/A	3	SC and RJ	Among different meiofaunal taxa, *Saccocirrus* egest microfibers without physical injury
34	Tarragô and Ozorio (2017)	EI	M	1	RS	Storm surges change the meiofauna distribution along with a beach profile
35	Santos and Venekey (2018)	DT/T	M/N	4	ES[a]	Higher meiofauna and nematodes densities during summer as well as higher nematode densities in high tidal level
36	Santos et al. (2019a, b)	EI	N	7	RJ	Nematode biological indexes did not indicate the ecological status of sandy beaches because the nematode distribution respond to a different sorting coefficient
37	Baia and Venekey (2019)	DT	M/N	1	PA	Genera diversity and richness increased toward the low tide mark and lower richness is found in the upper layer
38	Mello et al. (2019)	DT	K	2	SC, PR	The distribution pattern of *Franciscideres kalenesos* was related to the surf zones in intermediate/dissipative beaches
39	Herranz et al. (2019)	DT	K	1	RJ	*Cateria styx* adaptations for mechanical adhesion, as the enigmatic dorsal organ, through friction and interlocking, in an interstitial sandy beach habitat is inferred to be adhesive

(continued)

Table 3.1 (continued)

ID	Author	Focus	Fauna	N° beaches	State	Main finding
40	Tilbert et al. (2019)	DT/T	Ta	4	RN	Granulometry was the most important environmental factor in the meiofaunal spatial structure in tropical estuaries and *Batillipes pennaki* and *B. dandarae* were associated with medium sand
41	Santos et al. (2021)	EI	M/N	3	PA	Tardigrades, copepodes, and some nematodes species are more prone to suffer the effects of recreational activities in sandy beaches. However, meiofauna density and richness show a very fast recover after cessing recreational activities
42	Herranz et al. (2021a)	DT	K	2	RJ, PR	This study showed the neuroanatomy of aberrant kinorhynchs with an elongated, worm-like body plan and less distinct segmentation, including *Cateria styx* and *Franciscideres kalenesos*
43	Herranz et al. (2021b)	DT	K	2	RJ, PR	This study provided a comprehensive and comparative myoanatomical study distantly related kinorhynchs worm-like species, including *Cateria styx* and *Franciscideres kalenesos*
44	Brustolin et al. (2022)	DT	N	1	SP	This study deals with metacommunity concept and indicates the existence of high connectivity among sandy beach locations in within the same beach

Adapted and updated from Maria et al. (2016)
DT distribution pattern, *T* temporal distribution, *BI* biological interaction, *P* pollution, *EI* environmental impact, *M* meiofauna, *N* nematodes, *G* gastrotrichs, *A* interstitial annelids (polychaeta), *K* kinorhyncha, *Ta* Tardigrada; State: *PA* Pará, *PE* Pernambuco, *PR* Paraná, *SC* Santa Catarina, *SP* São Paulo, *RJ* Rio de Janeiro, *RN* Rio Grande do Norte, *RS* Rio Grande do Sul
[a]Volcanic sandy beaches located on an island

interstitial environment such as hydrodynamic energy and organic matter input (McLachlan and Defeo 2018). These conditions are more important for large-scale patterns of distribution whereas biological interactions have more influence over fine-scale distribution patterns (McLachlan and Defeo 2018).

Meiofaunal spatial variability has been studied at three spatial scales: macro-, meso-, and microscale. Macroscale distribution deals with studies of patterns for different latitudes and types of beaches, whereas mesoscale distribution implies along- and across-shore distribution patterns. Microscale studies, in turn, investigate changes in meters or centimeters. Nevertheless, for meiofauna, comparative studies across scales are lacking, and it is still unclear which of these scales of variation is the most important in structuring meiofaunal communities. At the larger scales (meso and macro), the meiofauna distribution is mainly driven by differences

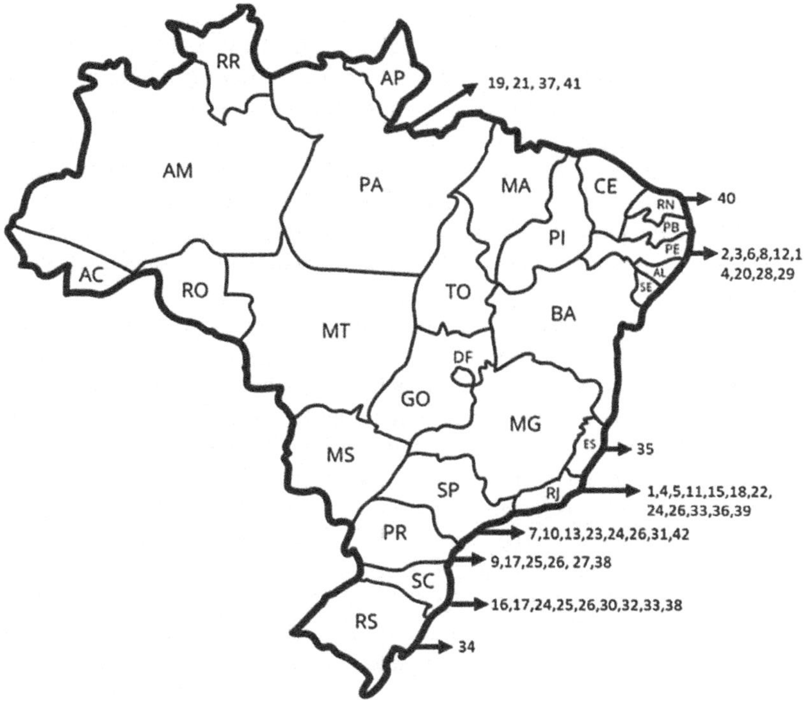

Fig. 3.1 Geographical distribution of the meiofauna ecological studies along the Brazilian coast. Numbers refer to the studies indicated in Table 3.1

in granulometry (Maria et al. 2016). However, despite the large extension of the Brazilian coastline, no study has investigated the influence of the latitude and the associated environmental variables (e.g., changes in tides and temperatures) on the distribution of the meiofauna. Such a study would be challenging since the morphodynamics of the beaches varies along the coast with differences in wave and tide regimes. For instance, as presented in Chap. 1, northern coast beaches are tide-dominated or tide-modified and heavily influenced by the Amazon river, while southeastern beaches are wave-dominated, being embayed or pocketed due to the presence of many bedrock headlands; coral and rock reefs occur on the sublittoral of northeastern beaches, reducing the wave energy that reaches the beach face (Short and Klein 2016). While the existence of a full range of beach types along the Brazilian coast makes the latitudinal comparison more difficult, it also provides an excellent scenario to develop studies aiming at comprehending different factors driving large-scale shifts in meiofauna community metrics (i.e., richness, abundance, and biomass).

Higher meiofauna density (ca. 2000 ind/cm^2) is usually found in exposed tropical and subtropical sandy beaches (Corgosinho et al. 2003; Rosa Filho et al. 2011). There is an expected trend of high diversity with increasing sediment grain size;

however, it may not apply to all meiofauna groups. It is reported that each nematode genus prefers a certain grain size (Fonseca et al. 2014) and sorting coefficient (Santos et al. 2019a). Higher species diversity of gastrotrichs is reported for beaches characterized by medium and fine sand on the São Paulo coast (Todaro and Rocha 2005), whereas higher annelids diversity is found in reflective beaches (i.e., coarse sediment) due to the inability of some interstitial species to live in compacted sediments (Di Domenico et al. 2009). Along the Brazilian coast, nematodes, copepods, and annelids are the most abundant meiofaunal groups. Kinorhynchs may also reach high numbers ecosystems composed of muddy sediments (Herranz et al. 2019; Mello et al. 2019).

In the mesoscale (i.e., within beaches), the distribution of meiofauna organisms is compartmentalized in zones on the across-shore gradient (McLachlan and Defeo 2018). Meiofaunal assemblages usually exhibit a patchy distribution in the interdial region with high density and diversity in the middle (Gomes and Rosa Filho 2009), following the intermediate disturbance hypothesis (Gheskiere et al. 2004). This across-shore distribution is also affected by anthropic influence. Higher diversity of nematodes, for example, is usually recorded in the lower beach level under pristine conditions (Venekey et al. 2014a, b). However, under different degrees of organic pollution, higher diversity may be observed on upper beach levels (Maria et al. 2013a) likely due to the influence of polluted waters.

The microscale distribution is commonly examined within the sediment column. Meiofaunal organisms can penetrate as deep as the oxygenized condition exists. Therefore the vertical distribution of meiofauna is driven by the balance between oxygenation and water drainage, sometimes even penetrating the groundwater level in reflective sandy beaches (McLachlan and Defeo 2018). Biological interactions among different nematode species are very pronounced on a scale of a few centimeters and may also influence microscale patterns (Maria et al. 2012). Some studies included sediment stratification to understand this fine-scale of distribution; yet, studies assessing this scale usually stratified the sediment in thick layers of 15 cm (Esteves et al. 1998), 10 cm (Albuquerque et al. 2007; Martins et al. 2015), or 2–5 cm (Baia and Venekey 2019), compromising the evidence of biological interactions, which are expected to occur in very fine scales of 1 cm (or less) deep (Maria et al. 2012). Meiofauna densities vary within the sediment and may be higher in the superficial layers in sheltered conditions (Baia and Venekey 2019) or deeper layers in more exposed sandy beaches (Martins et al. 2015), indicating the importance of hydrodynamics in regulating meiofauna densities.

3.3 Temporal Patterns

The temporal variability of the Brazilian sandy beach meiofauna has gained much less attention than the spatial patterns of distribution (Table 3.1). Temporal variation of the meiofauna is also expected to occur at multiple scales, from hours to days, seasons, and years. Environmental variables such as tides, temperature, and salinity also vary at multiple temporal scales and partly explain meiofauna temporal patterns of distribution (Esteves et al. 1998; Maria et al. 2013a). Nevertheless, temporal patterns are not consistent across the country. For tropical beaches on the northeastern coast, higher richness and abundance were found during dry seasons when low levels of salinity are observed (Gomes and Rosa Filho 2009). On the other hand, the species richness and abundance of meiofauna were higher during the rainy seasons on the southeastern coast, when temperatures were relatively higher (Albuquerque et al. 2007).

Interestingly, a study that assessed spatial and temporal variations showed that changes in the distribution patterns of interstitial annelids are related to shifts in the morphodynamic state of the sandy beaches between summer and winter (Di Domenico et al. 2009). Beaches classified as intermediate showed different interstitial annelids' associations between seasons. Species commonly found in dissipative beaches were observed during summer (e.g., *Protodriloides* sp. and *Hesionidens* sp.), while associations typical of reflective beaches were recorded during winter (e.g., *Saccocirrus* sp. and protodrilids). These associations changed seasonally related to the relative tide range and rainfall volume (Di Domenico et al. 2009).

3.4 Biodiversity

3.4.1 Copepoda

Among meiofaunal groups present on sandy beaches, Copepoda Harpacticoida may be considered one of the most abundant (McIntyre 1968), and its contribution to meiofauna abundance varies from 4% to 95% of the total meiofauna present in the sediment (Esteves et al. 1998; Albuquerque et al. 2007; Kihara et al. 2011). The first studies with this group on Brazilian beaches focused on taxonomic descriptions (Carvalho 1952; Jakobi 1953a, b, 1954, 1956; Jakobi and Nogueira 1960; Rouch 1962). Subsequent investigations continued tackling taxonomic aspects but also included ecological components such as the abundance and diversity of harpacticoids (Wandeness 1998; Corgosinho 2012). These studies were performed mainly in southeastern Brazil, with the highest number of studies done on the coast of the state of São Paulo (Masunari 1987; Alvarez 1988; Kihara 2003; Kihara and Rocha 2007; Kihara et al. 2011; Rocha et al. 2011; Björnberg and Kihara 2013; Björnberg 2010, 2014), and some in the northeastern coast (Wandeness et al. 1998; Silva 2006).

Very few studies along Brazilian beaches provided data on the composition of Copepoda Harpacticoida assemblages for sandy environments (Table 3.2). Up until now, a reduced number of taxa composed of a few genera have been registered in this environment. Among these few genera, *Arenopontia* – a typically interstitial taxon – may be categorized as a common genus of sandy beaches (Table 3.2). Overall, Copepoda Harpacticoida shows higher diversity in dissipative beach types

Table 3.2 Studies in Brazilian sandy beach environments showing the species composition of Copepoda Harpacticoida

Study	State	Type environment	Sampling information	Faunistic composition[a]
Wandeness et al. (1998)	RJ	Exposed Sandy beach – reflective	Total area (m^2) = 0.45	*Arenopontia acantha* accepted as *Neoleptastacus acanthus* (Chappuis 1954)
			One station/ intertidal	*Paraleptastacus spinicauda* accepted as *Paraleptastacus spinicauda us* (Scott and Scott 1895)
			5 strata	*Leptastacus aberrans* accepted as *Archileptastacus aberrans* (Chappuis 1954)
			12 sampling (1 per month)	*Eoschizopera* sp.
				Laophontina sp.
Wandeness et al. (1998)	PE	Estuarine sandbank	Total area (m^2) = 0.04	*Paraleptastacus spinicauda* accepted as *Paraleptastacus spinicaudus* (Scott and Scott 1895)
			One station/ intertidal	*Leptastacus rostratus* accepted as *Schizothrix rostratus* (Nicholls 1939)
			Without strata	*Paralaophonte (Paralaophonte) asellopsiformis* (Lang 1965)
			Only one sampling	*Laophonte cornuta* (Philippi 1840)
				Halectinosoma finmarchicus (Scott 1903) *nomen dubium* (according to Clément and Moore 2000)
				Harpacticus spinilosus (Lang 1965)
				Paramphiascella robinsonii (Scott 1902)
Silva (2006)	PE	Exposed Sandy beach – dissipative	Total area (m^2) = 0.15	*Arenopontia* (*Neoleptastacus*) *indica* accepted as *Neoleptastacus indicus* (Rao 1967)
			One station/ intertidal	*Arenopontia* (*Arenopontia*) *intermedia* accepted *Onycophontia intermedia* as (Rouch 1962)
			4, 5, or 8 strata	*Arenosetella germanica germanica* (Kunz 1937)
			3 sampling	*Noodtiella hoodensis* (Mielke 1979)
				Pseudoleptomesochrella bisetosa (Lindgren 1975)
				Attheyella (*Canthosella*) sp.[b]
				Ceuthonectes sp.[b]
				Diarthrodella sp.
				Elaphoidella sp.[b]
				Eoschizopera (*Eoschizopera*) sp. accepted as *Eoschizopera* sp.
				Oniscopsis sp.
				Schizoperopsis sp. accepted as *Schizopera* sp.
				Schizoperopsis (*Psammoschizoperopsis*) sp.

(continued)

Table 3.2 (continued)

Study	State	Type environment	Sampling information	Faunistic composition[a]
Kihara et al. (2011)[c]	SP	Sandy beaches (dissipative and reflective)	Qualitative sampling	*Afrolaophonte*
				Amphiascus
			7 sandy beaches	*Arenopontia*
				Arenosetella
				Filexilia

[a]Taxonomic status based on WORMS (Walter and Boxshall 2020)
[b]Freshwater taxa based on WORMS (Walter and Boxshall 2020)
[c]Genera mentioned to sandy beaches. Rocha et al. (2011) showed a complete list of Harpacticoida in different coastal environments from the northern coast of the state of São Paulo

compared to reflective ones. However, the number of observations is still very incipient to firmly establish a relationship between diversity and the morphodynamic state.

3.4.2 Nematoda

Nematodes are among the most abundant taxa on Brazilian sandy beaches, representing nearly 80% of total meiofauna in many studies (e.g., Wandeness et al. 1997; Souza-Santos et al. 2003; Pinto and Santos 2006). They may also be very diverse, even compared to other habitats, such as estuaries (Venekey et al. 2010). More than 50 nematode families were already registered along the Brazilian sandy beaches, and more than 70 genera are exclusively found in this habitat (some examples are given in Fig. 3.2a–h) (Venekey et al. 2010). However, few studies identify nematodes at low taxonomic levels (e.g., Bezerra et al. 1997; Netto et al. 2007; Maria et al. 2008, 2013c; Venekey et al. 2014a).

The taxonomic descriptions of marine nematodes from Brazilian sandy beaches began in the 1950s with the work of Gerlach on the southeastern and northeastern coasts (Gerlach 1956a, b, c, 1957). Only 50 years later, new taxon descriptions were published (Fonsêca-Genevois et al. 2009). One of the most abundant and diverse nematode taxa on Brazilian sandy beaches is the Chromadoridae family, which is commonly dominant in sediments of exposed sandy beaches (Urban-Malinga et al. 2004; Maria et al. 2013a, b, c). About 25 Chromadoridae genera are known to occur in Brazil. Prainha beach, on Bertioga (state of São Paulo), and Maragogi (state of Alagoas), are the type locality of *Hypodontolaimus pumilio* and *Prochromadorella spinosa*, respectively. Xyalidae family is also common on the Brazilian coast, with more than 30 species recorded (Venekey et al. 2014b). Among them, 13 have a Brazilian beach as type locality (Table 3.3).

Metachromadora (Desmodoridae) is the most common nematode genus found in ecological surveys, being registered along most beaches of the Brazilian coast except in the state of Pernambuco (Venekey et al. 2010). Four species have beaches of the southeastern coast as type locality: *Metachromadora pneumatica* and *M.*

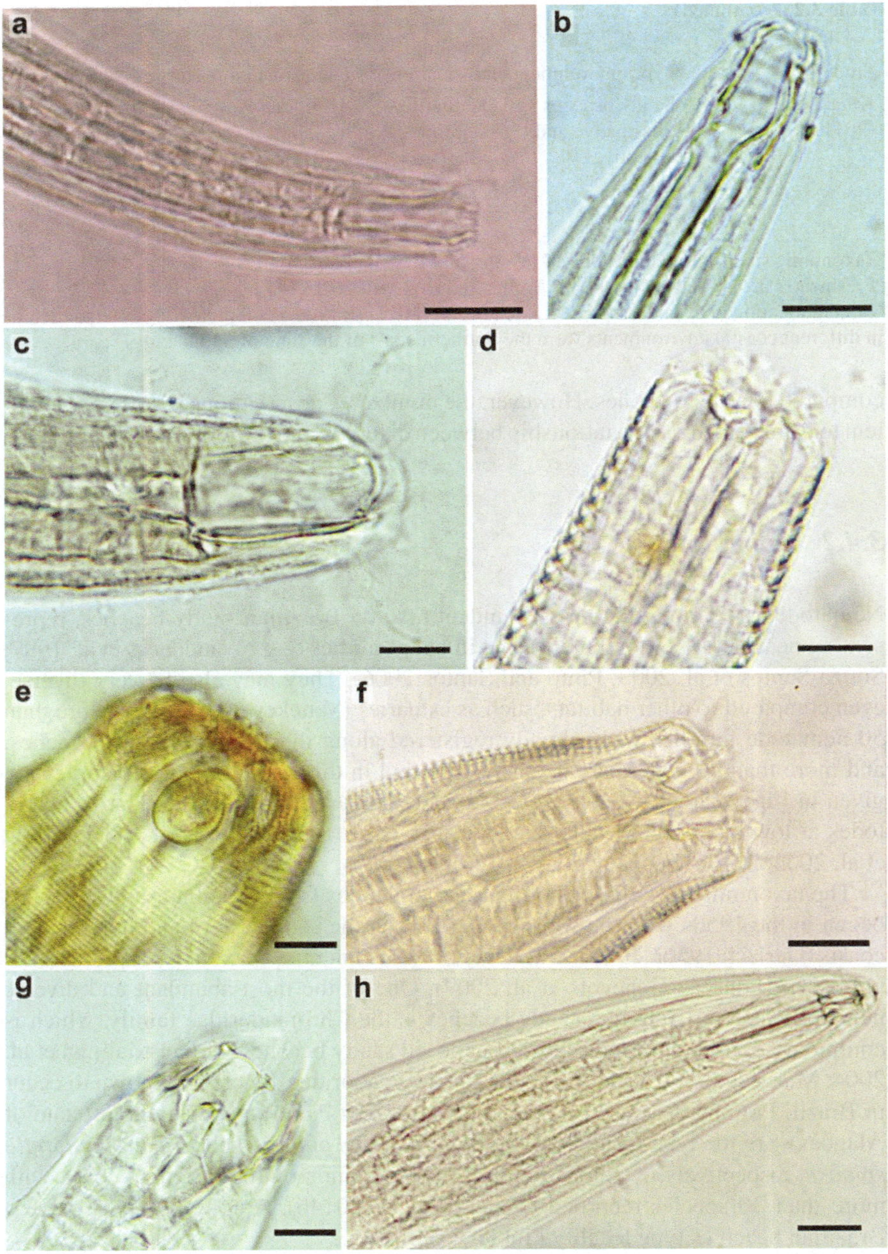

Fig. 3.2 Nematodes found at Brazilian sandy beaches: (**a**) *Ascolaimus*, scale 100 µm; (**b**) *Calyptronema*, scale 100 µm; (**c**) *Bathylaimus*, scale 100 µm; (**d**) *Euchromadora*, scale 100 µm; (**e**) *Pseudocromadora*, scale 100 µm; (**f**) *Parancanthoncus*, scale 150 µm; (**g**) *Enoplolaimus*, scale 100 µm; (**h**) *Trissonchulus*, scale 200 µm. (Photos: A.M. Esteves and T.F. Maria)

Table 3.3 Species of Xyalidae nematodes described in Brazil

Species	State[a]
Daptonema heterum (Gerlach 1957)[b]	SP
Elzalia floresi (Gerlach 1957)	SP
Metadesmolaimus tersus (Gerlach 1957)	SP
Paramonhystera stricta (Gerlach 1957)	SP
Pseudosteineria pavo (Gerlach 1957)[c]	SP
Pseudosteineria scopae (Gerlach 1956a, b)	PE
Rhynchonema cemae (Bezerra Smol and Vincx 2014)	PE
Rhynchonema veronicae (Bezerra Smol and Vincx 2014)	PE
Steineria ericia (Gerlach 1956a, b)	SP
Steineria marcorum (Gerlach 1956a, b)	SP
Steineria tripartita (Gerlach 1957)	SP
Theristus acribus (Gerlach 1957)	SP
T. stranus (Gerlach 1957)	SP

SP São Paulo, *PE* Pernambuco
[a]According to Venekey et al. (2014b)
[b]Mentioned as *Trichotheristus heterus* in Venekey et al. (2014b)
[c]Mentioned as *Steineria pavo* in Table 1 from Venekey et al. (2014b)

spectans occur in the state of São Paulo, whereas *M. prepapillata* and *M. verae* were described in the state of Rio de Janeiro.

3.4.3 *Annelida*

Annelida is a phylum with more than 17,000 described species with vast morphological diversity and ecological ubiquity (Weigert and Bleidorn 2016). Meiofaunal forms represent more than 400 species classified into 25 families of annelids (Worsaae 2020, 2021). While most of these families have both macrofaunal and meiofaunal representatives, 11 are considered entirely meiofaunal or interstitial (Worsaae and Kristensen 2005). Meiofaunal annelids are morphologically divergent, small, and comparatively simple forms, so they were initially grouped within the so-called Archiannelida (Worsaae and Kristensen 2005). Nowadays, based on phylogenomic studies, we know that these families are scattered throughout the annelid phylogenetic tree, plausibly evolving in convergence as small organisms either by secondary miniaturization or progenesis, depending on the group (Laumer et al. 2015; Struck et al. 2015).

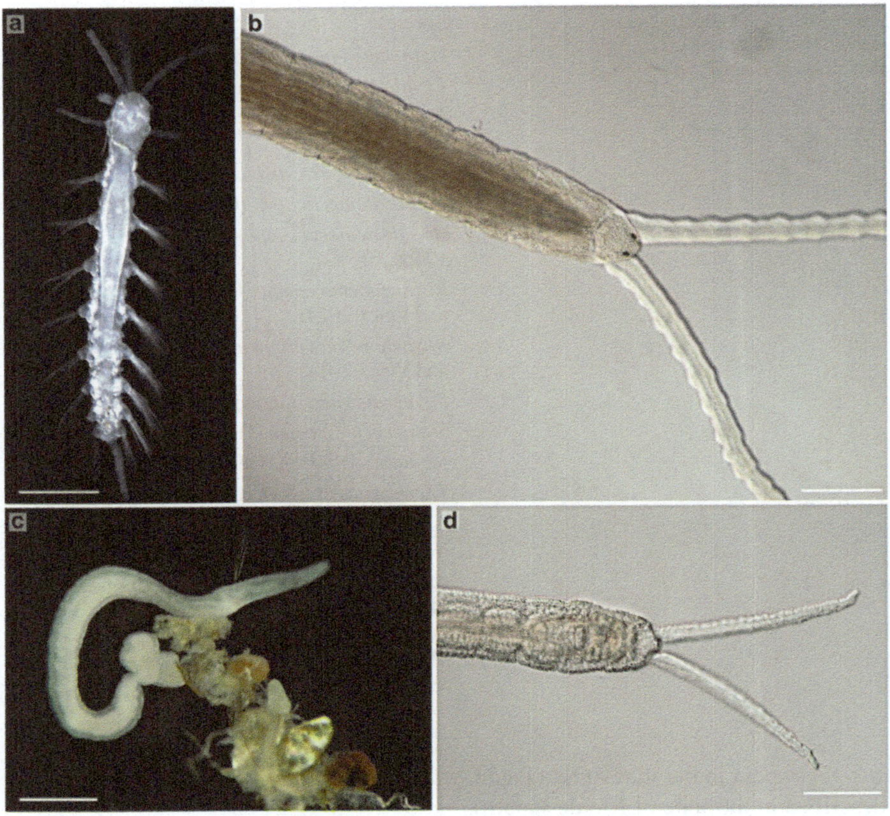

Fig. 3.3 Annelids and a nemertean from Brazilian sandy beaches: (**a**) *Nerilla mediterranea* (Annelida: Nerillidae), scale bar: 250 μm; (**b**) *Saccocirrus pussicus* (Annelida), scale 250 μm; (**c**) *Ototyphlonemertes* (Nemertea, see Sect. 3.4.7), scale 1 mm; (**d**) *Claudrilus cordeiroi* (Annelida), scale 100 μm. (Photos: M. Di Domenico; F. Gallucci; G. Fonseca; A.E. Migotto)

Among the exclusively interstitial families, Dinophilidae, Diurodrilidae, Nerillidae, Polygordiidae, Protodrilidae, Protodriloidae, and Saccocirridae were recorded in Brazil (some examples are given in Fig. 3.3a, b, and d). In total, 12 species have been recorded from these groups, placing the Brazilian coastline as a well-known diverse site regarding interstitial annelids families after European Countries (Italy, German, England, Spain, and France) and the USA. This diversity is linked to the sampling effort performed in the state of São Paulo, which started during the middle of the twentieth century with the studies conducted by Prof. Eveline D-B.R. Marcus and is still performed today, favored by the presence of marine research centers along the state coast. Currently, Brazil hosts the type locality of one genus (*Pharhyngocirrus*) and four nominal species of interstitial annelids (Marcus 1946, 1948; Siewing 1954; Westheide 1974; Santos and Silva 1992/93; Di Domenico et al. 2013, 2014a, b), most of them described at state of São Paulo, both by resident or visiting researchers. In contrast, the North and Northeast region of Brazil remain unexplored regarding sandy beach meiofaunal annelids.

The studies on meiofaunal annelids on the Brazilian sandy beaches have focused almost exclusively on taxonomic aspects (Marcus 1946, 1947a, b, 1948; Siewing 1954; Westheide 1974; Santos and Silva 1992/93; Di Domenico et al. 2013, 2014a, b). Records are from the coast of the states of São Paulo (Santos, Ilhabela, and Ubatuba), Rio de Janeiro (Rio de Janeiro city and Ilha Grande), Bahia (Ilhéus), Santa Catarina (Balneário Camboriú, Navegantes), and Paraná (Guaratuba and Matinhos).

Nerillidae is the most species-rich family of meiofaunal annelids, but only two species are officially recorded in Brazil: *Nerilla mediterranea* (Fig. 3.3a) and *Mesonerilla ecuadoriensis* (Marcus 1947a, b). *Nerilla mediterranea* was reported in Santos, state of São Paulo (Marcus 1947a, b), and at Praia Vermelha, state of Rio de Janeiro (P.J.P. Santos, unpublished data). *Mesonerilla ecuadoriensis* was first described for the Galápagos Islands (Schmidt and Westheide 1977) and was reported to Praia Vermelha, state of Rio de Janeiro (P.J.P. Santos, unpublished data). *Nerilla mediterranea* was recorded on the Itassusse Island subtidal zone at the Barequeçaba Beach, São Sebastião, state of São Paulo.

The highest diversity of Protodrilidae is known from the Eastern Atlantic and Mediterranean Sea (18 species, 51% of the 35 described species). In contrast, only seven species are known from the Western Atlantic: *Claudrilus corderoi* (Fig. 3.2d), from Ilhabela, state of São Paulo (Brazil) and possibly Dominica (Caribbean Sea), *Protodrilus gelderi* from Massachusetts Bay (USA), *P. smithsoni*, *P. hochbergi*, and *Claudrilus draco* (Martínez et al. 2013) from Panama and Belize, *Claudrilus ovarium*, and *Protodrilus pythonius* (Di Domenico et al. 2013) from Santa Catarina to Rio de Janeiro states in Brazil. A few records to the genus level are also reported from the Western Atlantic (Kirsteuer 1967; Di Domenico et al. 2009; Ruebush 1939). Despite its turbulence, the swash zone of reflective beaches seems to provide an ideal habitat for several adapted interstitial groups, including several protodrilid species (Di Domenico et al. 2009; McLachlan 1985). The presence of adhesive glands, as well as special epidermal glandular or skeletal structures, foraging strategies, and escape motility are the most common adaptations to such extremely turbulent environments (Delamare-Deboutteville 1960; Boaden 1995; Giere 2009; Jouin 1970; McLachlan 1988).

Saccocirridae is represented by two species belonging to the genus *Saccoccirus*. They are an important component of the interstitial communities in many exposed sandy beaches, where they are often found associated with nemerteans and feed on suspended particles while drifting along with the wave action (Di Domenico et al. 2014a, Di Domenico et al. 2019) (Fig. 3.3c) and proseriates (see Sect. 3.4.7). *Saccoccirus* species are non-selective and have been found to ingest microscopic fibers on several beaches of the world, including several sandy beaches on the Brazilian coast (Gusmão et al. 2016). The two species found in Brazil inhabit different environments in terms of grain size and hydrodynamics. *Pharyngocirrus gabriellae* is found in intertidal or subtidal regions in sheltered coasts. In contrast, *Saccocirrus pussicus* (Fig. 3.2b) is found in intertidal zones of exposed beaches (Di

Domenico et al. 2014a) and has distribution correlated with environmental factors such as waves, beach slope, and content of carbonates (Di Domenico et al. 2014b). Both genera are described as small, dependent on labile material, and capable of feeding by grabbing suspended material with their palps or directly with their mouths. It is suggested that *Pharyngocirrus gabriellae* is predominantly herbivorous while *Saccocirrus pussicus* is primarily carnivorous, but there is probably some overlapping in both diets (Jumars et al. 2015).

3.4.4 *Tardigrada*

The phylum Tardigrada is composed of microinvertebrates (from 70 to no more than 1000 μm) commonly found in terrestrial habitats (mainly in mosses and lichens) but also observed in marine and freshwater environments (Kaczmarek et al. 2015; Nelson et al. 2015). Currently, there are 30 families, 142 genera, 1298 species, and 34 subspecies assigned to Tardigrada (Degma et al. 2019). However, only ~220 species and subspecies are marine, representing less than 20% of all known tardigrade taxa (Fontoura et al. 2017).

Worldwide, marine tardigrades have been considerably more studied on the continental shelf than on the coast, with only a few records on sandy beaches. At least 32 species have been recorded on the Brazilian coast, 19 from sandy beaches, and 13 from continental margins (da Rocha et al. 2013; Santos et al. 2017, 2019b; Gomes-Júnior et al. 2018, 2020; Barros 2020). Regarding sandy beaches species, approximately half belong to the genus *Battilipes* (Fig. 3.4a). Most were not originally described from samples collected in Brazil (contrasting a general trend that marine tardigrades have small range size distributions; Kaczmarek et al. 2015; Miller and Perry 2016). Records of tardigrades were performed on few beaches along the Brazilian coastline, the majority located in the Northeastern coast (states of Alagoas, Ceará, Pernambuco, Rio Grande do Norte), with few in the Southeastern coast (states of São Paulo and Rio de Janeiro).

The first studies with marine Brazilian tardigrades were done by Marcus (1946) and Bois-Reymond Marcus (1952) that described the new species *Batillipes pennaki* and *Orzeliscus belopus* (also a new genus) from samples collected from the states of São Paulo and Rio de Janeiro, respectively. Subsequently, both species were recorded worldwide in many distinct biogeographic ecoregions and were considered cosmopolitan (Kaczmarek et al. 2015). However, a more integrative taxonomy approach detected intraspecific variations among populations of both species, which are nowadays considered a species complex (Santos et al. 2019b).

Most studies on Brazilian sandy beach tardigrades are taxonomical (Marcus 1946, 1952; Höfling-Epiphanio 1972; da Rocha et al. 2009, 2013; Moura et al. 2009; Santos et al. 2017; Barros 2020), while few of them investigated spatial and temporal distribution (Victor-Castro et al. 1999; da Rocha et al. 2002, 2004; Verçosa et al. 2009; Tilbert et al. 2019).

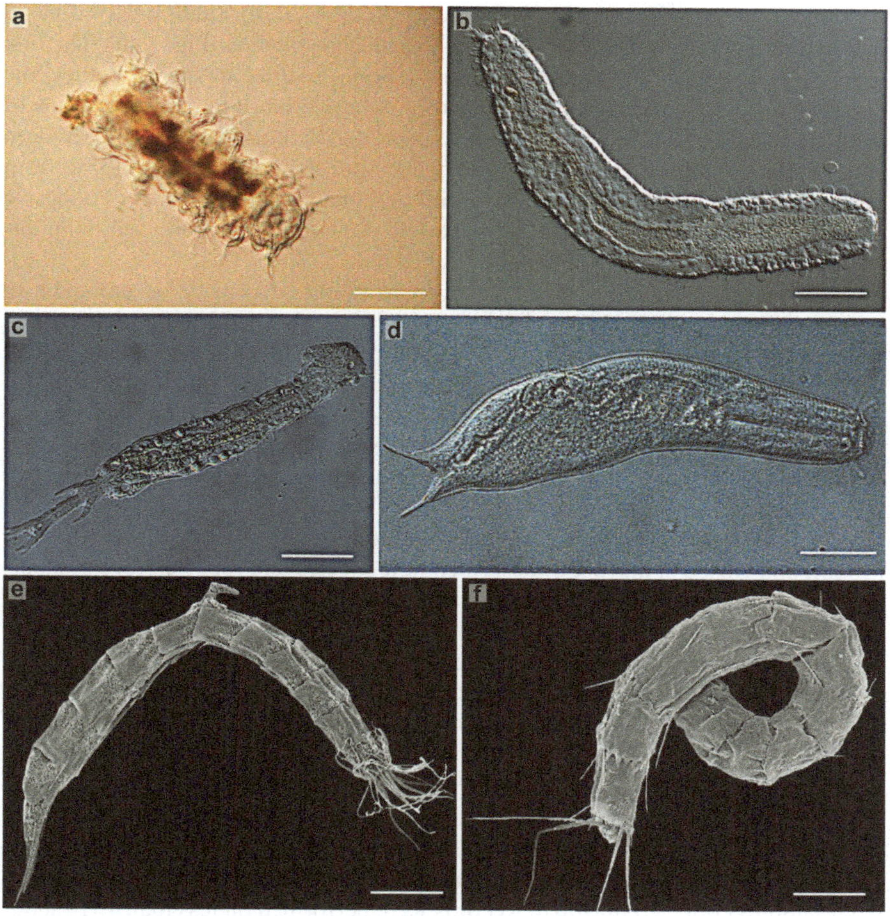

Fig. 3.4 Tardigrades, gastrotrichs, and kinorhynchs from Brazilian sandy beaches: (**a**) *Batillipes* sp. (Tardigrada), scale bar: 50 μm; (**b**) *Crasiella fonseci* (Macrodasyida), scale bar: 30 μm; (**c**) *Dendrodasys* sp. (Macrodasyida), scale bar: 50 μm; (**d**) *Aspidiophorus* sp. (Chaetonotida), scale bar: 50 μm; (**e**) *Cateria styx* with its enigmatic dorsal organ (Kinorhyncha), scale bar: 60 μm; (**f**) *Franciscideres kalenesos* (Kinorhyncha), scale bar: 50 μm. (Photos: M. Di Domenico; F. Gallucci; G. Fonseca; A.E. Migotto; A.R.S. Garraffoni; SEM photos (**e**) and (**f**) gently provided by Martin V. Sørensen)

3.4.5 Gastrotricha

Gastrotricha (80–3500 μm in total length) are commonly found in marine and freshwater benthic interstitial environments. They are recognized due to the presence of locomotory ventral cilia, complex anatomy, and life cycle with a predominance of hermaphroditism (Balsamo et al. 2014; Kieneke and Schmidt-Rhaesa 2015; Todaro et al. 2019a).

The taxon comprises almost 860 species (Todaro 2020), traditionally divided into two orders: Macrodasyida (Fig. 3.4b, c) and Chaetonotida (Fig. 3.4d). The first order has more than 370 species (99% are marine) with strap-shaped bodies, the presence of pharyngeal pores, and numerous adhesive tubes in distinct regions of the body (Kånneby and Hochberg 2015; Kieneke and Schmidt-Rhaesa 2015; Todaro et al. 2019a). Within Chaetonotida, nearly 500 species were described (35% from marine habitats) with tenpin-shaped bodies and one pair (rarely two) of adhesive tubules, limited to the posterior end (Balsamo et al. 2014; Kieneke and Schmidt-Rhaesa 2015).

Until two decades ago, marine gastrotrichs could be considered unknown in Brazil, with Marcus (1952) and Forneris (1985) only mentioning undescribed Gastrotricha species on the coast of the state of São Paulo. Todaro and Rocha (2004, 2005) were the first to conduct extensive samples along the Northern coast of the state of São Paulo and the southern part of the state of Rio de Janeiro. As a result, these authors reported almost 50 species (most of them undescribed) and described the first species endemic from Brazil, Macrodasys fornerisae. After these pioneers' studies, Todaro (2012, 2013), Araújo et al. (2014, 2016, 2022), Hochberg (2014), Garraffoni et al. (2017), Todaro et al. (2019b), Campos et al. (2020), and Bosco et al. (2020) described new species and reported recent occurrences for the coast of the states of São Paulo, Rio de Janeiro, Espírito Santo, and Bahia.

The studies on Brazilian marine gastrotrichs are almost exclusively taxonomic, with 24 species identified so far (Campos and Garraffoni 2019; Campos et al. 2020; Araújo et al. 2022). The Chaetonotida is the richest taxon in the number of species (16 species), followed by Macrodasyida (10 species). However, the marine gastrotrich richness is underestimated as more than 75% of the total number of species collected were reported as new species but have not been formally described yet (Todaro and Rocha 2004, 2005; Garraffoni et al. 2017; Campos and Garraffoni 2019). Only eight species, all macrodasyids, were initially described from samples collected in Brazil. Records were performed on less than 40 beaches along the Brazilian coastline, with approximately 70% located in São Paulo and the remaining 30% found in Rio de Janeiro, Espírito Santo, and Bahia (Campos and Garraffoni 2019). Ecological investigations on Brazilian gastrotrichs are limited to the studies done by Garraffoni et al. (2016) and Guidi et al. (2022). Garraffoni et al. (2016) attempted to understand the influence of environmental variables (e.g., sediment parameters, geographical position, location, and across-shore zone) on the gastrotrichs distribution over Brazilian sandy beaches. They found that the diversity patterns of Gastrotricha were explained by sediment textures, tidal zones, and localities. Guidi et al. (2020), in turn, investigated the spermatozoon morphology, spermatogenesis, and reproductive system of Acanthodasys australis (Bosco et al. 2020).

3.4.6 *Kinorhyncha*

The phylum Kinorhyncha is composed exclusively of marine and free-living metazoans. Kinorhynchs are segmented and spined animals, less than 1 mm long, and distributed worldwide (Sørensen and Pardos 2008). The taxon constitutes ca. 300 species distributed in 29 genera and 11 families (Sørensen and Pardos 2020) inhabiting sand or mud, from the intertidal zone to abyssal depths throughout the world (Neuhaus and Higgins 2002).

The taxonomical history of the Brazilian kinorhynchs remains overlooked, and only eight studies have been published until now (Gerlach 1956a, b; Higgins 1968; Dal Zotto et al. 2013; Sørensen 2011, 2014; Neuhaus and Kegel 2015; Herranz et al. 2019, 2021a, b; Mello et al. 2019). The aberrant species, *Cateria styx* (Fig. 3.4e), belonging to the family Cateriidae was described from intertidal sands on a high-energy beach at Macaé (state of Rio de Janeiro). The taxonomic status of the genus *Cateria* was discussed by Higgins (1968) based on new material collected at Toninhas beach, Ubatuba, state of São Paulo, and from Cavaleiro beach near the type locality. In this same study, the author also designated a new lectotype and paralectotypes. Recently, Neuhaus and Kegel (2015) published a taxonomic revision of type and non-type specimens belonging to the genus *Cateria* and redescribed *C. styx* using light microscopy techniques for the first time.

More than 50 years after Gerlach's study, Sørensen (2011) published the first report of *Echinoderes* from Brazil with the material collected in the state of São Paulo. In 2013, Dal Zotto and co-authors described another aberrant kinorhynch, from Ilhabela and São Sebastião (state of São Paulo) and Guaratuba (state of Paraná), with morphological characters unknown for the entire taxon, called *Franciscideres kalenesos*. However, due to uncertainties and conflicting signals, the authors preferred to treat *Franciscideres* as *incertae sedis*. After, Sørensen et al. (2015), based on phylogenetic relationships obtained by molecular and morphological (apomorphy: the presence of a neck region with segment-like ring) data, designated the new family Franciscideridae to accommodate *Franciscideres* and *Gracilideres* (Yamasaki 2019).

Cateria has been an enigmatic kinorhynch genus due to its aberrant worm-like shape and extremely thin cuticle. However, the rare occurrence of the species, only found in sandy intertidal habitats, and the poor preservation of the type material have hampered detailed studies of the genus over time. Sixty years after the original description of *Cateria styx*, Herranz et al. (2019) present an extensive morphological and functional study based on new material collected from Cavaleiro beach, state of Rio de Janeiro. *Cateria styx* was found to display a complex array of cuticular structures (spines, spinoscalids, and extraordinarily complex cuticular ornamentation) that was interpreted as an adaptation for mechanical adhesion, through friction and interlocking, in an interstitial habitat; the enigmatic dorsal organ is a hydrostatic structure, which functions is inferred to be adhesive (Herranz et al.

2019, 2021a, b). Herranz et al. (2021b) provided a comprehensive and comparative myoanatomical study of *Cateria styx*, *Franciscideres kalenesos*. Both species showed a weak external segmentation of the trunk with a distinct segmental arrangement of the musculature (Herranz et al. 2021b). Intriguingly, the aberrant *Cateria styx* and *Franciscideres kalenesos* found at intermediate Brazilian sandy beaches showed the nervous system to be segmental and similar to those of non-aberrant kinorhynchs; hereby not mirroring their otherwise aberrant and posteriorly shifted myoanatomy (Herranz et al. 2021a).

Recently, the spatial abundance of *Franciscideres kalenesos* (Fig. 3.4f) was described and associated with the hydrodynamic zones in intermediate/dissipative beaches. Also, the distribution of *F. kalenesos* occurred in a narrow strip of the beach, located in a high turbulence zone. Within this zone, *F. kalenesos* was distributed in patches separated every 30 m. The observed zonation in sandy beaches and the patchy distribution in turbulent habitats like these have not previously been described for a species of this phylum (Mello et al. 2019).

3.4.7 Other Groups

Besides the groups presented before, sandy beaches harbor an enormous number of meiofaunal species and genera belonging to Platyhelminthes (Proseriata and Rhabdocoela), Acoela (formerly included in Platyhelminthes), Mollusca (Opisthobranchia), and Nemertea (e.g., Marcus 1946, 1947a, b, 1948, 1950, 1955, 1957, 1958; Marcus and Marcus 1951, 1954a, b, 1955, 1956, 1957; Corrêa 1948, 1949, 1950, 1953, 1954, 1957).

Brazil is one of the most diverse known countries in the world for its "microturbellarian" fauna, and they are abundant on sandy beaches (Braccini et al. 2016). Most marine free-living Platyhelminthes species belong to the meiofauna, excluding most Polycladida and few representatives of the Tricladida *Maricola* (Cannon 1986). Members of the Kalyptorhynchia (Rhabdocoela) (Fig. 3.5c) may occur in epibenthic habitats such as the periphyton (mixture of algae, cyanobacteria, heterotrophic microbes, and detritus), while all other taxa are almost exclusively found interstitially and are among the dominant meiofaunal groups in sandy beaches and brackish-water habitats (Giere 2009). Recently two species of *Typhlopolycystis* and one species of *Carcharodorhynchus* (*C. brasiliensi*) were described for sandy beaches in the state of São Paulo. *Carcharodorhynchus brasiliensis, particularly*, was found on several beaches in São Sebastião (Reygel et al. 2014). Proseriata, the second most diverse taxon of interstitial Platyhelminthes, is usually the most abundant animal group in terms of the number of individuals in sediments with medium to coarse-sand and low organic matter content (Curini-Galletti 2001). This taxon may be so abundant in the swash zone of exposed reflective beaches that they characterize the entire interstitial community, known after the dominant genus in the Mediterranean, as the "*Otoplana*-zone" (Remane 1933; McLachlan and Defeo 2018). The genus *Otoplana* is endemic to the Mediterranean-Lusitanian Region

(Scarpa et al. 2017), but the family Otoplanidae has representatives of the genus *Kata* occupying the same environment on Brazilian sandy beaches. Two species, *Kata leroda* and *Kata evelinae,* were found on several sandy beaches in Brazil (state of São Paulo; Curini-Galletti 2014). These species differed in habitat preference: *K. leroda* (Fig. 3.5b) was common along the continental coast of São Sebastião, while *K. evelinae* was only found on a high-energy beach on the outer Coast of Ilhabela (Curini-Galletti 2014). Finally, among the Platyhelminthes the Brazilian waters have maybe one of the most intriguing tricladids closely related to *Maricola,* named *Puiteca camica* (Fig. 3.5a). These worms are found occasionally on sandy beaches between the headlands and may be found in algae associated with rocky shores or sand sediments from intertidal and subtidal (MDD, MCG, personal communication). For a complete overview of Platyhelminthes on Brazilian sandy beaches, see Marcus (1946, 1948, 1949, 1950, 1951, 1952, 1954), Curini-Galletti (2014), Reygel et al. (2014), and Schockaert et al. (2019).

The Phylum Acoelomorpha (Philippe et al. 2011) was reported in the Brazilian sandy beaches at the order-level as Acoela by Hooge and Rocha (2006), Marcus (1948, 1949, 1950, 1951, 1952, 1954), and Marcus and Marcus (1955, 1957) (Fig. 3.5d). The records of the order Nemertodermatida are still absent on the Brazilian Coast. The majority of Acoela and Nemertodermatida are typical

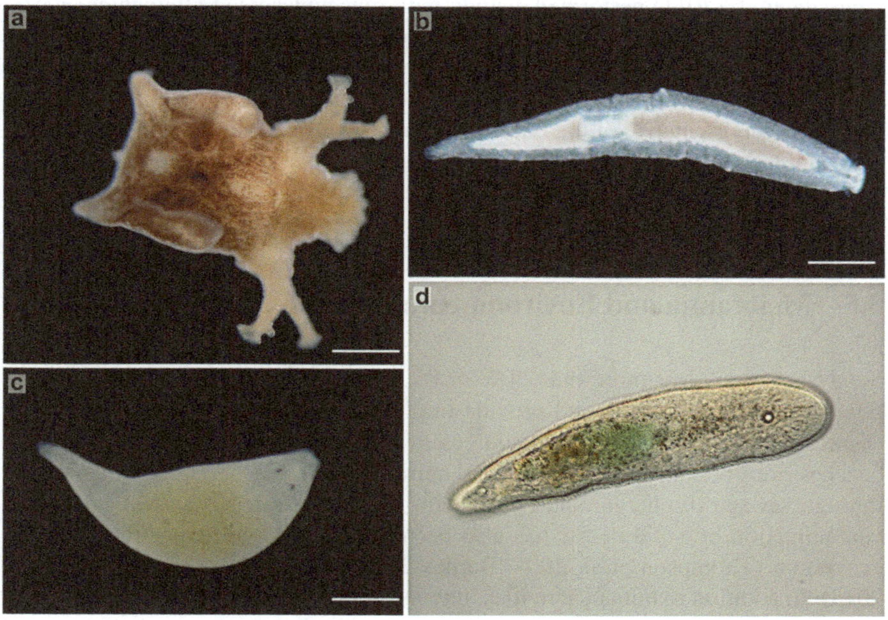

Fig. 3.5 Flatworms and related groups from Brazilian sandy beaches: (**a**) *Puiteca camica* (Tricladida, Platyhelminthes), scale bar: 200 μm; (**b**) *Kata leroda* (Otoplanidae, Platyhelminthes), scale bar: 50 μm; (**c**) unknown species (Rhadcoela, Platyhelminthes), scale bar: 50 μm; (**d**) unknown species (Acoelomorpha), scale bar: 40 μm. (Authors: M. Di Domenico; F. Gallucci; G. Fonseca; A.E. Migotto)

representatives of soft-bodied, interstitial meiofauna. Acoela is the most diverse group, with about 440 nominal species described worldwide (Jondelius and Jondelius 2020). In Brazil, there are 35 species described, 28 by Ernest Marcus and Eveline Du Bois-Reymond Marcus (from 1948 to 1957), and seven by Hooge and Rocha (2006). Hooge and Rocha (2006) sampled 31 sites along the sandy beaches around São Sebastião Coast and found 21 species of Acoela on different types of beaches.

Nemertea has more than 1300 species known (Norenburg and Stricker 2002; Strand and Sundberg 2010), and about 50 meiofaunal-sized species are known worldwide, occurring interstitially in coarse-sand sediments. The most characteristic interstitial nemerteans belong to the statocyst-bearing genus *Ototyphlonemertes*, with at least 33 nominal species known worldwide (Döhren and Bartolomaeus 2020) (Fig. 3.2c). *Ototyphlonemertes* is highly suggestive of the presence of cryptic species, which have already been demonstrated on molecular bases (Leasi et al. 2016). Envall and Norenburg (2001) established six "phylomorphs," small and morphologically homogeneous groups of morphotypes – Duplex, Pallida, Cirrula, Fila, Lactea, and Macintoshi – that could be diagnosed unambiguously, although one or more may be paraphyletic or even polyphyletic. In Brazil, Prof. Diva Diñiz Corrêa published several papers on the taxonomy of the meiofauna *Ototyphlonemertes* (Nemertea) commonly found on the Brazilian sandy beaches (Andrade et al. 2011). Andrade et al. (2011) performed a study with four widespread *Ototyphlonemertes* species from intertidal sand samples collected along the Brazilian Coast at eight sandy beaches from the states of Santa Catarina to Bahia: *O. erneba* (Pallida morph), *O. evelinae* (Duplex-morph), *O. parmula* (representative of the Fila morph), and *O. lactea* (representative of the Lactea morph). They found strong evidence of gene flow among *O. erneba* and *O. evelinae*, and their level of populational genetical structure is much lower than for the other two species.

3.5 Meiofauna and Environmental Impacts

Sandy beaches cover more than 30% of the world's ice-free coastlines. They are popular recreational spots and provide important habitats for wildlife (Schlacher et al. 2007). In Brazil, the main sources of anthropogenic disturbances in sandy beaches are (1) construction of coastal infrastructure, (2) organic contamination through sewage discharge, and (3) tourism (Amaral et al. 2016). More recently, contamination by solid debris has also been recognized as a major threat to this ecosystem (Thompson et al. 2004; Barnes 2005; Barnes et al. 2009; Suciu et al. 2017). In addition to human activities, threats posed by global climate change such as rising sea level, increased frequency and magnitude of extreme events, coastal erosion, rising sea temperatures, and ocean acidification are predicted to have consequences for sandy beaches globally (Amaral et al. 2016; Vousdoukas et al. 2020). Due to the high vulnerability of this ecosystem and the ubiquitous presence of meiofauna within its sand grains, studies worldwide have identified impacts on meiofauna, also suggesting these organisms as indicators of disturbance and environmental quality (Kennedy and Jacoby 1999; Gheskiere et al. 2005; Sun et al. 2014). Nevertheless, studies that have investigated the impacts of anthropogenic disturbances on meiofauna at Brazilian sandy beaches are extremely scarce.

The first two studies found in the peer-review literature are relatively recent (i.e., 2003), probably due to the somewhat recent development of meiofauna ecological studies in Brazil. Oliveira and Soares-Gomes (2003) investigated the effects of domestic sewage on the intertidal meiofauna on a sandy beach in the state of Rio de Janeiro, and Moellmann and Corbisier (2003) analyzed the effects of tourism on meiofauna on a sandy beach from the coast of the state of São Paulo. Whereas no effect of the organic pollution was observed by Oliveira and Soares-Gomes (2003), Moellmann and Corbisier (2003) suggested that trampling and recreational activities related to tourism probably affect the meiofauna, changing the vertical distribution and inducing downward migration in the sediment to avoid disturbance in areas of high human concentration. In both studies, however, some clear shortcomings regarding sampling design (which precludes the determination of a cause-effect relationship between human activities and meiofauna structure) as well as the low taxonomic resolution used (both have identified meiofauna to major zoological groups) demand caution when interpreting their results. Recently, Santos et al. (2021) evaluated the effects of recreational activities on meiofaunal communities and nematode assemblages on Amazonian sandy beaches (state of Pará). By sampling three beaches with distinct disturbances levels (Urban: Atalaia; Intermediary: Farol-Velho; and Protected: Corvinas) before, during, and after the tourist season, the authors found that the structure of the meiofaunal assemblages varied as a consequence of sediment compaction linked to the higher number of tourists.

Regarding urbanization, Felix et al. (2016) used an innovative approach to investigate the relationship between sandy beach visual attractiveness (classified in lower and higher rate units) and meiofauna biodiversity in urbanized and semi urbanized sandy beaches in Brazil and Uruguay. Lower-rated beach units exhibited signs of lower environmental quality, indicated by lower nematode densities and higher oligochaete densities. Interestingly, sandy beach infrastructure (i.e., kiosks, lifeguards, and many other leisure services) and sidewalk, conspicuous in lower-rated units and absent in higher-rated units, explained almost the same variability of meiofauna as grain size and sorting.

When it comes to marine debris, Gusmão et al. (2016) reported the ingestion of synthetic fibers by the polychaete *Saccocirrus* at sandy beaches in the states of Santa Catarina and Rio de Janeiro and suggested that their non-selective microphagous suspension-feeding behavior makes them more prone to ingest these microplastics. Laboratory observations showed that microfibers were rapidly egested with no apparent physical injury.

Finally, Santos et al. (2019a) analyzed the nematode assemblages from seven sandy beaches under different types of pressure in the state of Rio de Janeiro. Despite the high human density, most urban and touristic beaches did not show distinct values for the diversity measures when compared to the most preserved beaches.

3.6 Final Remarks

Sandy beaches are central economic assets, attracting more tourists and recreational users than any other coastal ecosystem. They harbor a rich and dense community of benthic fauna that plays multiple roles, such as nutrient regeneration, provision of food for fishes and seabirds, and trophic subsidies to terrestrial consumers. Meiofaunal organisms, due to their high abundance and short life-cycles, are essential for coastal food web and nutrient cycling. Moreover, they can be important indicators of environmental changes in sandy beach ecosystems.

Despite the late start, Brazilian sandy beach meiofauna studies have increased considerably over the past decades, and much of this important biodiversity has been unveiled. Nematodes are thus far the most well-known meiofaunal taxon on Brazilian sandy shores, but we are also discovering many species from different groups such as Copepoda, Annelida, and Gastrotricha.

The increasing threats associated with human activities and climate change claim an urgent need to understand their impacts on Brazilian sandy beaches. In this context, future studies should take advantage of meiofaunal organisms to better understand and preserve the biodiversity and functioning of Brazilian sandy beaches. Ideally, the focus should be given to broader studies with clear cause-effect hypotheses that consider proper variability so we can more realistically estimate the risks posed by anthropogenic disturbances on sandy beach ecosystems. That could be achieved through collaborative studies within the increasing number of meiofauna research groups, which are currently distributed from the northern to the southern coasts of Brazil, as well as with experts from other research areas if we want to estimate how impacts on meiofauna can be translated to effects on the ecosystem functions they provide.

References

Albuquerque EF, Pinto APB, Perez ADQ et al (2007) Spatial and temporal changes in interstitial meiofauna on a sandy ocean beach of South America. Braz J Oceanogr 55:121–131

Alvarez MPJ (1988) Harpacticoid copepods from Una do Prelado River (São Paulo, Brazil): genus *Schizopera*. Hydrobiologia 167–168:435–444

Amaral ACZ, Corte GN, Denadai MR et al (2016) Brazilian sandy beaches: characteristics, ecosystem services, impacts, knowledge and priorities. Braz J Oceanogr 64:5–16

Andrade SCS, Norenburg JL, Solferini VN (2011) Worms without borders: genetic diversity patterns in four Brazilian *Ototyphlonemertes* species (Nemertea, Hoplonemertea). Mar Biol 158:2109–2124

Araújo TQ, Balsamo M, Garraffoni ARS (2014) A new species of Pseudostomodella (Gastrotricha, Thaumastodermatidae) from Brazil. Mar Biodivers 44(3):243–248

Araujo TQ, Wieloch AH, Vidigal THD et al (2016) *Pseudostomella dolichopoda* Todaro, 2012 and *P. cataphracta* Ruppert, 1970 (Gastrotricha: Thaumastodermatidae): new records from Brazil and USA and an updated key to the genus. Check List 12(6):1986

Araújo TQ, Wieloch AH, Hochberg R, Garraffoni AR (2022) Description of Xenotrichula tropicalis sp. nov. (Gastrotricha: Chaetonotida) and new records of Xenotrichulidae species from Brazil and USA. Ann Zool 72:167–185

Baia E, Venekey V (2019) Distribution patterns of meiofauna on a tropical macrotidal sandy beach, with special focus on nematodes (Caixa d'Água, Amazon Coast, Brazil). Braz J Oceanogr 67:e19230

Balsamo M, Grilli P, Guidi L et al (2014) Gastrotricha: biology, ecology and systematics. Families Dasydytidae, Dichaeturidae, Neogosseidae, Proichthydiidae. In: Dumont H (ed) Identification guides to the plankton and benthos of inland waters, vol 24. Backuys Publishers, Weikersheim, pp 1–187

Barnes DKA (2005) Remote islands reveal rapid rise of Southern Hemisphere sea debris. ScientificWorldJournal 5:915–921

Barnes KA, Galgani F, Thompson RC et al (2009) Accumulation and fragmentation of plastic debris in global environments. Philos Trans R Soc Lond B Biol Sci 364:1985–1998

Barros RC (2020) Tardigrades research in Brazil: an overview and updated checklist. Arq Zool 51(1):1–11

Bezerra TNC, Fonsêca-Genevois V, Genevois B (1996) Distribuição horizontal e vertical da meiofauna em uma região tropical intermareal (Istmo de Olinda Pernambuco Brasil). Trab Oceanogr Univ Fed Pernamb 24:249–264

Bezerra TNC, Genevois B, Fonsêca-Genevois V (1997) Influência da granulometria na distribuição e adaptação da meiofauna na praia arenosa do Istmo de Olinda PE. In: Silva SA, Grohmann PA, Esteves AM (eds) Ecologia de praias arenosas do litoral brasileiro, vol 3. Programa de Pós-Graduação em Ecologia da UFRJ, Rio de Janeiro, pp 107–116

Bezerra TN, Smol N, Vincx M (2014) Two new species of Rhynchonema Cobb 1920 from a Brazilian sandy beach. Marine Biodivers 44 343–365

Björnberg TKS (2010) *Quinquelaophonte varians* n. sp. (Copepoda, Harpactiocida, Crustacea) and notes on its developmental stages. Pan Am J Aquat Sci 5(1):62–77

Björnberg TKS (2014) Three new species of benthonic Harpacticoida (Copepoda, Crustacea) from São Sebastião Channel. Nauplius 22(2):75–90

Björnberg TKS, Kihara TC (2013) On Tetragonicipitidae (Crustacea, Copepoda) from the Channel of São Sebastião, Brazil, with description of their nauplii and two new species of *Phyllopodopsyllus*. Zootaxa 3718(6):501–529

Boaden PJS (1995) Where Turbellaria? Concerning knowledge and ignorance of marine turbellarian ecology. Hydrobiologia 305(1–3):91–99

Bosco I, Lourenco AP, Guidi L et al (2020) Integrative description of a new species of Acanthodasys Remane, 1927 (Gastrotricha, Macrodasyida, Thaumastodermatidae) based on four distinct morphological techniques and molecular data. Zool Anz 286:31–42

Braccini JAL, Amaral SV, Leal-Zanchet AM (2016) Microturbellarians (Platyhelminthes and Acoelomorpha) in Brazil: invisible organisms? Braz J Biol 76(2):476–494

Brustolin MC, Fonseca G, Gallucci F (2022) Habitat attributes dictate the roles of dispersal and environmental filtering on metacommunity assembly at coastal soft-bottom ecosystems. Estuar Coast 45(2):592–602

Campos A, Garraffoni ARS (2019) A synopsis of knowledge, zoogeography and an online interactive map of Brazilian marine gastrotrichs. PeerJ 7:e7898

Campos A, Todaro MA, Garraffoni ARS (2020) A new species of *Paraturbanella* Remane, 1927 (Gastrotricha, Macrodasyida) from the Brazilian coast, and the molecular phylogeny of Turbanellidae Remane, 1926. Diversity 12:42

Cannon LRG (1986) Turbellaria of the world: a guide to families & genera. Queensland Museum, Brisbane, 136 pp

Carvalho JP (1952) Nova espécie do gênero Paraleptastacus Wilson, 1932 (Copepoda, Fam. Canthocamptidae). Pap Avulsos Dep Zool 11(4):37–40

Chappuis PA (1954) Nouveaux harpacticoides de la nappe phreatique de l`Adige. Mem Mus Civ Stor Nat Verona 4:157–162

Corgosinho PHC (2012) Talpacoxa brandini gen. et sp. nov. A new Nannopodidae Brady, 1880 (Copepoda: Harpacticoida) from submersed sands of Pontal do Sul (Paraná, Brazil). J Nat Hist 45–46:2865–2879

Corgosinho PHC, Metri R, Baptista C et al (2003) Abundance and diversity of the sublittoral meiofauna on two sandy beaches under different hydrodynamic conditions at Ilha do Mel (PR, Brazil). Lundiana 4:89–94

Corrêa DD (1948) Ototyphlonemertes from the Brazilian coast. Commun Zool Mus Hist Nat Mont 2:1–12

Corrêa DD (1949) Ecological studies of Brazilian Ototyphlonemertes. Commun Zool Mus Hist Nat Mont 3:1–7

Corrêa DD (1950) Sobre ototyphlonemertes do Brasil. Bol Fac Filos Ciênc Letras USP (Zool) 15:203–234

Corrêa DD (1953) Sobre a neurofisiologia locomotora de Hoplonemertinos e a taxonomia de Ototyphlonemertes. An Acad Bras Cienc 25:545–555

Corrêa DD (1954) Nemertinos do litoral Brasileiro. Bol Fac Filos Ciênc Letras USP (Zool) 19:1–122

Clément M, Moore CG (2000) A revision of the genus Halectinosoma (Copepoda: Harpacticoida: Ectinosomatidae): the H herdmani (Scott & Scott) group of species. Zool J Linn Soc 128(3):237–267

Corrêa DD (1957) Nemertinos do litoral Brasileiro – VI. An Acad Bras Cienc 29(2):251–271

Curini-Galletti M (2001) The Proseriata. In: Littlewood DTJ, Bray RA (eds) Interrelationships of the Platyhelminthes. Taylor and Francis, London/New York, pp 41–48

Curini-Galletti M (2014) Contribution to the knowledge of the Proseriata (Platyhelminthes: Rhabditophora) from southeast Brazil. Mar Biodivers 44(3):287–312

Curini-Galletti M, Artois T, Delogu V et al (2012) Patterns of diversity in soft-bodied meiofauna: dispersal ability and body size matter. PLoS One 7(3):e33801

Curvelo RR, Corbisier TN (2000) The meiofauna associated with Sargassum cymosum at Lazaro Beach, Ubatuba, São Paulo. Rev Bras Oceanogr 48:119–130

da Rocha CMC, Fonsêca-Genevois V, Victor-Castro FJ (2002) Distribuição espaço-temporal de *Batillipes pennaki* Marcus, 1946 (Tardigrada, Heterotardigrada) na margem sul da ilha de Itamaracá (Pernambuco – Brasil). Trop Oceanogr 28:35–46

da Rocha CMC, Fonsêca-Genevois V, Castro FJV et al (2004) Environmental interactions of Batillipes pennaki (Tardigrada, Heterotardigrada) in a tropical sandy beach (Itamaracá, Pernambuco, Brazil). Meiofauna Marina 13:79–86

da Rocha CMC, Verçosa MM, Santos ECL et al (2009) Marine tardigrades from the coast of Pernambuco, Brazil. Meiofauna Marina 17:97–101

da Rocha CMC, Santos ECL, Gomes EL Jr et al (2013) New records of marine tardigrades from Brazil. J Limnol 72(S1):102–107

Dal Zotto M, Di Domenico M, Garraffoni ARS et al (2013) *Franciscideres* gen. nov. – a new, highly aberrant kinorhynch genus from Brazil, with an analysis of its phylogenetic position. Syst Biodivers 11:303–321

Degma P, Bertolani R, Guidetti R (2019) Actual checklist of Tardigrada species, 35th edn, 59 pp. Available from: http://www.tardigrada.net/register/taxonomy.htm. Accessed 20 Mar 2020

Delamare-Deboutteville C (1960) Biologie des eaux souterraines littorales et continentales (No. 574.091698). Paris

Di Domenico M, Lana PDC, Garraffoni ARS (2009) Distribution patterns of interstitial polychaetes in sandy beaches of southern Brazil. Mar Ecol 30:47–62

Di Domenico M, Martínez A, da Cunha Lana P et al (2013) *Protodrilus* (Protodrilidae, Annelida) from the southern and southeastern Brazilian coasts. Helgol Mar Res 67(4):733–748

Di Domenico M, Martínez A, Almeida TCM et al (2014a) Response of the meiofaunal annelid *Saccocirrus pussicus* (Saccocirridae) to sandy beach morphodynamics. Hydrobiologia 734:1–16

Di Domenico M, Martínez A, Lana P et al (2014b) Molecular and morphological phylogeny of Saccocirridae (Annelida) reveals two cosmopolitan clades with specific habitat preferences. Mol Phylogenet Evol 75:202–218

Di Domenico M, Martínez A, Worsaae K (2019) Saccocirridae (Annelida) from the Canary Islands with a description of *Saccocirrus slateri* sp. nov. Mar Biodivers 49:2125–2139

Döhren J, Bartolomaeus T (2020) Nemertea. In: Schmidt-Rhaesa A (ed) Guide to the identification of marine meiofauna. Friedrich Pfeil Publishing, pp 227–238

Envall M, Norenburg JL (2001) Morphology and systematics in mesopsammic nemerteans of the genus *Ototyphlonemertes* (Nemertea, Hoplonemertea, Ototyphlonemertidae). Hydrobiologia 456:145–163

Esteves AM, Bloise C, Nogueira CSR (1998) Variação espaço-temporal da meiofauna ao longo de um período quinzenal, em um ponto fixo da Praia Vermelha, Rio de Janeiro. In: Anais do Simpósio de Ecossistemas Brasileiros, vol 4. ACIESP, pp 179–193

Felix G, Marenzi RC, Polette M et al (2016) Landscape visual quality and meiofauna biodiversity on sandy beaches. Environ Manag 58:682–693

Fonseca G, Gallucci F (2016) The need of hypothesis-driven designs and conceptual models in impact assessment studies: an example from the free-living marine nematodes. Ecol Indic 71:79–86

Fonseca G, Maria TF, Kandratavicius N et al (2014) Testing for nematode–granulometry relationships. Mar Biodivers 44(3):435–443

Fonsêca-Genevois V, Smol N, Decraemer W et al (2009) Bernardius lineatus gen. n., sp. n. (Enoplida: Enchelidiidae), a remarkable nematode with ornamented cuticle from a Brazilian tropical sandy beach. J Nematol 11(2):275–287

Fontoura P, Bartels PJ, Jørgensen A et al (2017) A dichotomous key to the genera of the marine heterotardigrades (Tardigrada). Zootaxa 4294:1–45

Forneris L (1985) Gastrotricha. In: Manual de Técnicas Para a Preparação de Coleções Zoológicas, vol 11. Sociedade Brasileira de Zoologia, São Paulo

Garraffoni ARS, Di Domenico M et al (2016) Patterns of diversity in marine Gastrotricha from Southeastern Brazilian coast is predicted by sediment textures. Hydrobiologia 773(1):105–116

Garraffoni ARS, Di Domenico M, Hochberg R (2017) New records of marine Gastrotricha from São Sebastião Island (Brazil) and the description of a new species. Mar Biodivers 47(2):451–459

Gallucci F, De Castro IB, Perina FC, Abessa DM, de Paula Teixeira A (2015) Ecological effects of Irgarol 1051 and Diuron on a coastal meiobenthic community: A laboratory microcosm experiment. Ecol Indic 58:21–31

Gerlach SA (1956a) Über einen aberranten Vertreter der Kinorhynchen aus dem Küstengrundwasser. Kiel Meeresforsch 12:120–124

Gerlach SA (1956b) Brasilianische Meeres-nematoden 1. (Ergebnisse eines studienaufenthaltes na der Universität São Paulo). Bol Inst Oceanogr 5(1–2):3–69

Gerlach SA (1956c) Die Nematodenbeseiedlung des tropischen Brandungsstrandes von Pernambuco, Brasilianische Meeres Nematoden II. Kiel Meeresforsch 12(2):202–218

Gerlach SA (1957) Die Nematodenfauna des Sandstrandes na der küste von Mittelb (Brasilianische Meeres-Nematoden IV). Mitt Zool Mus Berl 33:411–459

Gheskiere T, Hoste E, Vanaverbeke J et al (2004) Horizontal zonation patterns and feeding structure of marine nematode assemblages on a macrotidal, ultradissipative sandy beach (De Panne, Belgium). J Sea Res 52:211–226

Gheskiere T, Vincx M, Weslawski JM, Scapini F, Degraer S (2005) Meiofauna as descriptor of tourism-induced changes at sandy beaches. Mar Environ Res 60(2):245–265

Gheskiere T, Vincx M, Greet P et al (2006) Are strandline meiofaunal assemblages affected by a once only mechanical beach cleaning? Experimental findings. Mar Environ Res 61:245–264

Giere O (2009) Meiobenthology. The microscopic motile fauna of aquatic sediments, 2nd edn. University of Hamburg, Springer-Verlag, Berlin, Heidelberg

Gomes TP, Rosa Filho JS (2009) Composição e variabilidade espaço-temporal da meiofauna de uma praia arenosa amazônica (Ajuruteua, Pará). Iheringia Sér Zool 99(2):210–216

Gomes-Júnior E, Santos E, da Rocha CMC et al (2018) A new species of *Ligiarctus* (Tardigrada, Heterotardigrada) from the Brazilian continental shelf, Southwestern Atlantic Ocean. Mar Biodivers 48(1):5–12

Gomes-Júnior E, Santos É, da Rocha C et al (2020) The deep-sea genus *Coronarctus* (Tardigrada, Arthrotardigrada) in Brazil, South-Western Atlantic Ocean, with the description of three new species. Diversity 12(2):63

Guidi L, Garraffoni ARS, Semprucci F, Balsamo M (2020) Spermatozoa ultrastructure, spermatogenesis and reproductive system of Acanthodasys australis (Gastrotricha, Macrodasyida). Zool Anz 286:108–116

Guidi L ,Balsamo M, Ferraguti M, Todaro MA (2021) Reproductive organs and spermatogenesis of the peculiar spermatozoa of the genus Kryptodasys (Gastrotricha Macrodasyida) with an appraisal of the occurrence and origin of the tail-less spermatozoa in Gastrotricha. J Zool Syst Evol Res 59(8):1673–1688

Gusmão F, Di Domenico M, Amaral ACZ et al (2016) In situ ingestion of microfibres by meiofauna from sandy beaches. Environ Pollut 216:584–590

Herranz M, Di Domenico M, Sørensen MV et al (2019) The enigmatic kinorhynch *Cateria styx* Gerlach, 1956–a sticky son of a beach. Zool Anz 282:10–30

Herranz M, Park T, Di Domenico M et al (2021a) Revisiting kinorhynch segmentation: variation of segmental patterns in the nervous system of three aberrant species. Front Zool 18(1):1–23

Herranz M, Worsaae K, Park T et al (2021b) Myoanatomy of three aberrant kinorhynch species: similar but different? Zoomorphology 140(2):193–215

Higgins RP (1968) Taxonomy and postembryonic development of the Cryptorhagae, a new suborder for the mesopsammic kinorhynch genus *Cateria*. Trans Am Microsc Soc 87(1):21–39

Hochberg R (2014) Crasiella fonseci a new species of Gastrotricha (Macrodasyida Planodasyidae) from São Paulo Brazil. Marine Biodivers 44:237–242

Höfling-Epiphanio E (1972) Ocorrência de *Batillipes mirus* Richters, 1909 e *B. tubernatis* Pollock, 1971 (tardigrada) no litoral brasileiro. Ciênc Cult 24:358–359

Hooge MD, Rocha CE (2006) Acoela (Acoelomorpha) from the northern beaches of the state of São Paulo, Brazil, and a systematic revision of the family Otocelididae. Zootaxa 1335(1):1–50

Jakobi H (1953a) Novos Laophontidae (Copepoda – Crustacea) da costa brasileira. Dusenia 4(1):47–60

Jakobi H (1953b) Neue Tegastiden (Harpacticoida – Copepoda) von der Kueste Santa Catarinas (Brasilien). Dusenia 4(2):173–180

Jakobi H (1954) Harpacticoida (Cop. Crust.) da microfauna do substrato areno-lodoso do “Mar de Dentro” (Ilha do Mel – Baía de Paranaguá – Brasil). Dusenia 5(5/6):209–232

Jakobi H (1956) Novas espécies de Harpacticoidea (Copepoda – Crustacea) provenientes de regiões de água salobra da costa São Paulo – Paraná. Dusenia 7(3):159–171

Jakobi H, Nogueira MH (1960) Descrição de uma nova espécie de Ectinosoma encontrada em águas salobras da Ilha de Santa Catarina (Copepoda – Harpacticoidea). Bol Univ Paraná (Zool) 1(5):1–10

Jondelius U, Jondelius Y (2020) Acoelomorpha. In: Schmidt-Rhaesa A (ed) Guide to the identification of marine meiofauna. Friedrich Pfeil Publishing, pp 44–53

Jouin C (1970) Recherches sur les Protodrilidae (Archiannélides). I Etude morphologique et systématique du genre Protodrilus. Cah Biol Mar 11(4):367–434

Jumars PA, Dorgan KM, Lindsay SM (2015) Diet of worms emended: an update of polychaete feeding guilds. Annu Rev Mar Sci 7:497–520

Kaczmarek Ł, Bartels PJ, Roszkowska M et al (2015) The zoogeography of marine tardigrades. Zootaxa 4037(1):1–189

Kånneby T, Hochberg R (2015) Phylum Gastrotricha. In: Thorp J, Rogers DC (eds) Ecology and general biology: thorp and Covich’s freshwater invertebrates. Academic Press, pp 211–223

Kennedy AD, Jacoby CA (1999) Biological indicators of marine environmental health: meiofauna–a neglected benthic component? Environ Monit Assess 54:47–68

Kieneke A, Schmidt-Rhaesa A (2015) Gastrotricha. In: Schmidt-Rhaesa A (ed) Handbook of zoology. Gastrotricha, Cycloneuralia and Gnathifera. Vol. 3. Gastrotricha and Gnathifera. De Gruyter, Berlin, pp 1–134

Kihara TC (2003) Diversidade dos copépodes harpacticóides da meiofauna marinha do litoral norte do estado de São Paulo. Tese de doutorado. Universidade de São Paulo, São Paulo

Kihara TC, Rocha CEF (2007) *Microarthridion corbisierae* sp. nov. (Harpacticoida, Tachidiidae), um novo copépode da meiofauna do litoral norte do estado de São Paulo, Brasil. Rev Bras Zool 24(4):907–918

Kihara TC, Rocha CEF, Sousa RM et al (2011) Meiofauna de praia arenosa. In: Amaral ACZ, Nallin SAH (eds) Biodiversidade e ecossistemas bentônicos marinhos do litoral norte de São Paulo, sudeste do Brasil. UNICAMP/IB, Campinas, pp 388–401

Kirsteuer E (1967) Bredin-Archbold-Smithsonian biological survey of Dominica, 3: Marine archiannelids from Dominica. Proc US Natl Mus 123:1–6

Kunz H (1937) Zur Kenntnis der Harpacticoiden des Küstengewässers der Kieler Förde (Sudien an marinen Copepoden I). Kiel Meeresforsch 2(1):95–115

Kotwicki L, Szymelfenig M, De Troch M et al (2005) Latitudinal biodiversity patterns of meiofauna from sandy littoral beaches. Biodivers Conserv 14(2):461–474

Lang K (1965) Copepoda Harpacticoidea from the Californian Pacific coast. Kungl Sv Vetens Handlinger serie 10(2):1–560

Laumer CE, Bekkouche N, Kerbl A et al (2015) Spiralian phylogeny informs the evolution of microscopic lineages. Curr Biol 25:2000–2006

Leasi F, Andrade SC, Norenburg J (2016) At least some meiofaunal species are not everywhere. Indication of geographic, ecological and geological barriers affecting the dispersion of species of *Ototyphlonemertes* (Nemertea, Hoplonemertea). Mol Ecol 25:1381–1397

Lindgren E W (1975) Six meiobenthic Harpacticoida (Crustacea Copepoda) from North Carolina beaches. Cah Biol Mar 16(4):445–473

Marcus ED-BR (1946) On a new archeannelid, Saccocirrus gabriellae, from Brazil. Comun Zool Mus Hist Nat Montev 37(2):1–11

Marcus ED-BR (1947a) Turbelários marinhos do Brasil. Bol Fac Filos Ciênc Letras USP (Zool) 12:99–215

Marcus ED-BR (1947b) Nerilla mediterranea from Brazil. Comun Zool Mus Hist Nat Montev 45(2):1–6

Marcus ED-BR (1948) Further archiannelids from Brazil. Comun Zool Mus Hist Nat Montev 2:1–22

Marcus ED-BR (1952) On the South American Malacopoda. Bol Fac Filos Cienc Letras USP (Zool) 5:189–204

Marcus ED-BR (1955) On Turbellaria and polygordius from the Brazilian coast. Bol Fac Filos Cienc Letras USP (Zool) 20:19–65

Marcus E (1949) Turbellaria Brasileiros (7). Bol Fac Filos Cienc Letr Univ Sao Paulo Zool 14:7–155

Marcus Er (1950) Turbellaria Brasileiros (8). Bol Fac Filos Cienc Letr Univ Sao Paulo Zool 15:5–192

Marcus E (1951) Turbellaria brasileiros (9). Bol Fac Filos Cienc Letr Univ Sao Paulo Zool 16(9):5–215

Marcus E (1954) Turbellaria Brasileiros-XI. Pap Avulsos Zool 11:419–489

Marcus ED-BR (1957) On Turbellaria. An Acad Bras Cienc 29:153–191

Marcus ED-BR (1958) On South American Turbellaria. Acad Brasil Ciênc 30(3):391–417

Marcus E, Marcus ED-BR (1951) Contributions to the natural history of Brazilian Turbellaria. Comun Zool Mus Hist Nat Montev 3(63):1–25

Marcus E, Marcus ED-BR (1954a) The nudibranch Rhodope from South America. Comun Zool Mus Hist Nat Montev 4(68):1–8

Marcus E, Marcus ED-BR (1954b) Über Philinoglossacea und Acochlidiacea. Kiel Meeresforsch 10:215–223

Marcus E, Marcus ED-BR (1955) Sea-hares and side gilled slugs from Brazil. Bol Inst Oceanogr 6:349

Marcus E, Marcus ED-BR (1956) On two sacoglossan slugs from Brazil. Am Mus Novit 1796:1–21
Marcus E, Marcus ED-BR (1957) Turbellaria. Expl Hydrobiol Lacs Kivu Edouard et Albert Res Sci 3(2):27–52
Maria TF, Esteves AM, Smol N et al (2008) Nematodes from sandy beaches of Guanabara Bay, Rio de Janeiro, Brazil. Biociênci 16(2):92–103
Maria TF, De Troch M, Vanaverbeke J et al (2012) The importance of biological interactions for the vertical distribution of nematodes in a temperate ultra-dissipative sandy beach. Estuar Coast Shelf Sci 97:114–126
Maria TF, Paiva PC, Vanreusel A et al (2013a) The relationship between sandy beach nematodes and environmental characteristics in two Brazilian sandy beaches (Baía de Guanabara, Rio de Janeiro). An Acad Bras Ciênc 85(1):257–270
Maria TF, Esteves AM, Vanaverbeke J et al (2013b) Is nematode colonisation in the presence of *Scolelepis* in tropical sandy-beach sediment similar to the colonisation process in temperate sandy beaches? Braz J Biol 73(1):19–28
Maria TF, Smol N, Esteves AM (2013c) Two new species of *Metachromadora* (Nematoda: Desmodoridae) from Guanabara Bay, Rio de Janeiro, Brazil, and a revised dichotomous key to the genus. J Mar Biol Assoc UK 94(1):105–114
Maria TF, Wandeness AP, Esteves AM (2016) State of the art of the meiofauna of Brazilian sandy beaches. Braz J Oceanogr 64(sp2):17–26
Martínez A, Di Domenico M, Jörger K et al (2013) Description of three new species of *Protodrilus* (Annelida, Protodrilidae) from Central America. Mar Biol Res 9(7):676–691
Martins MO, Almeida TCM, Di Domenico M (2015) Vertical distribution of meiofauna on reflective sandy beaches. Braz J Oceanogr 63(4):469–480
Masunari S (1987) *Parathalestris mourei*, a new species of seaweed-dwelling copepod (Harpacticoida, Thalestridae) from Santos Bay, Brazil. Crustaceana 54(1):104–112
Mcintyre AD (1968) The meiofauna and macrofauna of some tropical beaches. J Zool 156:377–392
McLachlan A (1985) The biomass of macro-and interstitial fauna on clean and wrack-covered beaches in Western Australia. Estuar Coastal Shelf Sci 21(4):587–599
McLachlan A (1988) Behavioural adaptations of sandy beach organisms: an ecological perspective. In: Chelazzi G, Vannini M (eds) Behavioral adaptation to intertidal life. Springer, Boston, pp 449–475
McLachlan A, Defeo O (2018) The ecology of sandy shores, 3rd edn. Elsevier, Amsterdam, 560 pp
Medeiros LRA (1987) Conhecimento sobre meiobentos no Brasil e relato de um caso da costa sudeste-sul brasileira: Síntese de conhecimentos. Publ Aciesp 54:348–379
Mello CL, Carvalho AL, Faria LC et al (2019) Spatial distribution pattern of the aberrant Franciscideres kalenesos (Kinorhyncha) on sandy beaches of Southern Brazil. Zool Anz 282:44–51
Mielke (1979) Interstitielle Fauna von Galapagos XXV Longipediidae Canuellidae Ectinosomatidae (Harpacticoida). Mikrofauna Meeresboden 77: 1–106
Miller WR, Perry ES (2016) The coastal marine Tardigrada of the Americas. Zootaxa 4126(3):375–396
Moellmann AM, Corbisier TN (2003) Does touristic flow affect meiofauna of sandy beaches? Preliminary results. J Coast Res 35:590–598
Moore CG, Bett BJ (1989) The use of meiofauna in marine pollution impact assessment. Zool J Linnean Soc 96(3):263–280
Moreno M, Ferrero TJ, Granelli V et al (2006) Across shore variability and trophodynamic features of meiofauna in a microtidal beach of the NW Mediterranean. Estuar Coast Shelf Sci 66(3–4):357–367
Moura JR, Verçosa MM, Santos E et al (2009) Ocorrência de *Parastygarctus sterreri* Renaut-Mornant, 1970 e *Halechiniscus perfectus* Schulz, 1955 (Tardigrada, Heterotardigrada) no arquipélago São Pedro São Paulo, RN, Brasil. Atlantica 31:213–216

Nelson DR, Guidetti R, Rebecchi L (2015) Phylum Tardigrada. In: Thorp JH, Rogers DG (eds) Ecology and general biology: thorp and Covich's freshwater invertebrates, vol 1, 4th edn. Academic Press, pp 347–380

Netto SA, Meneghel A (2014) Pulse of marine subsidies: the role of surf diatom Asterionellopsis glacialis accumulations in structuring the meiofauna of sandy beaches. Mar Biodivers 44(3):445–457

Netto SA, Pagliosa PR, Fonseca ALO et al (2007) Interações entre o microfitobentos, meiofauna e macrofauna (Praia Comprida, Santa Catarina). Braz J Ecol 11:78–82

Neuhaus B, Higgins RP (2002) Ultrastructure, biology and phylogenetic relationships of Kinorhyncha. Integr Comp Biol 42:619–632

Neuhaus B, Kegel A (2015) Redescription of *Cateria gerlachi* (Kinorhyncha, Cyclorhagida) from Sri Lanka and of *C. styx* from Brazil, with notes on *C. gerlachi* from India and *C. styx* from Chile, and the ground pattern of the genus. Zootaxa 3965(1):1–77

Nicholls AG (1939) Marine harpacticoids and cyclopoids from the shores of the St Lawrence. Nat Canad Quebec 66:241–316

Norenburg JL, Stricker SA (2002) Chapter 7: Phylum Nemertea. In: Young CM (ed) Atlas of marine invertebrate larvae. Academic Press, San Diego, pp 163–177

Oliveira EB, Soares-Gomes A (2003) Punctual impact of a domestic sewage on intertidal meiofauna of Charitas beach, Niterói, Rio de Janeiro, Brazil. J Coast Res 35:573–579

Philippi A (1840) Zoologische Bemerkungen (Fortsetzung). Arch Naturgesch 6(1): 181–195

Philippe H, Brinkmann H, Copley RR et al (2011) Acoelomorph flatworms are deuterostomes related to Xenoturbella. Nature 470:255–260

Pinto TKO, Santos PJP (2006) Meiofauna community variation in a Brazilian tropical sandy beach. Atlantica 28(2):117–127

Rao GC (1967) On the life-history of a new sand dwelling harpacticoid copepod. Crustaceana 13(2):129–136

Remane A (1933) Verteilung und Organisation der benthonischen Mikrofauna der Kieler Bucht. Wiss Meeresun 21:161–221

Reygel P, Schockaert E, Janssen T et al (2014) Two new species of Carcharodorhynchus Meixner, 1938 (Platyhelminthes: Rhabdocoela: Schizorhynchidae) from Brazil and Lanzarote. Mar Biodivers 44(3):279–285

Rocha CEF, Kihara TC, Sousa RM Jr et al (2011) Copepoda. In: Amaral ACZ, Nallin SAH (eds) Biodiversidade e ecossistemas bentônicos marinhos do litoral norte de São Paulo, sudeste do Brasil. UNICAMP/IB, Campinas, pp 189–202

Rodríguez JG, Lastra M, López J (2003) Meiofauna distribution along a gradient of sandy beaches in northern Spain. Estuar Coast Shelf Sci 58:63–69

Rosa Filho JS, Gomes TP, Almeida MF et al (2011) Benthic fauna of macrotidal sandy beaches along a small scale morphodynamic gradient on the Amazon coast (Algodoal Island, Brazil). J Coast Res 64:435–439

Rouch R (1962) Harpacticoides (crustacés copépodes) d'Amérique du Sud. Biol l'Am Aust 1:237–280

Ruebush TK (1939) The occurrence of the two rare genera, *Protohydra* and *Protodrilus*, on the east coast of North America. Science 90(2348):617–618

Santos PJP, Silva VMAP (1992/93) Macrochaeta westheidei n.sp., first record of Acrocirridae for the Brazilian coast (Annelida, Polychaeta). Nerítica 7:7–12

Santos TMT, Venekey V (2018) Meiofauna and free-living nematodes in volcanic sands of a remote South Atlantic, oceanic island (Trindade, Brazil). J Mar Biol Assoc UK 98(8):1919–1934

Santos E, da Rocha CMC, Gomes-Junior E et al (2017) Three new *Batillipes* species (Arthrotardigrada: Batillipedidae) from the Brazilian coast. Zootaxa 4243(3):483–502

Santos GH, Cardoso RS, Maria TF (2019a) Bioindicators or sediment relationships: evaluating ecological responses from sandy beach nematodes. Estuar Coast Shelf Sci 224:217–227

Santos E, Veiga P, Rubal M et al (2019b) *Batillipes pennaki* Marcus, 1946 (Arthrotardigrada: Batillipedidae): deciphering a species complex. Zootaxa 4648(3):549–567

Santos TMT, Petracco M, Venekey V (2021) Recreational activities trigger changes in meiofauna and free-living nematodes on Amazonian macrotidal sandy beaches. Mar Environ Res 167:105289

Scarpa F, Cossu P, Delogu V et al (2017) Molecular support for morphology-based family-rank taxa: the contrasting cases of two families of Proseriata (Platyhelminthes). Zool Scr 46:753–766

Schlacher TA, Dugan J, Schoeman DS, Lastra M, Jones A, Scapini F, Defeo O (2007) Sandy beaches at the brink. Divers Distrib 13(5):556–560

Schockaert ER, Moons P, Janssen T et al (2019) On the genus *Typhlopolycystis* Karling, 1956 (Platyhelminthes, Kalyptorhynchia, Polycystididae), with data on the five known species and the description of eleven new species. Zootaxa 4603(1):81

Schmidt R, Westheide W (1977) Interstitielle Fauna von Galapagos XVII Polygordiidae Saccocrridae Protodrilidae Nerillidae Dinophilidae (Polychaeta). Mikrofauna Meeresboden 62:1–38

Schratzberger M, Ingels J (2018) Meiofauna matters: the roles of meiofauna in benthic ecosystems. J Exp Mar Biol Ecol 502:12–25

Scott T, Scott A (1895) XLVII—On some new and rare British Copepoda. J Nat Hist 16(95) 353–362

Scott T (1902) Notes on gatherings of Crustacea collected by the Fishery Steamer "Garland" and the steam trawlers "Star of Peace" and "Star of Hope" of Aberdeen during the year 1901. Rep Fish Board Scotl 20(3):447–485

Scott T (1903) Notes on the Copepoda from the Arctic Seas collected in 1890 by the Rev Canon A M Norman. Ann Mag Nat Hist 11(7): 4–32

Short AD, Klein AHF (2016) Brazilian beach systems. Springer, Cham, 611p

Siewing R (1954) Zur Verbeitung von Pisionidens indica Aiyar and Alikunnhi. Kiel Meeresforsch 10:81–83

Silva APC (2006) Efeito da maré na distribuição vertical da meiofauna e de Copepoda Harpacticoida na praia arenosa de Maracaípe (Pernambuco, Brasil). Dissertação de Mestrado, Universidade Federal Pernambuco, Recife

Silva VMAP, Grohmann PA, Nogueira CSR (1991) Studies of meiofauna of Rio de Janeiro, Brazil. Coast Zone 91(3):2011–2022

Sørensen MV, Pardos F (2008) Kinorhynch systematics and biology—an introduction to the study of kinorhynchs inclusive identification keys to the genera. Meiofauna Marina 16:21–73

Sørensen MV (2011) Kinorhyncha. In: Amaral ACZ, Nallin SAH (eds) Biodiversidade e ecossistemas bentônicos marinhos do litoral norte de São Paulo, sudeste do Brasil. Instituto de Biologia, pp 106–110

Sørensen MV (2014) First account of echinoderid kinorhynchs from Brazil, with the description of three new species. Mar Biodivers 44(3):251–274

Sørensen MV, Dal Zotto M, Rho HS, Herranz M, Sánchez N, Pardos F, Yamasaki H (2015) Phylogeny of Kinorhyncha based on morphology and two molecular loci, PLoS One 10(7) e0133440

Sørensen MV, Pardos FM (2020) Kinorhyncha. In: Schmidt-Rhaesa A (ed) Guide to the identification of marine meiofauna. Friedrich Pfeil Publishing, pp 239–270

Souza-Santos LP, Ribeiro VSS, Santos PJP et al (2003) Seasonality of intertidal meiofauna on a tropical sandy beach in Tamandare Bay (Northeast Brazil). J Coast Res 35:369–377

Strand M, Sundberg P (2010) Stjarnmaskar-slemmaskar: Sipuncula-Nemertea. Nationalnyckeln till Sveriges flora och fauna. Art Databanken, 188 pp

Struck TH, Golombek A, Weigert A et al (2015) The evolution of annelids reveals two adaptive routes to the interstitial realm. Curr Biol 25:1993–1999

Suciu MC, Tavares DC, Costa LL et al (2017) Evaluation of environmental quality of sandy beaches in southeastern Brazil. Mar Pollut Bull 119(2):133–142

Sun X, Zhou H, Hua E, Xu S, Cong B, Zhang Z (2014) Meiofauna and its sedimentary environment as an integrated indication of anthropogenic disturbance to sandy beach ecosystems. Mar Pollut Bull 88(1-2):26–267

Tarragô LD, Ozorio CP (2017) Influence of storm surges on intertidal meiofauna of an exposed sandy beach. Braz J Oceanogr 65(4):709–714

Thompson RC, Olsen Y, Mitchell RP et al (2004) Lost at sea: where is all the plastic? Science 304:838–838

Tilbert S, Castro FJV, Tavares G et al (2019) Spatial variation of meiofaunal tardigrades in a small tropical estuary (~6°S; Brazil). Mar Freshw Res 70:1094–1104

Todaro MA (2012) A new marine gastrotrich from the State of São Paulo (Brazil), with a key to species of *Pseudostomella* (Gastrotricha, Thaumastodermatidae). Zookeys 223:39–51

Todaro MA (2013) A new non-naked species of *Ptychostomella* (Gastrotricha) from Brazil. Zookeys 289:13–24

Todaro MA (2020) Marine and freshwater Gastrotricha. In: Gastrotricha World Portal. Available at: http://www.gastrotricha.unimore.it/marine.htm. Accessed 2 Feb 2020

Todaro MA, Rocha CEF (2004) Diversity and distribution of marine Gastrotricha along the northern beaches of the state of São Paulo (Brazil), with description of a new species of *Macrodasys* (Macrodasyida, Macrodasyidae). J Nat Hist 38(13):1605–1634

Todaro MA, Rocha CEF (2005) Further data on marine gastrotrichs from the State of São Paulo and the first records from the State of Rio de Janeiro (Brazil). Meiofauna Marina 14:27–31

Todaro MA, Sibaja-Cordero JA, Segura-Bermúdez OA, Coto-Delgado G, Goebel-Otárola N, Barquero JD, Cullell-Delgado M, Dal Zotto M (2019a) An introduction to the study of Gastrotricha, with a taxonomic key to families and genera of the group. Diversity 11(7):117

Todaro MA, Dal Zotto M, Kånneby T et al (2019b) Integrated data analysis allows the establishment of a new, cosmopolitan genus of marine Macrodasyida (Gastrotricha). Sci Rep 9(1):7989

Urban-Malinga B, Kotwicki L, Gheskiere TL, Jankowska K, Opalinski K, Malinga M (2004) Composition and distribution of meiofauna including nematode genera in two contrasting Arctic beaches. Polar Biol 27:447–457

Venekey V, Fonseca-Genevois VG, Santos PJP (2010) Biodiversity of free-living marine nematodes on the coast of Brazil: a review. Zootaxa 2568:39–66

Venekey V, dos Santos PJP, Fonsêca-Genevois VG (2014a) The influence of tidal and rainfall cycles on intertidal nematodes: a case study in a tropical sandy beach. Braz J Oceanogr 62(4):247–256

Venekey V, Gheller PF, Maria TF et al (2014b) The state of the art of Xyalidae (Nematoda, Monhysterida) with reference to the Brazilian records. Mar Biodivers 44:367–390

Verçosa MM, Santos ÉCLD, Souza JRBD et al (2009) Distribuição de *Stygactus bradypus* Schulz, 1951 (Tardigrada, Heterotardigrada) na margem sul da ilha de Itamaracá (PE, Brasil). Atlantica 31(1):25–33

Victor-Castro FJ, Fonsêca-Genevois V, Lira L et al (1999) Efeito da granulometria e da topografia sobre a distribuição de *Batillipes pennaki* Marcus, 1946 em zona tropical típica: restinga do Paiva, Pernambuco, Brasil. Trab Oceanogr 27(2):89–102

Vousdoukas MI, Ranasinghe R, Mentaschi L et al (2020) Sandy coastlines under threat of erosion. Nat Clim Chang 10:260–263

Walter TC, Boxshall G (2020) World of Copepods database. *Filexilia* Conroy-Dalton & Huys, 1997. Accessed through: World Register of Marine Species at: http://www.marinespecies.org/aphia.php?p=taxdetails&id=115192. Accessed 20 Mar 2020

Wandeness AP (1998) Estrutura espaço-temporal dos Copepoda Harpacticoida em um perfil na regiao de Macaé, RJ. Dissertação de Mestrado, Universidade Federal Fluminense, Niterói

Wandeness AP, Esteves AM, Nogueira CRS et al (1997) Meiofauna da Zona entremarés da Praia dos Anjos, Arraial do Cabo, RJ. In: Absalão RS, Esteves AM (eds) Ecologia de praias arenosas do litoral brasileiro. UFRJ, Rio de Janeiro, pp 93–106

Wandeness AP, Esteves AM, Nogueira CSR et al (1998) Copepoda Harpacticoida meiobentônicos da Coroa do Avião, Pernambuco. Trab Oceanogr 26(2):99–104

Weigert A, Bleidorn C (2016) Current status of annelid phylogeny. Org Divers Evol 16:345–362

Westheide W (1974) Interstitielle polychaete aus brasilianischen Sandstränden. Mikrofauna des Meeresbodens 31:1–16

Worsaae K (2020) Annelida (excluding Clitellata and Sipuncula). In: Schmidt-Rhaesa A (ed) Guide to the identification of marine meiofauna. Friedrich Pfeil Publishing, pp 239–270
Worsaae K, Kerbl A, Domenico MD, Gonzalez BC, Bekkouche N, Martínez A (2021) Interstitial Annelida. Diversity 13(2):77 https://doi.org/10.3390/d13020077
Worsaae K, Kristensen RM (2005) Evolution of interstitial polychaeta (Annelida). In: Morphology, molecules, evolution and phylogeny in polychaeta and related taxa. Springer, Dordrecht, pp 319–340
Yamasaki H (2019) *Gracilideres mawatarii*, a new genus and species of Franciscideridae (Allomalorhagida: Kinorhyncha)–a kinorhynch with thin body cuticle, adapted to the interstitial environment. Zool Anz 282:176–188

Chapter 4
Benthic Invertebrate Macrofauna

Guilherme Nascimento Corte and Antonia Cecilia Zacagnini Amaral

4.1 Introduction

The benthic invertebrate macrofauna is probably the better-known component of sandy beach biodiversity. These organisms are defined as those with sizes larger than 0.5 mm and are usually characterized according to their position on the seafloor. Many sandy beach organisms live and feed upon the sediment surface and are known as epifauna. Yet, most species are found within the sediment, inhabiting tubes or galleries, or simply living among the sand grains. Those are known as infaunal organisms.

Sandy beach macrofauna includes representatives of all major taxa such as nemerteans, anthozoans, platyhelminthes, sipunculids, annelids (polychaetes, oligochaetes, and echiurans), mollusks, crustaceans, hemichordates, insects, and echinoderms (McLachlan and Brown 2006). Most of these organisms permanently inhabit sandy beach environments, while others are eventual foragers. Their occurrence varies in several spatial scales and is mainly influenced by environmental characteristics such as sediment type, waves, and tides. To cope with the harsh and dynamic environment they live in, sandy beach macrofauna presents important adaptations such as a high degree of mobility and the ability to burrow rapidly (McLachlan and Brown 2006).

Sandy beach benthic macrofauna is essential for the functioning of sandy beach ecosystems as they drive important ecological processes such as bioturbation and cycling of organic matter. Also, several species are important resources for subsistence and income of fishing communities. In this chapter, we present the main

G. Nascimento Corte (✉)
College of Science and Mathematics, University of Virgin Islands, Saint Thomas, USVI, USA
e-mail: guilherme.corte@uvi.edu

A. C. Z. Amaral
Departamento de Biologia Animal, Instituto de Biologia, Universidade Estadual de Campinas IB/UNICAMP, Campinas, SP, Brazil

A. C. Z. Amaral et al. (eds.), *Brazilian Sandy Beaches*, Brazilian Marine Biodiversity, https://doi.org/10.1007/978-3-031-30746-1_4

macrobenthic groups found on Brazilian sandy beaches and discuss important topics of their ecology such as adaptations to the dynamic beach environment and spatial patterns of distribution. We also review the current knowledge and scientific gaps related to this important component of Brazilian sandy shores.

4.2 Diversity and Ecology of Brazilian Sandy Beach Macrofauna

The macrofaunal diversity of Brazilian sandy beaches follows similar trends observed worldwide, with a higher number of species and individuals belonging to Annelida (Polychaeta), Mollusca, and Crustacea (Amaral et al. 2016). Here, we present the most frequent and abundant groups along the Brazilian coast as well as other groups with high ecological importance due to their positive effect on local biodiversity or use as ecological indicators.

4.2.1 Annelida (Polychaetes)

Polychaete worms (annelids) are sensitive to wave action and increase in sediment grain size, becoming more common as beach exposure decreases and sands become finer (McLachlan and Brown 2006). Therefore, polychaetes are very abundant on sheltered or dissipative beaches, where they usually dominate the intertidal and subtidal zones (McLachlan and Brown 2006; Amaral et al. 2016). Due to the ecological and taxonomical advances made over the last few decades, the knowledge about the diversity and ecology of polychaetes in Brazil is significant. So far, about 1400 species, belonging to almost 500 genera and 70 families, have been registered on the Brazilian coast (10% of the valid species worldwide) (Amaral et al. 2013; Lana et al. 2017). The families Capitellidae, Glyceridae, Nephytidae, Nereididae, Onuphidae, Opheliidae, Orbiniidae, and Spionidae are likely the most abundant on Brazilian sandy shores, while Arenicolidae and Chaetopteridae can be highlighted due to their ecological importance.

Sandy beach *arenicolids* (lugworms) are among the largest polychaetes to be found on sandy beaches, ranging from 10 to 40 cm in length (Steiner et al. 2016). They typically occur in the intertidal region and shallow-water zones occupying U-shaped burrows easily recognized as a funnel-like depression indicating the head end and small piles of sediment (worm castings) in the tail end of the burrow (Steiner et al. 2016). Arenicolids are characterized as subsurface deposit feeders and can eat as much as several hundred times their body weights per day (Taghon 1988; Jumars et al. 2015). They circulate large volumes of water through their burrow and oxygenate the sand via bioturbation (i.e., the ecological process by which benthic macrofauna transport sediment particles and solutes through feeding, digging, and/or

Fig. 4.1 Annelida "Polychaeta." (**a**) Arenicolidae: *Arenicola brasiliensis*, full animal (São Paulo, photo: Gabriel Monteiro); (**b**) the marks left by the presence of *Arenicola*, a funnel-shaped depression and the piles of sediments, can be observed on the surface of the sediment (São Paulo, photo: Cecilia Amaral); (**c**) Capitellidae: *Capitella* sp., full animal (São Paulo photo: Gabriel Monteiro); (**d**) Chaetopteridae, full animal; (**e**) upper ends of Chaetoperidae tubes; (**f**) Glyceridae, full animal (São Paulo, photos: Álvaro Migotto)

building galleries). The species *Arenicola brasiliensis* is the most common arenicolid on Brazilian sandy beaches (Fig. 4.1a, b).

Capitellidae worms are highly opportunistic and capable of rapid colonization of environments subjected to organic enrichment, especially sheltered areas with high percentages of fine sand and mud. They are characterized as motile subsurface deposit feeders (Jumars et al. 2015; Silva et al. 2017). The genera *Capitella* and *Heteromastus* are the most common capitellids on Brazilian sandy beaches, and the species *Capitella capitata* is a very relevant taxon in ecological monitoring studies (Gorman et al. 2017; Checon et al. 2018a, 2021a). Yet, molecular studies found that this "species" is a complex of many cryptic species (Livi et al. 2017; Silva et al. 2017). Silva et al. (2017) investigated the diversity and distribution of the *C. capitata* complex along the Brazilian coast and revealed the existence of four new species: *Capitella aracaensis*, *C. biota*, *C. neoaciculata*, and *C. nonatoi* (Fig. 4.1c).

Chaetopterids are tubicolous and commonly found in U-shaped tubes in sediments of sheltered beaches and sand flats. They are characterized as mucus-net suspension feeders and have a body divided into three distinct sections, including flap-like chaetigers for active pumping (Jumars et al. 2015; Mirza et al. 2020). Chaetopterids are not among the most abundant polychaetes on Brazilian sandy beaches; however, they deserve particular attention due to the symbiotic communities they host inside their tubes. Britayev et al. (2017) studied the species belonging to the genus *Chaetopterus* in Vietnam and found that several species coexist in their tubes, such as the porcellanid crabs *Polyonyx* spp. and *Eulenaios cometes*, the pinnotherid crab *Tetrias*, the nudibranch *Phestilla*, the polynoid polychaete *Ophthalmonoe pettiboneae*, and the carapid fish *Onuxodon fowleri*. Similarly, the species *Chaetopterus variopedatus* (Fig. 4.1d, e), the most common along the Brazilian coast, is usually found with the commensal crab *Polyonyx gibbesi* inside its tube (Eça et al. 2013). According to Eça et al. (2013), *C. variopedatus* is a good biomonitor for metal contamination studies and can be used in ecological monitoring programs.

Glycerids are predatory worms found in soft and mixed sediments and rubble at all water depths, including sandy beaches. They are characterized as discreetly motile, setting up a gallery of connected burrows through which they hunt overpassing epifauna taking advantage of the great reach of their everted pharynx (Jumars et al. 2015). Three genera are found in Brazil: *Glycera*, *Hemipodia*, and *Hemipodus* (Amaral et al. 2013). The studies on Brazilian glycerids done so far are almost exclusively taxonomic. Recently, however, Almeida et al. (2019) used stable isotope analyses to show that *Hemipodia californiensis* is a top predator in sandy beach ecosystems (Fig. 4.1f).

Nephtyid worms are generally characterized as active burrowers. They often burrow just beneath the sediment-water interface in lower areas of intertidal zones and are known as carnivores, although several species may be subsurface deposit feeders (Jumars et al. 2015). The studies on Brazilian nephtyids focus on taxonomic aspects (e.g., Rizzo and Amaral 2007; Franco and Rizzo 2016), with 13 species identified so far: *Nephtys laciniosa*, *N. acrochaeta*, *N. caeca*, *N. fluviatilis*, *N. hombergi*, *N. magellanica*, *N. simoni*, *N. squamosa*, *Aglaophamus macroura*, *A. uruguayi*, *A. juvenalis*, *A. dibranchis*, and *Inermonephtys palpata*.

Nereidids (Fig. 4.2a) are motile or discretely motile omnivorous polychaetes (Jumars et al. 2015) mostly found in shallow-water sediments. Their omnivorous feeding behavior is usually characterized by selective deposit feeding on sand combined with macrophagous carnivory, scavenging, and herbivory (Jumars et al. 2015). Many species can undergo extensive transformations during reproduction, called epitokal modifications, which include enlargement of parapodial lobes, replacement of chaetae for natatory chaetae in posterior chaetigers, development of modified sensory structures, and retinal transdifferentiation (Peixoto and Santos 2016). These

Fig. 4.2 Annelida "Polychaeta." (**a**) Nereididae: *Laeonereis culveri,* anterior region (photo: Gabriel Monteiro); (**b**) Onuphidae: *Diopatra* sp., anterior and median region (photo: Álvaro Migotto); (**c**) *Diopatra* sp., upper end of the tube (photo: Álvaro Migotto); (**d**) Opheliidae: *Armandia* sp., full animal (photo: Gabriel Monteiro); (**e**) Orbiniidae, anterior region (photos: Gabriel Monteiro); (**f**) Spionidae: *Scolelepis* sp. (photo: Álvaro Migotto)

modifications allow them to adopt a short pelagic existence to swarm and spawn gametes. On Brazilian sandy shores, the euryhaline species *Laeonereis culveri* is commonly found on beaches close to estuarine areas. The population dynamics and secondary production *of Laeonereis culveri* were studied by Omena and Amaral (2003) in Southeast Brazil, while Sampieri et al. (2020) used cytochrome oxidase I (COI) barcode sequences to differentiate North and South American Atlantic populations of *L. culveri*.

Onuphids are elongated worms and most common in muddy sand of sheltered sandy beaches (Amaral et al. 2006). *Diopatra* is the most species-rich genus of Onuphidae (~60 species) worldwide and is characterized as tubiculous (Fig. 4.2b, c). In Brazil, the species *D. cuprea* is the most abundant. A recent study by Seixas et al. (2020), however, showed that *D. cuprea* is a species complex composed by four species described as *D. hannelorae* sp. nov., *D. marinae* sp. nov., *D. pectiniconicum* sp. nov., and *D. victoriae* sp. nov.

Opheliids (Fig. 4.2d) are very motile fusiform worms characterized as subsurface deposit feeders. Opheliids are active excavators and some species are strong swimmers as well (Jumars et al. 2015). On Brazilian sandy beaches, the most common genera are *Thoracophelia* (=*Euzonus*), *Armandia*, and *Travisia* (Souza and Borzone 2007). The population dynamics of *Thoracophelia furcifera* was investigated by Souza and Borzone (2007) and Otegui et al. (2012) in southern Brazil. Both studies found that this species has a well-defined distribution in the upper intertidal zone and recruitment peaks during winter, reaching densities up to 3000 individuals per square meter.

Orbiniid polychaetes (Fig. 4.2e) are motile burrowers very common on muddy and sandy sheltered beaches. The worms of this family may have a long body – reaching up to 300 mm – and are characterized as subsurface deposit feeders. The genera *Orbinia*, *Scoloplos*, and *Haploscoloplos* are the most common on Brazilian sandy beaches. Despite their importance in sandy beach ecosystems, population dynamics (Chap. 6) and ecological aspects of Orbiniidae species are still largely overlooked in Brazil.

Spionidae is one of the richest families within Polychaeta with more than 500 species and 30 genera worldwide (Rocha et al. 2009). Among the genera belonging to the Spionidae family, *Scolelepis* (Fig. 4.2f) is the most abundant in beach environments (Rocha et al. 2009), commonly occurring in habitats with fine sediments and high organic matter content (Amaral and Morgado 1994; Rizzo and Amaral 2001). The wide distribution of *Scolelepis* species is mainly attributed to the morphological characteristics that allow them to adapt well to different sandy beach environments, such as the presence of an acuminate prostomium, which facilitates rapid excavation in the sandy substrate, and long palps capable of capturing food in the water column and on the surface of the sediment. Therefore, this species can act as a suspension or deposit feeder according to the presence or absence of water flow (Pardo and Amaral 2004; Rocha et al. 2009; Jumars et al. 2015). Seven *Scolelepis* species have been recorded along the Brazilian coast: *S. chilensis*, *S. goodbodyi*, *S. squamata*, *S. texana*, *S. lighti*, *S. quinquedentata*, and *S. gaucha* (Rocha et al. 2009). The population dynamics of *S. gaucha*, the most abundant polychaete species in the intertidal zone of subtropical and temperate beaches along the Atlantic coast of South America (Santos 1991), was investigated by Santos (1991, 1994), who found that beach morphodynamics was the main factor controlling mortality and growth of this species. Souza and Borzone (2000) studied the distribution and population dynamics of *S. squamata* in Paraná state, South Brazil, and observed that organisms were concentrated in a narrow band along the middle intertidal zone. Moreover, they found that *S. squamata* may reach densities of more than 20,000 individuals per square meter and reproduce continuously throughout the year. Continuous reproduction was also observed for *S. goodbodyi* by MacCord and Amaral (2007) and Leão et al. (2012) in Southeastern Brazil. The importance of *Scolelepis* species is highlighted by its use as an indicator organism by the Brazilian monitoring network of coastal benthic habitats (ReBentos – http://www.rebentos.org/) (Amaral et al. 2015). Besides *Scolelepis*, the genera *Dispio*, *Laonice*, and *Prionospio* are also very common on Brazilian sandy shores, commonly found co-occurring within patches of *Scolelepis* species (Amaral and Nallin 2011).

4.2.2 *Mollusca*

Mollusks are usually the second-most important group in number of species and individuals on sandy beaches. However, they can often be the most important group when it comes to biomass, especially in protected bays and dissipative beaches (McLachlan and Brown 2006). Given their high biomass and importance as a food resource, mollusks are likely the most studied group regarding their population biology on Brazilian sandy beaches (Chap. 6). The taxonomic classes Gastropoda and Bivalvia are by far the most diverse on sandy shores in Brazil and worldwide.

4.2.2.1 Gastropoda

Most of the gastropods found on Brazilian sandy beaches belong to the families Cerithiidae, Nassariidae, Olividae, and Terebridae (Fig. 4.3).

Cerithiidae is a large family of medium-sized marine gastropods with size ranging from 3 to 150 mm. Cerithiids are herbivores and detritivores that graze the seabed mainly in the lower intertidal and sublittoral. The species most frequent on Brazilian sandy beaches is *Cerithium atratum* (Fig. 4.3a), an herbivorous that occupies both consolidated and unconsolidated environments and is generally associated with vegetation (Denadai et al. 2004). This species is an important component in the food web of coastal ecosystems and shell fragments of *C. atratum* were observed in the stomach of crustaceans, gastropods, sea stars, and fishes (Houbrick 1974). The population biology of *C. atratum* was studied on the northern coast of the state of São Paulo, Southeastern Brazil by Amaral et al. (2003) and Denadai et al. (2004), which found that this species represented about 45% of the total number of mollusks in intertidal assemblages of tide-dominated sandy beaches on the coast of São Paulo state.

Nassariids are a family of marine gastropods that are very common in dissipative beaches and sheltered areas. These snails have rounded shells with a high spire and can be herbivorous, carnivorous, or scavengers (Lucena et al. 2012). Within the family Nassariidae, the species *Phrontis* (=*Nassarius*) *vibex* (Fig. 4.3b) is likely the most common and most studied on Brazilian sandy beaches. Boehs et al. (2004) and Denadai et al. (2005) showed that *P. vibex* is one of the most abundant gastropods in

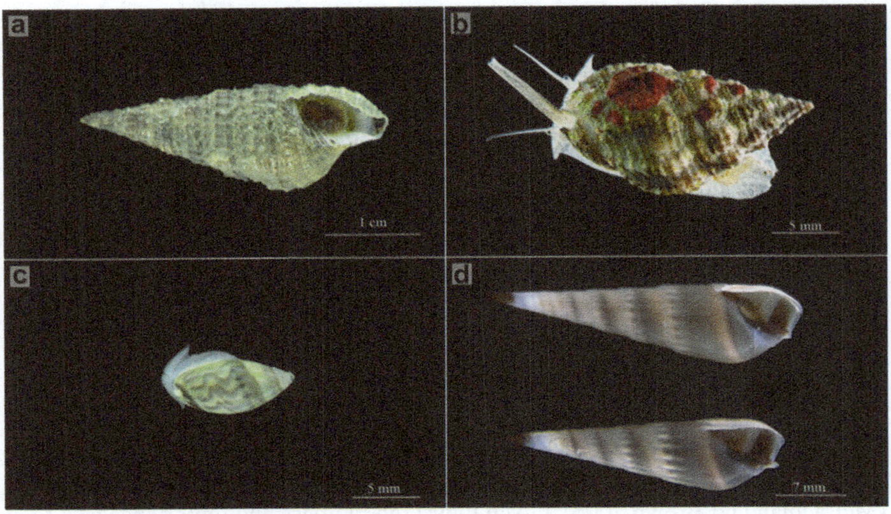

Fig. 4.3 Mollusca. Gastropoda. (**a**) Cerethiidae: *Cerithium atratum*; (**b**) Nassariidae: *Phrontis* (=*Nassarius*) *vibex*; (**c**) Olividae: *Olivella minuta*; (**d**) Terebridae: *Hastula cinerea*. (Photos: Álvaro Migotto)

sandy beach environments in South and Southeastern Brazil. The temporal and spatial variation in the egg-capsule deposition by *P. vibex* was investigated by Yokoyama and Amaral (2011a), who reported peaks during winter. The same authors studied the allometric growth of this species (Yokoyama and Amaral 2011b). The incidence of imposex (i.e., the presence of male characters in females, mainly related to Tributyltin (TBT) contamination) in different populations of *P. vibex* was examined by Cardoso et al. (2009, 2011) and Lima-Verde et al. (2010). The population dynamics (Chap. 6) of *P. vibex* were investigated by Cabrini and Cardoso (2012), who found a higher abundance during spring on a sheltered sandy beach in Southeast Brazil.

Individuals from the family *Olividae* are typically found semi-buried in the lower intertidal and subtidal zones of dissipative sandy beaches. They move by sub-superficial excavation and can be easily found by the trails they leave on the sediment (Fig. 4.4). The most common genera found on Brazilian sandy beaches are *Oliva*, *Olivella*, and *Olivancillaria*. The population biology of the species *Olivancillaria vesica* was studied by Caetano et al. (2003), who observed the highest population densities during winter and autumn in Southeast Brazil, while its feeding behavior and reproductive cycle and were investigated by Rocha-Barreira (2002, 2010, respectively). Yet, the most investigated olivid species on Brazilian shores is *Olivella minuta* (Fig. 4.3c), which had its distribution, growth, secondary production, reproduction, and diet investigated by Araujo and Rocha-Barreira (2012), Petracco et al. (2013), Corte et al. (2019a, b, 2020), and Checon et al. (2020, 2021b). For a long time, *O. minuta* was considered a carnivore/scavenger (Marcus and Marcus 1959; Arruda et al. 2003; McLachlan and Brown 2006). Yet, more recent studies have found that this species may shift its diet according to habitat

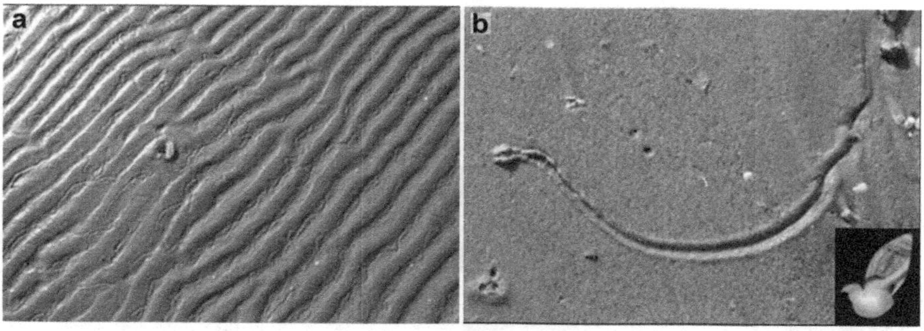

Fig. 4.4 Mollusca Gastropoda. (**a**, **b**) Olividae: *Olivella minuta*, traces left on the surface of the sediment. (Photos: Álvaro Migotto)

alterations and food availability, consuming both vegetal and animal organic matter (Corte et al. 2019a; Checon et al. 2020). The potential of *O. minuta* as an indicator of Tributyltin (TBT) contamination was assessed by Petracco et al. (2015).

Gastropods belonging to the family *Terebridae* are sand-dwelling carnivores that have high-spired shells with numerous whorls. *Hastula cinerea* (Fig. 4.3d) is likely the most common species of the Terebridae family on Brazilian sandy beaches, where it exhibits an aggregated pattern (tidal migration) and can reach densities of more than 100 individuals per square meter (Molina et al. 2013). Petracco et al. (2015) evaluated the potential of *H. cinerea* as a monitor of TBT contamination and suggested that this species has good potential based on its wide geographical distribution, common occurrence, easy collection, and association with TBT-contaminated sediments. Ragagnin and Turra (2021) used this species to show continued and widespread TBT contamination on Brazilian shores even after its international ban.

4.2.2.2 Bivalvia

Sandy beach bivalves typically occupy the lower zone of the intertidal area or occur below the tidal limit. Most of them are filter feeders but some species may feed on organic matter deposits on the sediment. Several families of bivalves can be registered on Brazilian sandy beaches. Among them, the families Donacidade, Tellinidae, and Veneridae are among the most common.

Bivalves belonging to the family *Donacidae* are adapted to strong hydrodynamism and can surf up and down the shore (McLachlan and Brown 2006). *Donax* species (Fig. 4.5a) are mostly filter feeders and present a wedge-shaped shell, which facilitates excavation and burial in the sediment. Their excavation is also facilitated by their foot with well-developed elevator muscles which allow a fast burrowing in the lower mid tidal and upper sublittoral zones of beaches of all morphodynamic types (Cardoso and Veloso 2003; Passos and Domaneschi 2004). Species from the genus *Donax* are among the most typical of Brazilian shores, where four species can be found: *D. striatus* and *D. denticulatus* are more commonly recorded on the

Fig. 4.5 Mollusca Bivalvia. (**a**) Donacidae: *Donax* sp.; (**b**) Tellinidae: *Tellina* sp.; (**c**) Tellinidae: *Macoma* sp.; (**d**) Tellinidae: *Macoma* sp., siphon marks on the sediment surface; (**e**) Veneridae: *Tivela mactroides*; (**f**) Veneridae: *Anomalocardia flexuosa*. (Photos: Álvaro Migotto)

Northeastern coast, while *D. hanleyanus* and *D. gemmula* are more frequently registered in the Southeastern region (Barroso et al. 2013). Many studies have investigated the ecological aspects of *Donax* species in Brazil, such as Cardoso and Veloso (2003), Petracco et al. (2019), and Almeida et al. (2019).

In contrast to donacids, bivalves from the *Tellinidae* family do not surf and are more easily found in sandy beach environments not subjected to strong hydrodynamism (McLachlan and Brown 2006). Therefore, they are more typical of sheltered and dissipative beaches, where they are considered major components of trophic chains due to their high number of species and abundance (Cardoso et al. 2015). Tellinids are usually deposit feeders and have ovate shells with slightly unequal valves. Most of the tellinids are distributed in the midlittoral and sublittoral zones of sandy beaches (Trevallion 1971; Denadai et al. 2001). The genera *Tellina* (Fig. 4.5b), *Eurytellina*, *Macoma* (Fig. 4.5c, d), and *Strigilla* are very common on Brazilian sandy beaches. The species *Eurytellina lineata* (*Tellina lineata*) and *T. versicolor* are widely distributed along the Brazilian coast, occurring from the Northeast to the South coast (Rios et al. 1994). Cardoso et al. (2015) investigated their population dynamics on a sheltered beach in Southeastern Brazil and found that *T. lineata* had higher density, biomass, and secondary turnover rates than *T. versicolor*, likely due to a difference in capacities to use the available resources.

Veneridae is a large family of medium-sized, suspension-feeder bivalves that displays a wide variety in shape and color. Similar to tellinids, they are more commonly found on sheltered or dissipative beaches. Many species from the family Veneridae are used as food resources along the Brazilian coast. Among them, the venerids *Tivela mactroides* (Fig. 4.5e), *Anomalocardia flexuosa (brasiliana)* (Fig. 4.5f), and *Iphigenia brasiliana* (*brasiliensis*) are some of the most common in Brazil, having value as economic and food resources for human populations (Amaral et al. 2016). Ecological aspects of *T. mactroides* have been extensively studied in Southeastern Brazil by Denadai et al. (2005, 2015a, b), Turra et al. (2014, 2016, 2018), and Rech et al. (2021). This species has an ontogenic distribution with young individuals mostly occurring in the subtidal and lower intertidal levels while bigger individuals dominate the upper intertidal zones of the beach (Denadai et al. 2005). It is likely that *T. mactroides* recruitment occurs in the subtidal/surf zone due to lower mortality rates related to stronger desiccation and individuals migrate to the intertidal after a growth period in the sublittoral (Denadai et al. 2005; Turra et al. 2014). The clam *Anomalocardia flexuosa* is one of the most conspicuous mollusks along the whole Brazilian coast and has been studied in the Southern (e.g., Boehs et al. 2008; Pezzuto et al. 2010), Southeastern (Mattos and Cardoso 2012; Corte et al. 2014, 2015, 2017), and Northeastern coast (e.g., Barreira and Araújo 2005; Boehs et al. 2008; Rodrigues et al. 2013; Oliveira et al. 2019).

4.2.3 *Crustacea*

The high mobility and resistant exoskeleton of crustaceans allow them to cope with the harsh swash climate of reflective beaches better than polychaetes and mollusks. Therefore, crustaceans are common across all beach types but are especially dominant in reflective ones. The exoskeleton also protects crustaceans against desiccation, which enables crustaceans to be the dominant macrofaunal group in upper zones of sandy beaches. Within Crustacea, the orders Amphipoda, Isopoda, and Decapoda host the most well-known groups of Brazilian sandy beach biodiversity.

Most *amphipods* inhabiting sandy beaches belong to the family Talitridae – the amphipod family most adapted to live in terrestrial environments (Serejo 2004). These amphipods are herbivorous or omnivorous and typical of the intertidal and upper zones of sandy beaches, where they are generally associated with wrack (although they can migrate up and down the shore in search of food) (McLachlan and Brown 2006). Talitrid amphipods exert a strong role in accelerating decomposition processes on sandy beaches and are also important prey of birds and fish (Veloso et al. 2008) (Fig. 4.6a). One particularly well-studied trait is their

Fig. 4.6 Arthropoda Crustacea. (**a**) Amphipoda: Talitridae (photo: Álvaro Migotto); (**b**) Decapoda: *Callichirus major* (photo: Cecflia Amaral); (**c**) *Albunea* (photo: Álvaro Migotto); (**d**) *Ocypode quadrata* (photo: Gabriel Monteiro); (**e**) *Calilnectes danae* (photo: Gabriel Monteiro); (**f**) Paguridae

orientation behavior (see the topic *Adaptations of sandy beach macrofauna* further in this chapter), which allows them to return to the optimal beach zone from a distant point (Scapini 2006, 2014). On Brazilian sandy beaches, the genera *Atlantorchestoidea* (*Pseudorchestoidea*), *Talorchestia,* and *Platorchestia* are the most abundant. The species *Atlantorchestoidea brasiliensis* is one of the most important species in macrofaunal assemblages of exposed sandy beaches in Brazil (Cardoso and Veloso 1996, 2001), as well as one of the most studied. Ecological investigations have focused on its population dynamics (Cardoso and Veloso 1996), reproduction (Cardoso and Veloso 2001), behavior strategies (e.g., Cardoso 2002), and influence of human disturbances (Veloso et al. 2008).

Similar to amphipods, *isopods* occur mostly in the upper zones of sandy beaches; however, they are also abundant in the midlittoral and sublittoral (McLachlan and

Brown 2006). Isopods are among the most abundant species of sandy beach macrofauna, generally characterized as scavengers, and have an important role in the energy flow of the beach ecosystem as they consume large amounts of organic matter and are consumed by top predators such as fish and birds (Petracco et al. 2010). The family Cirolanidae is the most conspicuous isopod taxon on sandy beach shores and includes the genus *Excirolana*, the most ubiquitous taxon on sandy beaches of the Americas (McLachlan and Brown 2006; Petracco et al. 2010). This genus is ovoviviparous and has no larval dispersal, recruiting directly to the benthos as juveniles. Two Excirolana species are very common in Brazil: *Excirolana armata* and *E. braziliensis*. The former is abundant on beaches of dissipative and intermediate morphodynamic states, while the second occurs in a wider range of morphodynamic beach states (Petracco et al. 2010). Studies have indicated that both species show latitudinal patterns, increasing in individual size and growth rates toward subtropical and temperate beaches, following an inverse relationship with mean water temperature (Cardoso and Defeo 2003, 2004; Petracco et al. 2010).

Decapods can be very abundant in the intertidal and subtidal zones of sandy beaches. A few species, such as the ghost shrimps *Callichirus major* and *Sergio mirim* (family Callianassidae) (Fig. 4.6b), present low-mobility. *Callichirus major*, popularly known as "corrupto" in Brazil, is a cryptic, solitary shrimp that inhabits deep burrows on sandy beaches, generally in the lower midlittoral or below mean water level of sheltered shores and dissipative beaches (Rodrigues and Shimizu 1997). This species is widely used as bait for artisanal fisheries along the Brazilian coast and has been studied in the Northeast (e.g., Botter-Carvalho et al. 2007), Southeast (e.g., Laurino et al. 2020a), and South (e.g., Souza et al. 1998). Similarly, *Sergio mirim* lives in shallow sublittoral zones of more sheltered shores and dissipative beaches and is found from the southern part of the state of Bahia (northeastern Brazilian) to the north of Argentina. The Pinnotherid crab *Pinnixa patagoniensis* is typically found associated with *S. mirim* burrows (Alves and Pezzuto 1998).

Most Brazilian decapods, however, are highly motile. Hermit crabs, mainly belonging to the genus *Clibanarius*, *Isocheles* (family Diogenidae), and *Pagurus* (family Paguridae), may be commonly found in the lower intertidal and subtidal zones of dissipative and tide-dominated Brazilian beaches and have been studied by Turra and Leite (1999), Sant'Anna et al. (2012a, b), Souza et al. (2015), Gorman et al. (2016), Turra et al. (2019), and Danin et al. (2020). Decapods may also be very abundant in the intertidal and subtidal zones of more turbulent beaches, such as species belonging to the genera *Albunea* (Fig. 4.6c), *Lepidopa*, *Hippa*, and *Emerita*, all present in considerable numbers on Brazilian sandy shores. The mole crab *Emerita brasiliensis*, popularly known as tatuí or tatuíra, inhabits the midlittoral zone of reflective and dissipative beaches of southeastern and southern regions of Brazil, where it usually attains high biomass and production (Petracco et al. 2017). Given its high ecological importance in bioturbation and as a consumer, several studies have investigated the ecology of *E. brasiliensis* in Brazil (e.g., Defeo and Cardoso 2004; Celentano and Defeo 2006; Celentano et al. 2010; Petracco et al. 2003, 2017).

Probably the most conspicuous decapods on Brazilian shores are crabs of the family Ocypodidae. Along the Brazilian coast (and also worldwide), ocypodid crabs occupy burrows in the upper zones of the shore. Ocypode crabs can act as scavengers or predators and defend their burrows with agonistic displays. They usually respond predictably to human impacts (Chap. 9), mainly by reduced burrow abundances and size, and have been suggested as useful bioindicators of anthropic impacts (Costa and Zalmon 2019a; Costa et al. 2019). The most common species is *Ocypode quadrata* (Fig. 4.6d), popularly known in Brazil as "Maria-farinha" or "Caranguejo fantasma," which has been the focus of many ecological investigations such as Turra et al. (2005), Pombo and Turra (2013, 2017, 2019), Pombo et al. (2017a, b), Campagnoli et al. (2018), Costa and Zalmon (2019a, b), Costa et al. (2019, 2021, 2022), Barboza et al. (2021), Souza et al. (2021), and Arueira et al. (2022). Other common brachyuran on Brazilian sandy beaches are blue crabs of the genera *Callinectes* (Fig. 4.5e) and the speckled swimming crab *Arenaeus cribarius*, scavengers that inhabit the subtidal zones of dissipative/sheltered sandy beaches.

4.2.4 Insecta

Several insect species are common inhabitants of sandy beach environments, while others are eventual sandy beach foragers. Among the insect fauna eventually found on sandy beaches, beetles and ants are the most common representatives. Their presence is strongly linked to vegetation in the dune area and the occurrence of stranded organic matter that can be used as a food resource (McLachlan and Brown 2006).

Regarding insects that truly inhabit sandy beaches, the order Coleoptera is likely the most important group (Corte et al. 2022). Sandy beach coleopterans have been recognized as important indicators of human impacts on sandy beach ecosystems, and the most abundant families on Brazilian shores are Tenebrionidae, Staphylinidae (rove beetles), and Cicindelinae (tiger beetles). In Southeast Brazil, Veloso et al. (2006) showed that the density of the tenebrionid *Phaleria testacea* is lower at urbanized beaches when compared to protected ones. Similarly, Costa and Zalmon (2019a, b) found that the abundance of the tiger beetle *Cylindera nivea* was negatively related to the urbanization level, and no individuals were found on high-impacted beaches. Also, the rove beetles of the genus *Bledius* (Fig. 4.7a, b) are used as indicator organisms by the Brazilian monitoring network of coastal benthic habitats (ReBentos – http://www.rebentos.org/). Four *Bledius* species have been registered along the Brazilian coast: *B. bonariensis*, *B.caribbeanus*, *B. fernandezi*, and *B. hermani*. Gandara-Martins et al. (2010) analyzed the relationship between sandy beach characteristics and the occurrence of three *Bledius* species on 12 sandy beaches of Paraná state. They *B. hermani* was the dominant species and most abundant and frequent on beaches with coarse sand and more reflective to intermediate morphodynamic. On the other hand, *B. bonariensis* had higher densities at more dissipative beaches. Furthermore, the highest densities of *Bledius* were recorded on beaches located in a national park, suggesting that the group is affected by human interference.

Fig. 4.7 Arthropoda Insecta. (**a**) Coleoptera: *Bledius* sp.; (**b**) *Bledius* sp., races left on the surface of the sediment. (Photos: Edilson Caron)

4.2.5 Echinodermata

Echinoderms are most commonly associated with rocky shores and are not the most typical group of sandy beach biodiversity; yet, they can be very abundant in the sublittoral zone of dissipative shond. *Leodia sexiesperforata* are commonly found along the Brazilian coast from the Northeast to the South (Tavares and Borzone 2006). Echinoderms are likely the least-studied major macrofaunal group on Brazilian beaches; however, some studies have been performed. Tavares and Borzone (2006) investigated the reproductive cycle of *Mellita quinquiesperforata* in an intermediate-dissipative and a reflective beach in South Brazil. The authors found that both populations exhibited a main reproductive period during spring/summer, but the spawning period was shorter in the former environment. Besides echinoids, asteroids (e.g., *Luidia senegalensis*; *Astropecten marginatus*) (Fig. 4.8b, c), holothurians (e.g., *Protankyra benedeni*) (Fig. 4.8d), and ophiuroids (e.g., *Hemipholis cordifera* and *Microphiopholis atra*) (Fig. 4.8e, f) can also be found in significant numbers in Brazilian sandy beaches (Alitto et al. 2018).

Fig. 4.8 Echinodermata. (**a**) Echinoidea: *Mellita quinquiesperforata* (photo: Álvaro Migotto); (**b**) Asteroidea: *Luidia senegalensis* (photo: Álvaro Migotto); (**c**) *Astropecten marginatus* (photo: Álvaro Migotto); (**d**) Holothuroidea: *Protankyra benedeni* (photo: Jéssica Prata); (**e**) Ophiroidea: *Hemipholis cordifera* (photo: Álvaro Migotto); (**f**) *Microphiopholis atra* (photo: Renata Alitto)

4.3 Adaptations of Sandy Beach Macrofauna

Sandy beach species need to respond quickly to the environmental changes constantly observed on sandy beaches such as the instability of the sediment, action of waves, and constant variation of the tides. Therefore, they display behavioral, morphological, and physiological adaptations that enable them to cope with such a dynamic environment.

4.3.1 Locomotion

Due to the intense dynamics observed in sandy beaches and the soft-bottom nature of the sedimentary bed, sessile animals are rare in this ecosystem. Thus, various forms of locomotion, such as surfing, crawling, jumping, and running are recorded in sandy beach macrofauna (Corte et al. 2021). The form and intensity of locomotion vary among groups and may also be influenced by environmental characteristics (Laurino et al. 2020b; Checon et al. 2021b). While most forms of locomotion demand high energy consumption, surfing is a means of transport with little energy expenditure and is usually associated with tidal migrations (up and down the intertidal region) or movement of the waves. In this form of transport, the animal moves with the help of the winds or water using part of its body as a sail. This form of locomotion is common mainly in mollusks, such as bivalves of the genus *Donax* and gastropods of the genus *Bulla*, which can extend their feet and siphons to capture the energy of the waves and winds (McLachlan and Brown 2006).

4.3.2 Excavation

Being able to bury quickly can be considered a matter of life and death for the inhabitants of sandy beaches. By burying themselves in the sand, they avoid their displacement by the action of waves, minimize water loss and desiccation, and also escape from predators. Excavation is such a fundamental activity among macrofaunal organisms that different adaptations have appeared.

Many mollusks and crustaceans have smooth bodies without ornamentation, which reduces resistance and facilitates penetration into the sediment. Bivalves, like representatives of the genera *Donax* and *Tellina*, have a wedge-shaped shell, which facilitates the downward penetration into the sediment. Polychaetes, in turn, excavate the sediment using the proboscis (eversible part of the pharynx) and parapodia (lateral projections of the body) as anchoring and excavation structures (Jumars et al. 2015).

Despite having rigid bodies due to their protective exoskeletons, crustaceans can excavate more quickly than most soft-bodied organisms and are probably the most diverse group in terms of excavation. Members of the Hippidae family such as the species *Emerita brasiliensis* perform the excavation "backward," using the last pair of legs to throw sand forward; individuals from the Talitridae family begin the excavation using their head; while *Ocypode* crabs perform lateral excavation by moving sand with their legs (Fig. 4.5d).

The sand dollar *Mellita quinquiesperforata* excavates and crawls by the sequential action of the spines (i.e., metachronal waves). This species has drainage channels that lead from the central region of the disc into the lunules and ambital notches, helping to relieve the excess water pressure on the oral surface and maintain its stability in strong currents (Fig. 4.9a–f).

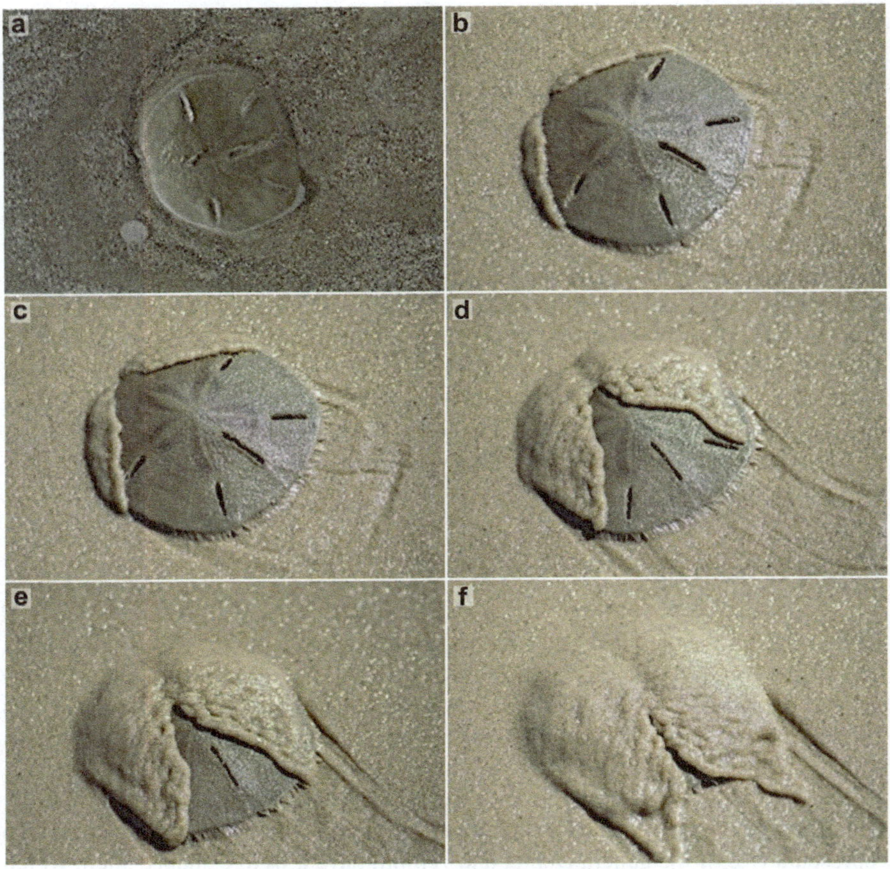

Fig. 4.9 Echinodermata. (**a–f**) Echinoidea: *Mellita quinquiesperforata*, excavation. (Photos: Álvaro Migotto)

4.3.3 Rhythmic Behavior

In general, rhythms can be defined as periodic changes in the behavior of species and are controlled by endogenous and environmental factors. The cyclical environmental changes that occur on sandy beaches (wave, tidal, lunar cycles, deposition and erosion of sediments, etc.) demand certain rhythmic behavior from organisms inhabiting these environments to perform their fundamental activities (e.g., foraging, reproduction, migration) always under adequate conditions.

Several species of macrofauna that can be easily found on the Brazilian coast, such as the crustaceans *Excirolana armata*, *Emerita brasiliensis*, and *Ocypode quadrata*, and the bivalve *Donax hanleyanus*, show rhythmic behavior following the rise and fall of the tide (Pombo et al. 2017b; Laurino et al. 2020b). Pombo et al. (2017a) showed that even species that inhabit the upper regions of the beach, such as the ghost crab *Ocypode quadrata*, may be significantly affected by tidal variations.

To start their activities, sandy beach macrobenthos mainly uses environmental clues, such as mechanical disturbance in the sediment, water content in the sediment, and low-frequency sounds (which are used as evidence of waves and swash) (McLachlan and Brown 2006). However, they may also rely on endogenous stimuli, especially in crustaceans which seem to have an internal clock that helps control these activities (Jelassi et al. 2015).

4.3.4 Orientation

As discussed in the previous topic, rhythmic activities are fundamental for many marine organisms and are triggered by environmental and endogenous elements. Several organisms rise to the surface and migrate to other regions of the beach according to the stimulus received; however, it is important not only to know when to start moving but also where to go.

Environmental stimuli can be divided into non-directional and directional (McLachlan and Brown 2006). Non-directional clues act as a trigger for the start of the activity, such as the higher water content in the sediment or the low-frequency sounds resulting from the action of the waves. Directional stimuli, in turn, are orientation cues that indicate where the animals should go. Examples of these stimuli are light exposure and intensity, the slope of the beach face, and water currents. Their effectiveness as orientation clues has been proven for many sandy beach crustacean species, especially sand-hoppers and beach-hoppers (Amphipoda, Talitridae) (Scapini et al. 2019). Bessa et al. (2017) performed field orientation experiments in Southeast Brazil using the talitrid *Atlantorchestoidea brasiliensis* and found that they use landscape cues and sun compass (i.e., solar orientation) to orient seaward. Moreover, they also found that the precision of orientation is highest in beaches without human access.

4.3.5 Feeding, Breathing, and Reproduction

The available resources on sandy beaches constantly change with environmental conditions, which results in the dominance of opportunistic species (which feed on a large variety of resources) and a small number of specialist species. Due to the lack of plants in the intertidal and subtidal zones, most of the beach fauna is formed by suspension feeders (that feed on resources available in the water column) and deposit feeders (that feed on debris and particulate organic matter deposited in the substrate). Some species, such as polychaetes of the genus *Scolelepis*, can change between suspension- and deposit-feeding modes depending on environmental conditions. In general, carnivorous species also vary their behavior between predator and scavenger according to food availability.

As sandy beach ecosystems lie at the interface between the land and the sea, species may obtain oxygen from the air or the water. Species around and above the drift line are usually terrestrial and air-breathing forms. On the other hand, species inhabiting the submerged zones of sandy beaches are marines and obtain dissolved oxygen from the water. Organisms inhabiting the intertidal zone are usually marines, but they are constantly exposed to both air and water. Consequently, they face regular periods of low-oxygen availability. To cope with these harsher conditions, some species, such as the crustacean *Emerita* spp., decrease their activities during low-oxygen periods, while other species, such as the polychaete *Arenicola*, have extracellular hemoglobin that can transport about 40 times more oxygen than human hemoglobin. Interestingly, *Arenicola*'s hemoglobin is compatible with all human blood types, and researchers have been investigating its use for human health (Zal and Rousselot 2014).

In addition to oxygen availability, variations in temperature and salinity can be major obstacles for inhabitants of sandy beaches. Changes in the salinity of the interstitial water may affect the behavior of sandy beach species, including their excavation capability (Laurino et al. 2020b). For this reason, many sandy beach organisms show greater tolerances than the range of values commonly found in their habitats. Examples of euryhaline (i.e., tolerant to salinity variations) and eurythermal (i.e., tolerant to thermal variations) species commonly found throughout the Brazilian coast are the bivalve *Anomalocardia flexuosa* and the polychaete *Laeonereis culveri*.

Sandy beach macrobenthos also show high reproductive variability in response to variations in environmental factors. Defeo and Cardoso (2002, 2004), for example, showed that the mole crab *Emerita brasiliensis* may shift from continuous to seasonal reproductive and recruitment events from subtropical to temperate beaches. Similarly, Corte et al. (2014) and Corte (2015) found that the reproductive cycle of the clam *Anomalocardia flexuosa* may differ on beaches with different morphodynamics and food availability. The availability of hard substrates may also strongly influence the reproduction of benthic species. Egg-laying sandy beach species, for example, may travel long distances to find a suitable place for oviposition or even defer their reproduction in the absence of suitable substrate. An alternative to overcome these difficulties is to attach their egg capsules to living organisms such as algae, bivalves, or gastropods. In Southeast Brazil, Corte et al. (2018) showed that the gastropod *Olivella minuta* may attach up to 55 egg capsules to the shell of living *Tivela mactroids* individuals.

4.4 Spatial Patterns of Sandy Beach Macrofauna

The number of species and individuals of sandy beach macrofaunal communities is characteristically heterogeneous in space, and variations in the spatial patterns of beach macrofauna are observed at different scales, from different latitudes to a few centimeters within the same beach. Understanding these variations is a fundamental step to predicting how environmental changes will affect sandy beach biodiversity and developing efficient conservation and management plans.

4.4.1 Macroscale Patterns

Macroscale variations in the distribution of sandy beach macrofauna are mainly determined by the environmental (morphodynamics) characteristics observed in different types of beaches and latitudinal differences (Defeo and McLachlan 2005).

One of the main paradigms in sandy beach ecology is that species richness, abundance, and biomass of macrofauna increase from reflective to dissipative beaches (McLachlan and Brown 2006; Amaral et al. 2016; Corte et al. 2022). This pattern is a result of the benign environmental characteristics observed in dissipative beaches, such as smaller sediment size, lower beach slope, and reduced wave action, which allow the occurrence of soft-bodied species and/or slow burrowers (Defeo and McLachlan 2005). These environmental features also allow species commonly found in the sublittoral to be recorded in the lower intertidal zone of dissipative beaches. As a consequence, vast populations and a large diversity of species may be found on dissipative beaches (Checon et al. 2018b; Corte et al. 2022). On the other hand, only terrestrial or highly mobile species, adapted to vigorous swash, can establish populations on reflective beaches. As a result, in extreme cases, less than one individual of macrofauna is found per square meter on reflective beaches, while this value can reach thousands of individuals at dissipative/sheltered beaches (McLachlan and Brown 2006).

Many studies have compared macrofaunal assemblages between sandy beaches of different morphodynamic states in Brazil (e.g., Nucci et al. 2001; Turra et al. 2003; Denadai et al. 2005). In Southeast Brazil, Checon et al. (2018b) investigated the macrofauna of several beaches and found that dissipative beaches had higher species richness than reflective beaches. Differences in assemblages were mainly related to sediment type, beach slope, and beach width, confirming the influence of morphodynamic features on sandy beach macrofauna. The authors also found that beaches closely located had more similar biodiversity than beaches far from each other, demonstrating the importance of spatial variables in structuring sandy beach macrofauna.

Morphodynamic features can also strongly influence the population attributes of sandy beach macrofauna (Chap. 6). Petracco et al. (2017) assessed the morphodynamic effects on production and turnover rate (P/B ratio) of the mole crab *Emerita brasiliensis* and found that production and P/B ratio increased toward intermediate/dissipative conditions due to higher growth rate and higher frequency of recruits. The morphodynamic influence on sandy beach populations, however, seems to vary between species. Corte et al. (2017) compared studies investigating population attributes of the clam *Anomalocardia flexuosa* and found that populations from areas with low hydrodynamic conditions (i.e., tidal flats and dissipative beaches) showed lower abundance, growth, recruitment, and turnover rate when compared to populations inhabiting intermediate beaches.

In addition to the type of beach, latitudinal differences also influence the macroscale variations of sandy beach macrofaunal communities. It is hypothesized that, as observed in terrestrial ecosystems, sandy beach communities become more diverse toward the tropics, mainly due to a larger group of species existing in tropical regions (Soares 2003). Yet, an inverse pattern is observed concerning the biomass of individuals: the closer to the equator, the smaller the size of individuals, mainly due to the lower primary productivity found in tropical regions (Defeo and McLachlan 2005). In a comprehensive study examining macrofaunal communities from 263 beaches on the Atlantic and Pacific coasts of South America, Defeo et al. (2017) found that macrofaunal richness decreased from tropical to temperate beaches in the Pacific and followed a parabolic trend in the Atlantic, with the highest biodiversity found at tropical and mid latitudinal bands. Latitudinal trends were mostly explained by differences in beach slope, tidal range, and chlorophyll *a*.

Latitudinal patterns have also been reported for sandy beach populations. Cardoso and Defeo (2004), for example, analyzed 11 populations of the isopod *Excirolana braziliensis* distributed from tropical (9°N) to temperate (39°S) sandy beaches in Atlantic and Pacific Oceans. The authors found strong latitudinal patterns, with breeding and recruitment patterns shifting from continuous on tropical beaches to seasonal on temperate beaches. Also, females became sexually mature at smaller sizes and had lower individual fecundity on tropical beaches. Results were consistent on both Atlantic and Pacific coasts of South America, thereby reinforcing the influence of latitude.

4.4.2 Mesoscale Patterns

Mesoscale patterns refer to variations that occur within the same beach. The patterns of distribution of macrofauna in this scale are predominantly influenced by the exposure/submersion gradient that is evident with tidal fluctuations. During periods of low tide, areas with different environmental characteristics (especially the content of water retained in the sediment) can be defined from the low tide water line to the supralittoral. These environmental differences establish a distribution pattern known as zonation, with species with different tolerances inhabiting different areas.

According to McLachlan and Jaramillo (1995), beaches can present four zonation patterns: (1) without apparent zonation (there is no significant difference between the upper and lower regions of the beach at low tide); (2) with two zones, one where the animals that remove oxygen from the air are located and another inhabited by animals that remove oxygen from the water (Brown's Zones); (3) three zones (sub-terrestrial fringe, midlittoral fringe, and sublittoral fringe), based on the distribution of crustaceans proposed by Dahl (1952); and (4) four zones (dry, retention, resurgence, and saturation), according to the amount of water retained in the sediment during low tide (Salvat 1964). Although they have different numbers of divisions, the division schemes of Dahl and Salvat are quite similar (Fig. 4.10), and sandy beach researchers usually recognize the existence of three zones: supralittoral, midlittoral (intertidal), and sublittoral. It is important to highlight that, despite being characterized as separated units, these zones form the core of the "littoral active zone" (LAZ), a land-sea interface environment characterized by the interchange of material between the sea and the land through wind- and wave-driven transport.

The supralittoral zone corresponds to the upper zones of Dahl and Salvat and is the most terrestrial area of the beach environment (strongly associated with the backshore). In this area, there is a high thermal variation, which increases the risk of desiccation for macrofaunal species. Additionally, due to the limited time in which this zone is submerged (only during exceptionally high tides), food resources from the marine environment are scarce, and species need to have high mobility to search for resources. Because of these characteristics, animals with high resistance to water loss and mobility, such as crustaceans (e.g., *Ocypode* spp. and Talitrids), and insects (e.g., the coleopterans *Phaleria* spp. and *Bledius* spp.) are the most common in this zone (Fig. 4.11).

The midlittoral (intertidal) is equivalent to the middle zone of Dahl and the zones of retention and resurgence of Salvat. It is characterized by an intermediate submersion time, which gives this region less thermal stress when compared to the

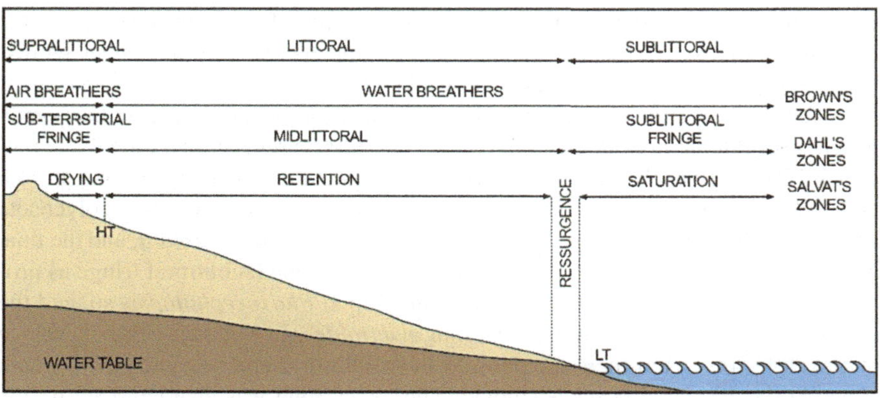

Fig. 4.10 Schemes of macrofauna zonation on sandy beaches. Water table corresponds to the upper level of an underground surface in which the soil or rocks are permanently saturated with water. HT: high tide line; LT: low tide line. (Modified from McLachlan and Brown (2006). Licensed by Elsevier and Copyright Clearance Center)

Fig. 4.11 Common macrofaunal species on Brazilian beaches and their zone of occurrence

supralittoral. However, the movement of water is more intense, which results in greater mechanical stress. The typical species of this zone have high mobility and/or are well adapted to the changing environmental conditions such as the crustaceans *Excirolana* (Isopoda) and *Emerita* (Hippidae), the gastropods *Hastula cinerea* and *Olivela minuta*, the bivalve *Donax* spp., and the polychaete *Scolelepis* sp. (Spionidae) (Fig. 4.11).

The sublittoral corresponds to the lower zone in the Dahl scheme and the Salvat saturation zone. It remains completely submerged most of the time (only its upper part may be exposed during extremely low tides) and is the zone with less thermal variation. However, it may be the zone with higher mechanical stress because of the constant wave action (surf zone). In reflective beaches, the strong wave energy precludes most of the macrofaunal species from inhabiting the sublittoral zone. On the other hand, this zone is where we can find the greatest diversity of species on dissipative and intermediate beaches (Corte et al. 2022).

A few investigations have focused on the zonation patterns of Brazilian sandy beach macrofauna. In South Brazil, Neves et al. (2007) registered a three-zone pattern in Cassino beach. The supralittoral zone was dominated by the ghost crab *Ocypode quadrata*; the midlittoral was characterized by the polychaeta *Thoracophelia (Euzonus) furciferus* and the isopod *Excirolana armata*; and the final zone, formed by the lower midlittoral and inner surf zone (sublittoral fringe as proposed by Dahl), was characterized by the amphipod *Phoxocephalopsis* sp. and the clams *Donax hanleyanus* and *Mesodesma mactroides*. In Southeast Brazil, Veloso et al. (2003) restricted their sampling to the midlittoral and recognized two main zones in the 15 sandy beaches studied. The lower zone was characterized by the mole crab *Emerita braziliensis*, the surf clam *Donax hanleyanus*, and the polychaetes *Scolelepis squamata* and *Hemipodia californiensis*, while the upper zone was characterized by the isopopd *Excirolana braziliensis*, the amphipod

Atlantorchestoidea brasiliensis and the coleopteran *Phaleria testacea* (Fig. 4.11). In a comprehensive study performed on 90 beach sites in Southeast Brazil, Corte et al. (2022) found that the shallow sublittoral and surf zones harbor the largest number of species and biomass among all sandy beach zones.

The zonation pattern, however, is not fixed and may be influenced by morphodynamic features. Alves and Pezzuto (2009) investigated the zonation pattern of benthic macrofauna of three sandy beaches with different morphodynamic characteristics in South Brazil, one dissipative, one intermediate, and one reflective, and found that macrobenthic zonation presented marked differences across the morphodynamic spectrum, with a higher number of species in lower zones of the dissipative beach. Additionally, they observed expansion in size and an increase in the overlapping of zones from reflective to dissipative conditions. These differences were mainly related to distinctive across-shore gradients in sediment moisture levels, sediment instability, and mean grain size throughout the morphodynamic spectrum.

4.4.3 Microscale Patterns

Microscale patterns correspond to variations from millimeters to meters. They are generally a result of environmental variations occurring in very small spatial scales (e.g., freshwater supply and water content in the sediment) and biotic interactions (Chap. 7). As observed worldwide, the number of ecological investigations at this spatial scale on Brazilian shores is much lower when compared to studies on macro and mesotidal scales; however, a few studies can be found. On the North coast of São Paulo, Southeast Brazil, Camargo (2009) showed that the abundance of gastropod *Olivella minuta* follows the ripple marks on the sediment (centimeter scale), with a higher number of individuals at sites with higher water content. By using field experiments, Laurino et al. (2020b) investigated the influence of freshwater on the vertical displacement of intertidal macrofauna and found that the isopod *Excironala armata* remains closer to the surface during freshwater floods. Also, Laurino and Turra (2021) assessed the macrofaunal spatial changes across few-meters gradients of freshwater influence and found that salinity reductions decreased the overall macrobenthic abundance and richness. This effect was higher for polychaetes, which had 85% fewer individuals near freshwater sources. Conversely, salinity reductions did not affect crustacea abundance across the gradient.

4.5 Ecological and Economic Importance

Sandy beach macrofauna plays an important role in coastal ecosystem functioning. They participate in both marine and terrestrial food chains and are important prey items for birds, fishes, and other invertebrates. Besides the trophic importance of macrofauna as basal food web organisms and primary consumers, many other species are important food resources for coastal species and also human populations.

Marine invertebrate fisheries on sandy shores have critical socioeconomic importance, especially in developing countries like Brazil (Defeo 2003). Sandy beach macrofaunal species have long been harvested by human populations, being the target of artisanal, recreational, or commercial fisheries[1] (Defeo 2003; McLachlan and Brown 2006). On Brazilian shores, clams and crabs are the groups most exploited by human populations. The venerid clams *Anomalocardia flexuosa* and *Tivella mactroids*, for example, are important fishery resources collected along the Brazilian coast (Corte et al. 2015, 2017; Amaral et al. 2016). Pezzuto and Silva (2015) investigated the exploitation of this species in the Pirajubaé Marine Extractive Reserve (State of Santa Catarina) and found that each fisherman may collect more than 270 kg of live clams per day, adding up to almost 900 tons in 1 year. Denadai et al. (2015a) reported intense harvesting of *T. mactroids* on the North Coast of the state of São Paulo, where it is gathered by both residents and tourists. Other species, such as the blue crab *Callinectes* spp. and the tanaidacean *Callichirus major*, also have economic relevance and are used as a food resource or bait. It is important to note, however, that most coastal fisheries along Brazilian coast are not supported by management or regulations based on solid scientific knowledge and most stocks are already overexploited and only support small-scale artisanal fisheries (Corte et al. 2015).

The cycling of nutrients is also an important function performed by sandy beach macrofauna. Given their strong association with the sediment, with many species living in burrows or inside the sediment, sandy beach macrofauna drives important processes such as sediment reworking, bio-irrigation, and organic matter decomposition, thereby modulating nutrient exchange through the sediment-water interface (Defeo et al. 2009). In Brazil, Pennafirme et al. (2019, 2020) assessed bioturbation by the polychaete *Laeonereis acuta* and found that it elevated bacterial metabolic activity and increased the carbohydrate consumption rate by the bacterial consortium.

Finally, macrobenthos is usually regarded as effective indicators for ecosystem health assessment and monitoring of human-induced impacts because of their widespread distribution, high biomass, and specific ecological requirements (Borja et al. 2016). Several species such as the ghost crab *Ocypode quadrata*, the tiger beetle *Cylindera nivea*, and the polychaete *Scolelepis squamata* (Amaral et al. 2015) respond to anthropogenic disturbances and have been used in impact assessments along the Brazilian coast (Costa et al. 2019) (Chap. 9).

[1]As defined by McLachlan and Brown (2006) a recreational fishery is the collection of the fishery resource for bait or food without sale or dependence on the resource. Artisanal fisheries are performed by individuals or small groups using traditional methods with focus on subsistence or sale. Finally, a commercial fishery is performed by corporate or collective organizations and focuses only on the sale of the resource.

4.6 Brazilian Sandy Beach Macrofauna: Current Knowledge and Scientific Gaps

A comprehensive review of the knowledge about Brazilian sandy beach macrofauna was performed by Amaral et al. (2016), which found a total of 126 studies on the ecology of sandy beach macrofauna published between 1976 and 2014. Although the investigation of Brazilian sandy beach macrofauna began in the 1970s, it was only in the late 1990s that the number of studies increased (Amaral et al. 2016). Ever since, studies on Brazilian sandy beach macrofauna have been published every year in international scientific journals, with higher production after 2000 (Fig. 4.12). The vast majority of studies (96%) focused on describing population and community patterns, as well as explaining their relation with environmental characteristics, particularly sediment type, salinity, and waves (Amaral et al. 2016). The species *Emerita brasiliensis*, *Excirolana brasiliensis*, *Atlantorchestoidea brasiliensis*, and *Ocypode quadrata* were the most studied crustaceans. Regarding mollusks, the most examined species are the bivalves *Anomalocardia brasliana*, *Tivela mactroides*, and *Donax hanleyanus*, as well as the gastropod *Olivella minuta*. Species belonging to the genus *Scolelepis* and *Laeonereis* were the most investigated within Polychaeta.

Despite the increased number of studies over the last decades, the amount of research conducted in each region of Brazil is quite unbalanced. While approximately 60% of the studies were conducted in the southeastern region, only 2% were done in the northern region (Northeast: 12%; South: 25%) (Amaral et al. 2016). The

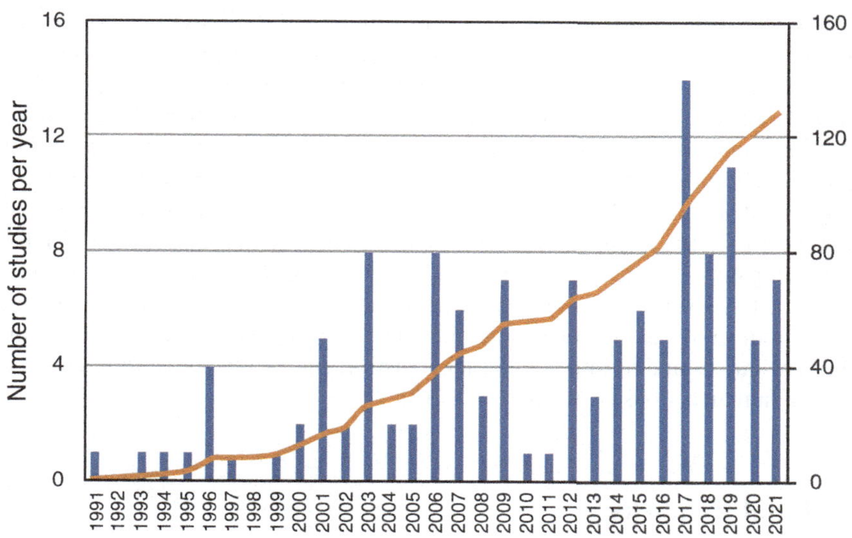

Fig. 4.12 Number of studies carried out on Brazilian sandy beach macrofauna. Only studies listed on the ISI Web of Science® were considered. Keywords: Brazil* AND beach* AND (macrofauna OR macrobenthos)

great heterogeneity of Brazilian sandy beaches (Chap. 1) and the limited knowledge about their biodiversity highlight the urgent need for further studies on a larger spatial scale to gain a better understanding of the Brazilian sandy beach biodiversity and the processes driving these ecosystems. This situation is perhaps most dire for the North region of Brazil. This area has unique environmental features such as a macrotidal regimen and strong influence of the rivers of the Amazon floodplain, however, to our knowledge, only Rosa Filho et al. (2009, 2011), Souza et al. (2021), and Santos et al. (2022) investigated the macrobenthic diversity of local sandy shores.

Another concern about sandy beach studies in Brazil relies on their duration, as only 5% have lasted more than 2 years. It is necessary to apply long-term studies to obtain consistent data to unveil the population dynamics of sandy beach species and serve as a baseline for predictive modeling of the responses of macrofaunal communities under different scenarios of global and regional changes (Amaral et al. 2016, http://www.rebentos.org). The understanding of responses of sandy beach macrofauna to environmental changes would also benefit from field and laboratory experiments, an approach that has proved to contribute significantly to a better understanding of these phenomena and their consequences (e.g., Laurino et al. , 2020b). It is also important to investigate the population biology of a larger number of species, given that most population studies examined a relatively small set of species (Chap. 6).

A relevant challenge that needs to be overcome in sandy beach research, including the research conducted in Brazil, is the necessity of performing ecological investigations that consider the sandy beach ecosystem as a whole. Most studies on the ecology of sandy beach, macrobenthic fauna inclusive, were restricted to the upper shore or intertidal zone, a fact that can result in fragmented information (Corte et al. 2019b, 2022). In this regard, it is important to perform investigations along the whole Littoral Active Zone, considering the supralitoral, the midlittoral, and sublittoral, as well as to conduct multidisciplinary studies that focus on the interdependence of adjacent ecosystems such as sandy beaches, dunes, rocky shores, and mangroves. A well-developed initiative can be found in the project Biota-Araçá, funded by the São Paulo Research Foundation (FAPESP), which presents the biodiversity of Araçá Bay, an area with 534.500 m^2 in Southeast Brazil comprising four sandy beaches, mangroves, intertidal mudflats, and rocky reefs. By investigating the ecological, geographical, historical, and socioeconomic characteristics of the bay, the project demonstrated how the advancement of scientific knowledge on biodiversity and its socioeconomic relevance is essential to the preservation of these resources and to improve the legal instruments fundamental to reduce the impact of human actions (Amaral et al. 2018).

4.7 Final Remarks

Brazilian sandy beach macrofauna plays a key role in supporting coastal ecosystems and human populations. Nevertheless, our knowledge about this important component of coastal biodiversity is sparse and insufficient to correctly understand its diversity and importance. This situation is further hindered by Brazilian historical and socioeconomic characteristics that endanger all its natural ecosystems, including sandy beaches. To gain a better understanding of Brazilian sandy beaches, including macrofaunal biodiversity, systematic and long-term studies are required on multiple scales. Only through this approach, we will be able to develop effective management and conservation plans that successfully protect sandy beach ecosystems in the long term and ensure their adequate functioning.

Acknowledgments This work was supported by the Fundação de Amparo à Pesquisa do Estado de São Paulo (FAPESP) (GNC, 2017/17071-9) and Conselho Nacional de Desenvolvimento Científico e Tecnológico (CNPq) (ACZA, 301551/2019-7; GNC, 165320/2020-6).

References

Alitto RAS, Bueno ML, Guilherme PDB et al (2018) Shallow-water brittle stars (Echinodermata: Ophiuroidea) from Araçá Bay (Southeastern Brazil), with spatial distribution considerations. Zootaxa 4405(1):1–66

Almeida TC, Rocha PF, Zalmon IR et al (2019) Is there an indication of the origin of nutrient supply in different morphological structures of macrofauna at two different Brazilian southeastern sandy beaches? Comparison by C and N stable isotopes. Environ Sci Pollut Res 26(32):33023–33029

Alves EDS, Pezzuto PR (1998) Population dynamics of Pinnixa patagoniensis Rathbun, 1918 (Brachyura: Pinnotheridae) a symbiotic crab of *Sergio mirim* (Thalassinidea: Callianassidae) in Cassino Beach, Southern Brazil. Mar Ecol 19(1):37–51

Alves EDS, Pezzuto PR (2009) Effect of morphodynamics on annual average zonation pattern of benthic macrofauna of exposed sandy beaches in Santa Catarina, Brazil. Braz J Oceanogr 57(3):189–203

Amaral ACZ, Morgado EH (1994) Alteraciones en la fauna de anélidos poliquetos de Araçá, São Sebastião (SP – Brasil). Rev Acad Colomb Cienc Exact Fis Nat 19:147–152

Amaral ACZ, Nallin SAH (2011) Biodiversidades e ecossistemas bentônicos marinhos do litoral norte de São Paulo, Sudeste do Brasil. Universidade Estadual de Campinas, Campinas

Amaral ACZ, Denadai MR, Turra A (2003) Intertidal macrofauna in Brazilian subtropical tide-dominated sandy beaches. J Coast Res SI35:446–455

Amaral ACZ, Elias R, Bone D et al. (2013) III Latin American symposium of polychaetes (III SILPOLY, Isla Margarita, Venezuela). Pan-Am J Aquat Sci 8:1–2

Amaral ACZ, Rizzo AE, Arruda EP (2006) Manual de identificação dos invertebrados marinhos da região sudeste-sul do Brasil. EdUSP, São Paulo

Amaral ACZ, Yokoyama LQ, Rocha MB et al (2015) Monitoramento de populações de *Scolelepis* (Polychaeta: Spionidae). In: Turra A, Denadai MR (eds) Protocolos Para o Monitoramento de Habitats Bentônicos Costeiros. Instituto Oceanográfico, São Paulo, pp 224–232

Amaral ACZ, Corte GN, Rosa Filho JS et al (2016) Brazilian sandy beaches: characteristics, ecosystem services, impacts, knowledge and priorities. Braz J Oceanogr 64:5–16

Amaral ACZ, Ciotti AM, Fonseca G (2018) Biodiversity and functioning of a subtropical coastal ecosystem: subsidies for integrated management. Ocean Coast Manag 164:1–156

Araujo PHV, Rocha-Barreira CA (2012) Population dynamic and secondary production of *Olivella minuta* (Gastropoda: Olividae) on Sandy Beach in Northeastern Brazil. Amici Molluscarum 20:7–15

Arruda EP, Domaneschi O, Amaral ACZ (2003) Mollusc feeding guilds on sandy beaches in São Paulo State, Brazil. Mar Biol 143(4):691–701

Arueira VF, Zalmon IR, Costa LL (2022) Is the ghost crab's feeding behavior a good early indicator of human pressure in sandy beaches? Reg Stud Mar Sci 53:102381

Barboza CAM, Mattos G, Soares-Gomes A et al (2021) Low densities of the ghost crab Ocypode quadrata related to large scale human modification of sandy shores. Front Mar Sci 8:589542

Barreira CAR, Araújo MLR (2005) Ciclo reprodutivo de *Anomalocardia brasiliana* (Gmelin, 1791) (Mollusca, Bivalvia, Veneridae) na Praia do Canto da Barra, Fortim, Ceará, Brasil. Bol Inst Pesca 31:9–20

Barroso CX, Rabay SG, Passos FD et al (2013) An extended geographical distribution of *Donax gemmula* Morrison, 1971 (Bivalvia: Donacidae): new record from the Brazilian Northeastern coast. Check List 9(5):1087–1090

Bessa F, Scapini F, Cabrini TMB et al (2017) Behavioural responses of talitrid amphipods to recreational pressures on oceanic tropical beaches with contrasting extension. J Exp Mar Biol Ecol 486:170–177

Boehs G, Absher TM, da Cruz-Kaled A (2004) Composition and distribution of benthic molluscs on intertidal flats of Paranaguá Bay (Paraná, Brazil). Sci Mar 68(4):537–543

Boehs G, Absher TM, da Cruz-Kaled AC (2008) Population ecology of *Anomalocardia brasiliana* (Gmelin, 1791) (Bivalvia, Veneridae) on Paranaguá bay, Paraná, Brazil. Bol Inst Pesca 34(2):259–270

Borja A, Elliott M, Andersen JH et al (2016) Overview of integrative assessment of marine systems: the ecosystem approach in practice. Front Mar Sci 3:20

Botter-Carvalho ML, Santos PJP, Carvalho PVVC (2007) Population dynamics of *Callichirus major* (Say, 1818) (Crustacea, Thalassinidea) on a beach in northeastern Brazil. Estuar Coast Shelf Sci 71:508–516

Britayev TA, Mekhova E, Deart Y et al (2017) Do syntopic host species harbour similar symbiotic communities? The case of Chaetopterus *spp.* (Annelida: Chaetopteridae). PeerJ 5:e2930

Cabrini TMB, Cardoso RS (2012) Population biology of Nassarius vibex (Say, 1822) on a sheltered beach in Southeastern Brazil. J Shellfish Res 31(3):809–815

Caetano CHS, Veloso VG, Cardoso RS (2003) Population biology and secondary production of *Olivancillaria vesica vesica* (Gmelin, 1791) (Gastropoda: Olividae) on a sandy beach in Southeastern Brazil. J Molluscan Stud 69(1):67–73

Camargo RM (2009) Distribuição espaço-temporal de Olivella minuta (LNK,1807) (Mollusca, Gastropoda, Olividae) na zona entremarés da Baía do Araçá, litoral norte do Estado de São Paulo. Master Thesis, Universidade de São Paulo.

Campagnoli ML, Pombo M, Turra A (2018) Ghost crab burrows simulation shows differential across-shore persistence. Crustaceana 91:821–830

Cardoso R (2002) Behavioural strategies and surface activity of the sandhopper *Pseudorchestoidea brasiliensis* (Amphipoda: Talitridae) on a Brazilian beach. Mar Biol 141(1):167–173

Cardoso RS, Defeo O (2003) Geographical patterns in reproductive biology of the Pan-American sandy beach isopod *Excirolana braziliensis*. Mar Biol 143(3):573–581

Cardoso RS, Defeo O (2004) Biogeographic patterns in life history traits of the Pan-American sandy beach isopod *Excirolana braziliensis*. Estuar Coast Shelf Sci 61(3):559–568

Cardoso RS, Veloso VG (1996) Population biology and secondary production of the sandhopper Pseudorchestoidea brasiliensis (Amphipoda: Talitridae) at Prainha Beach, Brazil. Mar Ecol Prog Ser 142:111–119

Cardoso RS, Veloso VG (2001) Embryonic development and reproductive strategy of *Pseudorchestoidea brasiliensis* (Amphipoda: Talitridae) at Prainha Beach, Brazil. J Nat Hist 35(2):201–211

Cardoso RS, Veloso V (2003) Population dynamics and secondary production of the wedge clam *Donax hanleyanus* (Bivalvia: Donacidae) on a high-energy, subtropical beach of Brazil. Mar Biol 142(1):153–162

Cardoso RS, Caetano CHS, Cabrini TMB (2009) Biphallia in imposexed females of marine gastropods: new record for *Nassarius vibex* from Brazil. Braz J Biol 69(1):223–224

Cardoso RS, Meireis F, Mattos G (2011) Crustaceans composition in sandy beaches of Sepetiba Bay, Rio de Janeiro, Brazil. Check List 7(6):778–781
Cardoso RS, Galhardo LB, Cabrini TM (2015) Population ecology and secondary production of congeneric bivalves on a sheltered beach in Southeastern Brazil. J Shellfish Res 34(3):931–938
Celentano E, Defeo O (2006) Habitat harshness and morphodynamics: life history traits of the mole crab *Emerita brasiliensis* in Uruguayan sandy beaches. Mar Biol 149(6):1453–1461
Celentano E, Gutiérrez NL, Defeo O (2010) Effects of morphodynamic and estuarine gradients on the demography and distribution of a sandy beach mole crab: implications for source–sink habitat dynamics. Mar Ecol Prog Ser 398:193–205
Checon HH, Vieira DC, Corte GN et al (2018a) Defining soft bottom habitats and potential indicator species as tools for monitoring coastal systems: a case study in a subtropical bay. Ocean Coast Manag 164:68–78
Checon HH, Corte GN, Shah Esmaeili YM et al (2018b) Nestedness patterns and the role of morphodynamics and spatial distance on sandy beach fauna: ecological hypotheses and conservation strategies. Sci Rep 8(1):1–10
Checon HH, Silva MO, Corte GN et al (2020) Full stomachs at empty tides: tidal cycle affects feeding activity and diet of the sandy beach gastropod *Olivella minuta*. J Molluscan Stud 86:219–227
Checon HH, Corte GN, Silva CF et al (2021a) Using the *Capitella complex* to investigate the effects of sympatric cryptic species distinction on ecological and monitoring studies in coastal areas. Mar Biodivers 51:48
Checon HH, Silva MO, Corte GN et al (2021b) Night underwater rides: the activity of a sandy beach gastropod is affected by interactive effects of light availability and water level. Mar Biol Res 17:523–528
Corte GN (2015) Reproductive cycle and parasitism in the clam *Anomalocardia brasiliana* (Bivalvia: Veneridae). Invertebr Reprod Dev 59:66–80
Corte GN, Yokoyama LQ, Amaral ACZ (2014) An attempt to extend the Habitat Harshness Hypothesis to tidal flats: a case study of *Anomalocardia brasiliana* (Bivalvia: Veneridae) reproductive biology. Estuar Coast Shelf Sci 150:136–141
Corte GN, Yokoyama LQ, Coleman RA et al (2015) Population dynamics of the harvested clam *Anomalocardia brasiliana* (Bivalvia: Veneridae) in Cidade Beach, south-east Brazil. J Mar Biol Assoc UK 95(6):1183
Corte GN, Coleman RA, Amaral ACZ (2017) Environmental influence on population dynamics of the bivalve *Anomalocardia brasiliana*. Estuar Coast Shelf Sci 187:241–248
Corte GN, Gonçalves-Souza T, Checon HH et al (2018) When time affects space: dispersal ability and extreme weather events determine metacommunity organization in marine sediments. Mar Environ Res 136:139–152
Corte GN, Yokoyama LQ, Checon HH et al (2019a) Spatial and temporal variation in the diet of the sandy beach gastropod *Olivella minuta*. Invertebr Biol 138(4):e12269
Corte GN, Yokoyama LQ, Denadai MR et al (2019b) Egg-capsule deposition of the marine gastropod *Olivella minuta* the importance of an interspecific relationship with the soft-bottom bivalve *Tivela mactroides*. J Molluscan Stud 85:126–132
Corte GN, Yokoyama LQ, Tardelli DT et al (2020) Spatial patterns of the gastropod *Olivella minuta* reveal the importance of tide-dominated beaches and the subtidal zone for sandy beach populations. Reg Stud Mar Sci 39:101454
Corte GN, Checon HH, Amaral ACZ (2021) Praias. In: Pereira RC, Soares-Gomes A (eds) Ecologia marinha, 1st edn. Interciência, Rio de Janeiro, pp 131–162
Corte GN, Checon HH, Shah Esmaeili Y et al (2022) Evaluation of the effects of urbanization and environmental features on sandy beach macrobenthos highlights the importance of submerged zones. Mar Pollut Bull 182:113962
Costa LL, Zalmon IR (2019a) Multiple metrics of the ghost crab Ocypode quadrata (Fabricius, 1787) for impact assessments on sandy beaches. Estuar Coast Shelf Sci 218:237–245

Costa LL, Zalmon IR (2019b) Sensitivity of macroinvertebrates to human impacts on sandy beaches: a case study with tiger beetles (Insecta, Cicindelidae). Estuar Coast Shelf Sci 220:142–151

Costa LL, Machado PM, Zalmon IR (2019) Do natural disturbances have significant effects on sandy beach macrofauna of Southeastern Brazil? Fortschr Zool 36:1–10

Costa LL, Soares-Gomes A, Zalmon IR (2021) Burrow occupation rates and spatial distribution within habitat of the ghost crab *Ocypode quadrata* (Fabricius, 1787): implications for impact assessments. Reg Stud Mar Sci 44:101699

Costa LL, Arueira VF, Ocaña FB et al (2022) Are ghost crabs (*Ocypode* spp.) smaller on human-disturbed sandy beaches? A global analysis. Hydrobiologia 849:3287–3298

Dahl E (1952) Some aspects of the ecology and zonation of the fauna on sandy beaches. Oikos 4:1–27

Danin APF, Pombo M, Martinelli-Lemos JM et al (2020) Population ecology of the hermit crab *Clibanarius symmetricus* (Anomura: Diogenidae) on an exposed beach of the Brazilian Amazon coast. Reg Stud Mar Sci 33:100944

Defeo O, Cardoso R (2002) Macroecology of population dynamics and life history traits of the mole crab Emerita brasiliensis in Atlantic sandy beaches of South America. Alemanha. Mar Ecol Prog Ser 239:169–179

Defeo O (2003) Marine invertebrate fisheries in sandy beaches: an overview. J Coast Res SI35:56–65

Defeo O, Cardoso RS (2004) Latitudinal patterns in abundance and life-history traits of the mole crab *Emerita brasiliensis* on South American sandy beaches. Divers Distrib 10(2):89–98

Defeo O, McLachlan A (2005) Patterns, processes and regulatory mechanisms in sandy beach macrofauna: a multi-scale analysis. Mar Ecol Prog Ser 295:1–20

Defeo O, McLachlan A, Schoeman DS et al (2009) Threats to sandy beach ecosystems: a review. Estuar Coast Shelf Sci 81(1):1–12

Defeo O, Barboza CAM, Barboza FR et al (2017) Aggregate patterns of macrofaunal diversity: an interocean comparison. Glob Ecol Biogeogr 26:823–834

Denadai MR, Amaral ACZ, Turra A (2001) Spatial distribution of molluscs on sandy intertidal substrates with rock fragments in south-eastern Brazil. Estuar Coast Shelf Sci 53(5):733–743

Denadai MR, Amaral ACZ, Turra A (2004) Biology of a tropical intertidal population of *Cerithium atratum* (Born, 1778) (Mollusca, Gastropoda). J Nat Hist 38:1695–1710

Denadai MR, Amaral ACZ, Turra A (2005) Structure of molluscan assemblages in sheltered intertidal unconsolidated environments. Braz Arch Biol Technol 48(5):825–839

Denadai MR, Pombo M, Bernadochi LC et al (2015a) Harvesting the beach clam *Tivela mactroides*: short-and long-term dynamics. Mar Coast Fish 7(1):103–115

Denadai MR, Le Sueur-Maluf L, Marques CG et al (2015b) Reproductive cycle of the trigonal clam *Tivela mactroides* (Bivalvia, Veneridae) in Caraguatatuba Bay, Southeastern Brazil. Mar Biol Res 11(8):847–858

Eça GF, Pedreira RMA, Hatje V (2013) Trace and major elements distribution and transfer within a benthic system: Polychaete *Chaetopterus variopedatus*, commensal crab *Polyonyx gibbesi*, worm tube, and sediments. Mar Pollut Bull 74:32–41

Franco NB, Rizzo AE (2016) Nephtyidae (Annelida: Polychaeta) from the Campos Basin, including two new species and a new record. Zootaxa 4114(3):291–308

Gandara-Martins AL, Borzone CA, Rosa LC, Caron E (2010) Ocorrência de três espécies do gênero Bledius leach, 1819 (Coleoptera, Ctaphylinidae, Oxytelinae) nas praias arenosas expostas do Paraná, Brasil. Braz J Aquat Sci Technol 14(2):23–30

Gorman D, Souza ECF, Leite FPP et al (2016) Olfactory selectivity in intertidal hermit crabs: aggregation behavior by *Pagurus criniticornis* (Decapoda, Anomura) in response to simulated predation on the gastropod *Cerithium atratum*. Hydrobiologia 772:31–43

Gorman D, Corte GN, Checon HH et al (2017) Optimizing coastal and marine spatial planning through the use of high-resolution benthic sensitivity models. Ecol Indic 82:23–31

Houbrick RS (1974). Growth studies on the genus *Cerithium* (Gastropoda; Prosobranchia) with notes on ecology and microhabitats. Nautilus, 88:14–27

Jelassi R, Bohli-Abderrazak D, Ayari A et al (2015) Endogenous activity rhythm in *Talitrus saltator*, Britorchestia brito (Crustacea, Amphipoda) and Tylos europaeus (Crustacea, Isopoda) from Barkoukech Beach (Tabarka, Tunisia). Biol Rhythm Res 46(6):873–886

Jumars PA, Dorgan KM, Lindsay SM (2015) Diet of worms emended: an update of polychaete feeding guilds. Annu Rev Mar Sci 7:497–520

Lana PC, Pagliosa P, Paiva PC et al (2017) Polychaetes in Brazil: people and places, past, present and future. Bol Inst Oceanogr Ven:24–50

Laurino IRA, Turra A (2021) The threat of freshwater input on sandy beaches: a small-scale approach to assess macrofaunal changes related to salinity reduction. Mar Environ Res 171:105459

Laurino IRA, Buchmann FS, Hernáez P (2020a) Spatial-temporal distribution of the burrowing shrimp Callichirus major (Say, 1818) (Decapoda, Callichiridae) in preserved populations of Southeastern Brazil. Thalassas 36:333–342

Laurino IRA, Checon HH, Corte GN et al (2020b) Flooding affects vertical displacement of intertidal macrofauna: a proxy for the potential impacts of environmental changes on sandy beaches. Estuar Coast Shelf Sci 245:106882

Leão LSD, Soares-Gomes A, Costa T et al (2012) Population dynamics and reproductive strategy of *Scolelepis goodbodyi* (Polychaeta: Spionidae) in a subtropical Atlantic beach. Fortschr Zool 29(3):195–202

Lima-Verde FB, de Castro ÍB, de Almeida Rocha-Barreira C (2010) Imposex occurrence in *Nassarius vibex* from South America: a potential bioindicator in estuarine environments. Mar Biodivers Rec 3:e30

Livi S, Tomassetti P, Vani D et al (2017) Genetic evidences of multiple phyletic lineages of *Capitella capitata* (Fabricius 1780) complex in the Mediterranean Region. J Mediterr Ecol 15:5–11

Lucena JMD, Meirelles CAOD, Matthews-Cascon H (2012) Feeding behavior of *Nassarius vibex* (Gastropoda: Nassariidae). Arq Ciênc Mar 45:60–67

MacCord FS, Amaral ACZ (2007) The reproductive cycle of Scolelepis goodbodyi (Polychaeta, Spionidae). Mar Biol 151(3):1009–1020

Marcus E, Marcus E (1959) Studies on Olividae. Bol Fac Filos Ciênc Letras USP (Zool) 22:99–188

Mattos G, Cardoso RS (2012) Population dynamics of two suspension-feeding bivalves on a sheltered beach in Southeastern Brazil. Helgol Mar Res 66(3):393–400

McLachlan A, Brown AC (2006) The ecology of sandy shores, 2nd edn. Elsevier, Amsterdam

McLachlan A, Jaramillo E (1995) Zonation on sandy beaches. Oceanogr Mar Biol 33:305–335

Mirza JD, Migotto ÁE, Yampolsky IV et al (2020) *Chaetopterus variopedatus* bioluminescence: a review of light emission within a species complex. Photochem Photobiol 96:768–778

Molina WF, de Lima Filho PA, Dantas VFR et al (2013) Morphological differences associated with colour morphs of the auger snail *Hastula cinerea* (Conoidea: Terebridae) from north-east Brazil. J Mar Biol Assoc UK 93(4):1035

Neves LPD, Silva PDSRD, Bemvenuti CE (2007) Zonation of benthic macrofauna on Cassino Beach, southernmost Brazil. Braz J Oceanogr 55(4):293–307

Nucci PR, Turra A, Morgado EH (2001) Diversity and distribution of crustaceans from 13 sheltered sandy beaches along São Sebastião Channel, south-eastern Brazil. J Mar Biol Assoc UK 81:475–484

Oliveira IBD, Lavander HD, Lima PCMD et al (2019) Effect of stocking density on the growth and survival of *Anomalocardia brasiliana* (Gmelin, 1791) (Bivalvia: Veneridae) post-larvae. Cienc Rural 49(12):e20190420

Omena E, Amaral A (2003) Sandy beach morphodynamic and the polychaete fauna in southeast Brazil. J Coast Res SI35:431–439

Otegui MB, Blankensteyn A, Pagliosa PR (2012) Population structure, growth and production of *Thoracophelia furcifera* (Polychaeta: Opheliidae) on a sandy beach in Southern Brazil. Helgol Mar Res 66(4):479–488

Pardo EV, Amaral ACZ (2004) Feeding behavior of *Scolelepis* sp. (Polychaeta: Spionidae). Braz J Oceanogr 52(1):74–79

Passos FD, Domaneschi O (2004) Biologia e anatomia funcional de *Donax gemmula* Morrison (Bivalvia, Donacidae) do litoral de São Paulo, Brasil. Rev Bras Zool 21(4):1017–1032

Peixoto AJM, Santos AJM (2016) Reproductive biology of *Perinereis anderssoni* (Polychaeta: Nereididae) in a subtropical Atlantic Beach. Invertebr Reprod Dev 60(3):201–211

Pennafirme S, Machado AS, Machado AC et al (2019) Monitoring bioturbation by a small marine polychaete using microcomputed tomography. Micron 121:77–83

Pennafirme S, Machado AC, Machado AS et al (2020) Unveiling the role of bioturbation on bacterial activity in metal-contaminated sediments. Sci Total Environ 744:140988

Petracco M, Veloso VG, Cardoso RS (2003) Population dynamics and secondary production of *Emerita brasiliensis* (Crustacea: Hippidae) at Prainha Beach, Brazil. Mar Ecol 24(3):231–245

Petracco M, Cardoso RS, Corbisier TN (2010) Population biology of *Excirolana armata* (Dana, 1853) (Isopoda, Cirolanidae) on an exposed sandy beach in Southeastern Brazil. Mar Ecol 31(2):330–340

Petracco M, Cardoso RS, Turra A (2013) Patterns of sandy-beach macrofauna production. J Mar Biol Assoc UK 93(7):1717

Petracco M, Camargo RM, Berenguel TA et al (2015) Evaluation of the use of *Olivella minuta* (Gastropoda, Olividae) and Hastula cinerea (Gastropoda, Terebridae) as TBT sentinels for sandy coastal habitats. Environ Monit Assess 187(7):440

Petracco M, Cardoso RS, Martinelli Filho JE et al (2017) Effects of beach morphodynamic features on production and P/B ratio of the crab Emerita brasiliensis Schmitt, 1935 (Decapoda: Hippidae) in sandy beaches of South America. J Mar Biol Assoc UK 97(6):1215–1221

Petracco M, Aviz D, Martinelli Filho JE et al (2019) Effects of physical features on production of three macrofaunal species in different sandy beach zones in South America. Estuar Coast Shelf Sci 218:23–30

Pezzuto PR, Silva D (2015) A pesca e o manejo do berbigão (*Anomalocardia brasiliana*) (Bivalvia: Veneridae) na Reserva Extrativista Marinha do Pirajubaé, SC, Brasil. Desenvolv Meio Ambiente 34:169–189

Pezzuto PR, Schio C, Almeida TC (2010) Efficiency and selectivity of the *Anomalocardia brasiliana* (Mollusca: Veneridae) hand dredge used in southern Brazil. J Mar Biol Assoc UK 90(7):1455

Pombo M, Turra A (2013) Issues to be considered in counting burrows as a measure of Atlantic ghost crab populations, an important bioindicator of sandy beaches. PLoS One 8(12):e83792

Pombo M, Turra A (2017) Variation in the body growth parameters of the ghost crab *Ocypode quadrata* from morphodynamically distinct sandy beaches. Braz J Oceanogr 65(4):656–665

Pombo M, Turra A (2019) The burrow resetting method, an easy and effective approach to improve indirect ghost-crab population assessments. Ecol Indic 104:422–428

Pombo M, Oliveira A, Xavier LY et al (2017a) Natural drivers of distribution of ghost crabs *Ocypode quadrata* and the implications of estimates from burrows. Mar Ecol Prog Ser 565:131–147

Pombo M, Campagnoli ML, Turra A (2017b) Continuous, video-recording assessment of daily activity cycle of the ghost crab *Ocypode quadrata* Fabricius, 1787 (Brachyura: Ocypodidae) in Southeastern Brazil. J Crustac Biol 38:133–139

Ragagnin MN, Turra A (2021) Imposex incidence in the sandy beach snail *Hastula cinerea* reveals continued and widespread tributyltin contamination after its international ban. Reg Stud Mar Sci 49:102118

Rech TF, Soto GAT, Turra A (2021) Species with insufficient data and red lists: the dilemma of the beach trigonal clam Tivela mactroides. J Nat Conserv 62:126024

Rios EC, Haimovici M, Peres JAA et al (1994) Seashells of Brazil. Editora FURG, Rio Grande

Rizzo AE, Amaral ACZ (2001) Environmental variables and intertidal beach annelids of Sao Sebastiao channel (State of Sao Paulo, Brazil). Rev Biol Trop 49(3–4):849–857

Rizzo AE, Amaral ACZ (2007) Nephtyidae (Annelida: Polychaeta) from São Paulo State, Brazil, including a new record for the Brazilian coast. Biota Neotrop 7(3):253–263

Rocha MB, Radashevsky V, Paiva PC (2009) Espécies de *Scolelepis* (Polychaeta, Spionidae) de praias do Estado do Rio de Janeiro, Brasil. Biota Neotrop 9(4):101–108

Rocha-Barreira CA (2002) Feeding behavior of *Olivancillaria vesica auricularia* (Lamarck, 1810) (Mollusca, Olividae). Thalassas 18:83–89

Rocha-Barreira CA (2010) Reproductive cycle of *Olivancillaria vesica auricularia* (Lamarck, 1910) (Mollusca: Gastropoda: Olividae) in Southern Brasil. Arq Ciênc Mar 43:110–118

Rodrigues SDA, Shimizu RM (1997) Autoecology of *Callichirus major* (Say, 1818). Oecologia Aust 3(1):155–170

Rodrigues AML, Borges-Azevedo CM, Costa RS et al (2013) Population structure of the bivalve *Anomalocardia brasiliana*, (Gmelin, 1791) in the semi-arid estuarine region of northeastern Brazil. Braz J Biol 73(4):819–833

Rosa Filho JS, Almeida MF, Silva DEA et al. (2009) Spatial and temporal changes in the benthic fauna of a macrotidal Amazon sandy beach, Ajuruteua, Brazil. J Coast Res SI(56):1823–1827.

Rosa Filho JS, Gomes TP, Almeida MF et al. (2011). Benthic fauna of macrotidal sandy beaches along a small-scale morphodynamic gradient on the Amazon coast (Algodoal Island, Brazil). J Coast Res 64:435–439.

Sampieri BR, Steiner TM, Baroni PC et al (2020) How oogenesis analysis combined with DNA barcode can help to elucidate taxonomic ambiguities: a polychaete study-based approach. Biota Neotrop 20(3):e20200959

Salvat B (1964) Les conditions hydrodynamiques interstitielles des sediments meubles intertidaux et la repartition verticale de la fauna endogee. Cah Rech Acad Sci Paris 259:1576–1579

Sant'Anna BS, Da Cruz Dominciano LC, Buozi SF et al (2012a) Is shell partitioning between the hermit crabs *Pagurus brevidactylus* and *Pagurus criniticornis* explained by interference and/or exploitation competition? Mar Biol Res 8:662–669

Sant'Anna BS, de Marchi MRR, dos Santos DM et al (2012b) Effects of tributyltin exposure in hermit crabs: *Clibanarius vittatus* as a model. Environ Toxicol Chem 31:632–638

Santos PJPD (1991) Morphodynamical influence of a temporary freshwater stream on the population dynamics of *Scolelepis gaucha* (Polychaeta: Spionidae) on a sandy beach in southern Brazil. Bull Mar Sci 48(3):657–664

Santos PJPD (1994) Population dynamics and production of *Scolelepis gaucha* (Polychaeta: Spionidae) on the sandy beaches of Southern Brazil. Mar Ecol Prog Ser 110:159–165

Santos TMT, Petracco M, Venekey V (2022) Effects of vehicle traffic and trampling on the macrobenthic community of Amazonian macrotidal sandy beaches. J Mar Biol Assoc UK 102:1–23

Scapini F (2006) Keynote papers on sandhopper orientation and navigation. Mar Freshw Behav Physiol 39(1):73–85

Scapini F (2014) Behaviour of mobile macrofauna is a key factor in beach ecology as response to rapid environmental changes. Estuar Coast Shelf Sci 150:36–44

Scapini F, Bessa F, Gambineri S et al (2019) Talitrid (Crustacea, Amphipoda) orientation as across scale bioindicator of sandy beaches environmental conditions: a meta-analytic approach. Estuar Coast Shelf Sci 220:25–37

Seixas VC, Steiner TM, Solé-Cava AM et al (2020) Hidden diversity within the *Diopatra cuprea* complex (Annelida: Onuphidae): morphological and genetics analyses reveal four new species in the south-west Atlantic. Zool J Linnean Soc 191:637–671

Serejo CS (2004) Cladistic revision of talitroidean amphipods (Crustacea, Gammaridea), with a proposal of a new classification. Zool Scr 33(6):551–586

Silva CF, Seixas VC, Barroso R et al (2017) Demystifying the *Capitella capitata* complex (Annelida, Capitellidae) diversity by morphological and molecular data along the Brazilian coast. PLoS One 12(5):e0177760

Soares AG (2003) Sandy beach morphodynamics and macrobenthic communities in temperate, subtropical and tropical regions: a macroecological approach. Thesis, University of Port Elizabeth, South Africa

Souza JR, Borzone CA (2000) Population dynamics and secondary production of *Scolelepis squamata* (Polychaeta: Spionidae) in an exposed sandy beach of southern Brazil. Bull Mar Sci 67(1):221–233

Souza JR, Borzone CA (2007) Population dynamics and secondary production of *Euzonus furciferus* Ehlers (Polychaeta, Opheliidae) in an exposed sandy beach of Southern Brazil. Rev Bras Zool 24(4):1139–1144

Souza JRB, Borzone CA, Brey T (1998) Population dynamics and secondary production of *Callichirus major* (Crustacea: Thalassinidea) on a southern Brazilian sandy beach. Arch Fish Mar Res 46:151–164

Souza ECF, Turra A, Leite FPP et al (2015) Intra-specific competition drives variation in the fundamental and realized niches of the hermit crab, *Pagurus criniticornis*. Bull Mar Sci 91:343–361

Souza DGC, Petracco M, Danin APF et al (2021) Population structure of and use of space by ghost crabs (Brachyura: Ocypodidae) on an equatorial, macrotidal sandy beach. Estuar Coast Shelf Sci 258:107376

Steiner TM, Nogueira JMM, Amaral ACZ (2016) Annelida. In: Fransozo A, Negreiros-Fransozo ML (eds) Zoologia dos invertebrados, 1st edn. Editora Roca, Rio de Janeiro, pp 296–372

Taghon GL (1988) The benefits and costs of deposit feeding in the polychaete *Abarenicola pacifica*. Limnol Oceanogr 33(5):1166–1175

Tavares YA, Borzone CA (2006) Reproductive cycle of *Mellita quinquiesperforata* (Leske) (Echinodermata, Echinoidea) in two contrasting beach environments. Rev Bras Zool 23(2):573–580

Trevallion A (1971) Studies on *Tellina tenuis* da Costa. III. Aspects of general biology and energy flow. J Exp Mar Biol Ecol 7(1):95–122

Turra A, Leite FPP (1999) Population structure and fecundity of the hermit crab *Clibanarius antillensis* Stimpson 1862 (Anomura, Diogenidae) in Southeastern Brazil. Bull Mar Sci 64:281–289

Turra A, Amaral ACZ, Denadai MR et al (2003) Intertidal macrofauna in brazilian subtropical sandy beach landscapes. J Coast Res 35:446–455

Turra A, Gonçalves MAO, Denadai MR et al (2005) Spatial distribution of the ghost crab *Ocypode quadrata* in low-energy tide-dominated sandy beaches. J Nat Hist 39:2163–2177

Turra A, Petracco M, Amaral ACZ et al (2014) Temporal variation in life-history traits of the clam *Tivela mactroides* (Bivalvia: Veneridae): density-dependent processes in sandy beaches. Estuar Coast Shelf Sci 150:157–164

Turra A, Xavier LY, Pombo M et al (2016) Assessment of recreational harvesting of the trigonal clam *Tivela mactroides*: socioeconomic aspects and environmental perception. Fish Res 174:58–67

Turra A, Corte GN, Amaral ACZ et al (2018) Non-linear curve adjustments widen biological interpretation of relative growth analyses of the clam *Tivela mactroides* (Bivalvia, Veneridae). PeerJ 6:e5070

Turra A, Fernandez WS, Ragagnin MN et al (2019) The effect of ocean acidification on the intertidal hermit crab Pagurus criniticornis is not modulated by cheliped amputation and sex. Mar Environ Res 1:104794

Veloso VG, Caetano CHS, Cardoso RS (2003) Composition, structure and zonation of intertidal macrofauna in relation to physical factors in microtidal sandy beaches at Río de Janeiro State, Brazil. Sci Mar 67(4):393–402

Veloso VG, Silva ES, Caetano CH et al (2006) Comparison between the macroinfauna of urbanized and protected beaches in Rio de Janeiro State, Brazil. Biol Conserv 127(4):510–515

Veloso VG, Neves G, Lozano M et al (2008) Responses of talitrid amphipods to a gradient of recreational pressure caused by beach urbanization. Mar Ecol 29:126–133

Yokoyama LQ, Amaral ACZ (2011a) Temporal variation in egg-capsule deposition by *Nassarius vibex* (Gastropoda: Nassariidae). Invertebr Reprod Dev 55(2):82–90

Yokoyama LQ, Amaral ACZ (2011b) Allometric growth of a common Nassariidae (Gastropoda) in south-east Brazil. J Mar Biol Assoc UK 91(5):1095–1105

Zal F, Rousselot M (2014) Extracellular hemoglobins from annelids, and their potential use in biotechnology. In: La Barre S, Kornprobst J-M (eds) Outstanding marine molecules. Wiley, Weinheim, pp 361–373

Chapter 5
Vertebrate Biodiversity

Yasmina Shah Esmaeili, Bruna Pagliani, Robson Henrique de Carvalho, and Leonardo Lopes Costa

5.1 Sandy Beach as a Habitat for Vertebrates

Sandy beaches comprise different habitats, from ever-immersed surf zones inhabited primarily by marine species to the intertidal-supralittoral interface where marine, semi-terrestrial, and terrestrial organisms coexist. This unique ecosystem forms the gateway between land and sea and, therefore, has a great diversity of vertebrates occupying a wide range of habitats in the sandy beach realm. Although some species are considered residents, most vertebrates are transient species on beaches, that is, do not spend their entire life in the ecosystem.

The use of sandy beach habitats by vertebrates is highly driven by physical factors; however, ecological interactions are also important determinants, particularly food availability (Fig. 5.1) and predation risk (Dugan et al. 2003; Yasué 2006). The surf zones of sandy beaches are the main areas of wave energy dissipation; this

Y. Shah Esmaeili (✉)
Departamento de Biologia Animal, Instituto de Biologia, Universidade Estadual de Campinas, Campinas, SP, Brazil

Departamento de Oceanografia Biológica, Instituto Oceanográfico, Universidade de São Paulo, São Paulo, SP, Brazil

B. Pagliani
Instituto de Biodiversidade e Sustentabilidade – NUPEM, Universidade Federal do Rio de Janeiro, Macae, RJ, Brazil

Laboratório de Ecologia Comportamental e Bioacústica, Universidade Federal de Juiz de Fora, Juiz de Fora, MG, Brazil

R. H. de Carvalho
Laboratório de Herpetologia – Répteis, Instituto de Ciências Biológicas, Universidade Federal de Juiz de Fora, Juiz de Fora, MG, Brazil

L. L. Costa
Laboratório de Ciências Ambientais, Universidade Estadual do Norte Fluminense Darcy Ribeiro, Campos dos Goytacazes, RJ, Brazil

A. C. Z. Amaral et al. (eds.), *Brazilian Sandy Beaches*, Brazilian Marine Biodiversity, https://doi.org/10.1007/978-3-031-30746-1_5

Fig. 5.1 The kelp gull (*Larus dominicanus*) feeding on a juvenile fish on a sandy beach on the Northern coast of São Paulo. (Photo: Bruna R. Debastiani)

turbulence contributes to the resuspension of sediment and infauna (i.e., organisms that live within the sediment of the seafloor), providing food and shelter for fishes, mainly juveniles (Olds et al. 2017), feeding grounds for birds and sharks (Costa et al. 2017a; Bornatowski et al. 2012), and nursery grounds for most of these potential flagship species,[1] targeted in conservation initiatives (Zacharias and Roff 2001).

In this chapter, we present the main groups of vertebrates that utilize Brazilian sandy beaches (fishes, birds, sea turtles, and cetaceans), as well as the main impacts that threaten their occurrence. We also discuss some conservation and management initiatives and the knowledge gaps that need to be fulfilled to better protect the vertebrate biodiversity of Brazilian beaches.

5.2 Sandy Beach Biodiversity

5.2.1 Fishes

Based on an extensive literature review ($n = 53$), we found that on average, 57 fish species are encountered in the Brazilian sandy beach surf zones, although some studies have reported up to 89 species in a single beach (Soeth et al. 2014). Carangidae are cited as one of the most abundant families (65% of the studies on the

[1] Charismatic species chosen to raise support and public awareness and contribute to biodiversity conservation in a given place or social context.

Brazilian coast), followed by Sciaenidae (38%), Mugilidae (34%), Atherinopsidae (31%), Clupeidae (28%), Engraulidae, and Gerreidae (21% each). According to a 2-year study performed by Araujo et al. (2018), the southeastern region has a higher number of species compared to the south and northeast, due to the higher primary productivity associated with upwelling zones in this region.

Some of the most frequently registered species in sandy beach surf zone studies in Brazil are the False hering (*Harengula clupeola*) (Fig. 5.2a), Brazilian silversides (*Atherinella brasiliensis*) (Fig. 5.2b), Florida pompano (*Trachinotus carolinus*) (Fig. 5.2c), Rio anchovy (*Anchoa januaria*), Atlantic sabretooth anchovy (*Lycengraulis grossidens*), White mullet (*Mugil curema*), Brazilian sardine (*Sardinella brasiliensis*), Silver mojarra (*Eucinostomus argenteus*) (Fig. 5.2d), Piquitinga anchovy (*Anchoa tricolor*), and Sand drum (*Umbrina coroides*).

Fig. 5.2 Fish species commonly found in the surf zone of Brazilian sandy beaches: (**a**) False hering (*Harengula clupeola*), (**b**) Brazilian silversides (*Atherinella brasiliensis*), (**c**) Florida pompano (*Trachinotus carolinus*), and (**d**) Silver mojarra (*Eucinostomus argenteus*), some of the most abundant surf zone fish species of Brazilian sandy beaches, (**e**) Longnose stingray (*Hypanus guttatus*). (Photos: Carolina Correia Siliprandi, source: Amaral et al. 2016), (**f**) Blacktip reefshark (*Carcharhinus melanopterus*), recorded by surf-BRUVs of the North Coast of São Paulo in a sandy beach surf zone. (Photo: Yasmina Shah Esmaeili)

Although the surf zones of sandy beaches have several resident species, they are also considered a nursery for the juvenile stages of several fish species that continue their lifespan in other habitats. Species such as flounders, rays, and several shark species (Fig. 5.2e, f) are known to reproduce and spawn in shallow-water environments such as sandy beaches (Yokota and Lessa 2006). Additionally, habitats such as the intertidal and tidal pools are home to fish species such as gobiids. Although often considered rare species, gobiid diversity in sandy beach environments may be underestimated by conventional beach seine sampling because of their cryptic habits (sand burrowing).

As surf zones constitute important feeding and nursery sites for fishes, food availability, and physical or human disturbances are decisive factors for fish distribution (Shah Esmaeili et al. 2022). Compared to low-hydrodynamic ecosystems, such as coastal lagoons, mangroves, and estuaries, sandy beaches have a low diversity of fish species and functional groups, theoretically attributed to the homogeneity of habitats and physical instability (Azevedo et al. 2017; Araújo et al. 2018). However, habitat complexity (e.g., the presence of banks of macroalgae and proximity to rocky shores) constitutes a source of increased complexity that promotes an increase in the richness and diversity of fish in the surf zone (Gaelzer and Zalmon 2008a; Andrades et al. 2014; Costa et al. 2017b; Olds et al. 2017).

Morphodynamic conditions play a crucial role in shaping fish communities of sandy beaches, and dissipative beaches (Chap. 1) are home to more species compared to intermediate and reflective beaches on the Brazilian coast (Félix et al. 2007; Oliveira and Pessanha 2014; Shah Esmaeili et al. 2022; Fig. 5.3), following

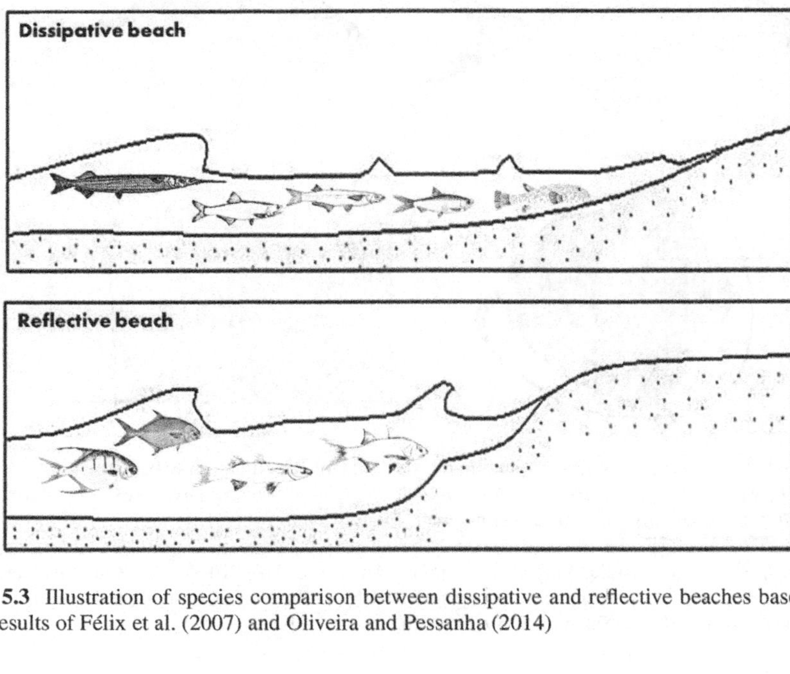

Fig. 5.3 Illustration of species comparison between dissipative and reflective beaches based on the results of Félix et al. (2007) and Oliveira and Pessanha (2014)

the global pattern (Lercari et al. 2010). Sheltered and moderately exposed beaches show similar ichthyofaunal composition, being inhabited mainly by clupeids and engraulids (Félix et al. 2007). Comparative studies showed that species such as *L. grossidens*, *R. bahiensis*, *A. brasiliensis*, *S. testudineus*, and *Hyporhamphus unifasciatus* occurred primarily in dissipative beaches (Oliveira and Pessanha 2014), while exposed beaches had a higher occurrence of species such as *T. carolinus*, *Trachinotus goodei*, *M. liza*, and *P. virginicus* (Félix et al. 2007; Oliveira and Pessanha 2014).

Only a few functional groups, such as benthic and zoobenthivorous species, were most abundant at sites with intermediate wave exposure, as shown by Shah Esmaeili et al. (2022), who surveyed fish communities at 77 sites along 150 km of São Paulo's coastline, corroborating the intermediate disturbance hypothesis (Connell 1978; Clark 1997).

Although salinity, turbidity, and wind speed influence fish communities, there is a high variability of responses (both negative and positive) reported in the literature (reviewed by Olds et al. 2018). Increasing turbidity, for example, can provide more efficient protection of juvenile fishes from visual predators, but it can be correlated with high physical harshness that in turn creates a less favorable habitat. Also, microhabitat features (e.g., the presence of gutters, runnels, and sandbars) probably interact with these environmental determinants to drive surf zone fish diversity, contradicting the idea that surf zones are homogeneous habitats (Borland et al. 2017). Additionally, seasonality can be a more decisive factor in structuring fish communities than spatial variations (Gondolo et al. 2011; Del Favero and Dias 2013, 2015; Dantas et al. 2016a; Costa et al. 2017b), with the seasonal species turnover resulting in higher fish diversity in warmer months (Costa et al. 2017b; Pereira et al. 2015). Short-term variation, for example, nictemeral, is also responsible for changes in species composition, as predators tend to approach shallow waters at night and during rising tides (Gaelzer and Zalmon 2008a,b; Dantas et al. 2016b).

5.2.2 *Birds*

Seabirds (e.g., gulls and terns) and shorebirds (e.g., plovers and sanderlings) generally comprise the bulk of birds frequenting intertidal zones of sandy beaches (Lunardi et al. 2012). Some shorebirds use upper beach zones (backshore) as nesting areas during the breeding season, and others rely on invertebrates as important feeding resources along the entire beach in both wintering and breeding sites. Sandy beaches can be important feeding sites for both resident and migrant species, especially when preferential adjacent low-energy ecosystems (e.g., mudflats, coastal lagoons, and mangroves) have suffered habitat degradation (Hubbard and Dugan 2003).

A review conducted by Vooren and Brusque (1999) found that a total of 148 species of sea and shorebirds have been registered on the Brazilian coast, making up 8.8% of all bird species recorded by Sick (1997) for Brazil. They are represented by

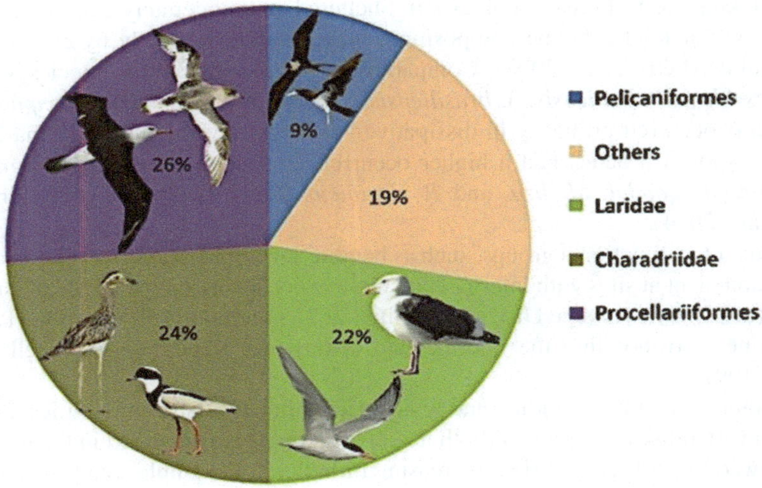

Fig. 5.4 Percentages of families of Brazilian shore and seabirds found on sandy beaches based on Vooren and Brusque (1999)

9 orders and 29 families. Of the total of 148 species, 81% belong to only three orders (Fig. 5.4): 26% to Procellariiformes (albatrosses – Diomedeidae; petrels – Procellariidae); 9% to Pelecaniformes (frigates – Fregatidae, boobies, and the like – Sulidae (Fig. 5.5a)); 24% to the Charadriiformes – family Charadriidae (torches, plovers, and the like); and 22% to Charadriiformes – family Laridae (seagulls – Stercorariidae, terns and the like – Sternidae), representing a total of 28% of the world species belonging to these three orders. Among the 87 Scolopacidae (sandpipers) species reported in the world, 26% occur in Brazil.

The coast of South America includes important wintering and stopover grounds for more than 2.9 million migratory Nearctic shorebirds. Many spend over half of their lives in these areas, including in Brazilian hotspots along one of the primary migration routes, known as the Atlantic route, that goes along the entire coast of Brazil from the state of Amapá to Rio Grande do Sul (ICMBio 2019). The migratory birds concentrate in humid areas with high food availability such as in the North along the coast of Amapá, the salt marches of Para, and the Reentrâncias Maranhenses State Protected Area in Maranhão; in the Northeast, along the coast of Icapuí (CE), Galinhos and Areia Branca (RN), Coroa do Avião (PE), the Piaçabuçu State Protected Area (AL), Mangue Seco, and Cacha-Prego (BA); in the South, at the Lagoado Peixe National Park (RS) (ICMBio 2019).

Therefore, there is evidence of the high importance of Brazil concerning the conservation of sea and shorebirds worldwide, including migrant species. An example is the Magellanic Penguin, *Spheniscus magellanicus*. Although they do not breed in Brazil, they are found along the Brazilian coast outside the breeding season, during the austral fall and winter (Baldassin et al. 2010).

Fig. 5.5 Birds species that feed and nest on Brazilian sandy beaches: (**a**) *Sula leucogaster* foraging in the sandy beach surf zone. (Photo: Danilo Rangel). (**b**) Eggs of *Charadrius collaris* in a nest on a sandy beach. (Photo: Julyana Madureira). (**c**) *Charadrius semipalmatus* feeding on intertidal zone of sandy beach. (Photos: Bruna Debastiani). (**d**) *Athene cunicularia* on sandy beach backshore. (Photo: Julyana Madureira). (**e**) *Passer domesticus* feeding on the remains of a dead bird on an urbanized sandy beach on the North coast of São Paulo. (Photo: Bruna Debastiani)

Six bird species nest on Brazilian sandy beaches: the four Charadriidae *Haematopus palliatus*, *Charadrius falklandicus*, *Charadrius collaris* (Fig. 5.5b), and *Charadrius wilsonia*, and the two Sternidae *Gelochelidon nilotica* and *Phaetusa simplex* (Vooren and Brusque 1999). Charadriidae and Scolopacidae that nest in North America constitute the most abundant migrant shorebirds present in the coastal zone of Brazil. The blowtorches, plovers, and other similar birds, from the Suborder Charadrii, occur mainly on oceanic beaches and/or estuaries, and contribute to 44% of the bird biodiversity of these habitats.

Along the Brazilian coast, the intertidal zone of beaches and mangroves are important as feeding grounds for both migratory and resident Charadriidae in the country (Mancini et al. 2017, 2018) (Fig. 5.5c). Birds feed on the infaunal invertebrates of these habitats, and the abundance of organisms that constitute their food usually has seasonal cycles. The peaks of the bird abundance align with the occurrence of their prey species (Lunardi et al. 2012) and the arrival of birds that overwinter in subtropical regions is adjusted to match these food peaks.

Several seagulls, terns, and species such as the Black skimmer (*Rynchops niger*) feed in the coastal areas, and land on the beaches in flocks to rest and groom their plumage (Escalante 1970; Vooren and Chiaradia 1990; Naves 1999). Seagulls are particularly tolerant and coexist with humans on beaches, feeding opportunistically on food scraps, or even ingesting litter (Yorio et al. 1998). However, other resident (e.g., Ciconiiformes) or migrant birds (Charadriiformes) usually require pristine beaches with an abundance of food to use this ecosystem as suitable habitat (Lunardi et al. 2012; Costa et al. 2017a).

Sandy beaches are used mainly by the above-mentioned marine birds, but in some cases, herons, terrestrial passerine species (e.g., *Pitangus sulphuratus*, *Machetornis rixosa*), and birds of preys (e.g., caracarinis, owls, and vultures) rely on opportunistic feeding opportunities in beach habitats (Fig. 5.5d, e).

Besides food availability, the presence of migratory shorebirds on sandy beaches is directly or indirectly related to hydrodynamics due to various elements such as the birds' thermal homeostasis (inundation risk), foraging (grain size related to the penetration of the beak on the sand), and perceived predation risk (beach slope influences the perception of predators) (Lunardi et al. 2012; Yasué 2006).

5.2.3 Sea Turtles

Of all seven species of sea turtles, five feed and/or nest on Brazilian sandy beaches: *Caretta caretta*, *Chelonia mydas*, *Lepidochelys olivacea*, *Dermochelys coriacea*, and *Eretmochelys imbricata* (Marcolvaldi and Marcovaldi 1999). The presence of these species in the feeding and nesting areas makes Brazilian coastal regions priority areas for the conservation of sea turtles worldwide (Wallace et al. 2011; Selig et al. 2014).

The Loggerhead turtle (*Caretta caretta*) has a circumglobal distribution and is highly migratory. In Brazil, the priority nesting areas are located in the north of the state of Bahia and the states of Rio de Janeiro, Espírito Santo, and Sergipe (Santos et al. 2011) (Fig. 5.6). Occurrences of individuals at different life stages are also recorded on the coast of several states in Brazil between the states of Pará and Rio Grande do Sul, both in coastal and oceanic waters (Santos et al. 2011).

Green turtles (*Chelonia mydas*) are distributed in tropical and subtropical regions where they feed and breed. Among the species of sea turtles found in Brazil, the green turtle is the only one that nesting on oceanic islands (De Padua Almeida et al. 2011) (Fig. 5.6). Their nesting season is from December to June (De Padua Almeida et al. 2011; Bellini et al. 2013) (Fig. 5.7a, b).

The Olive Ridley turtle (*L. olivacea*) has a circumglobal distribution and is a highly migratory species (Plotkin 2010). The priority nesting area in Brazil is located between the southern coast of the state of Alagoas and the northern coast of the state of Bahia with the highest density of nests in the state of Sergipe (De Castilhos et al. 2011) (Fig. 5.6). Juveniles and adults occur in coastal and oceanic

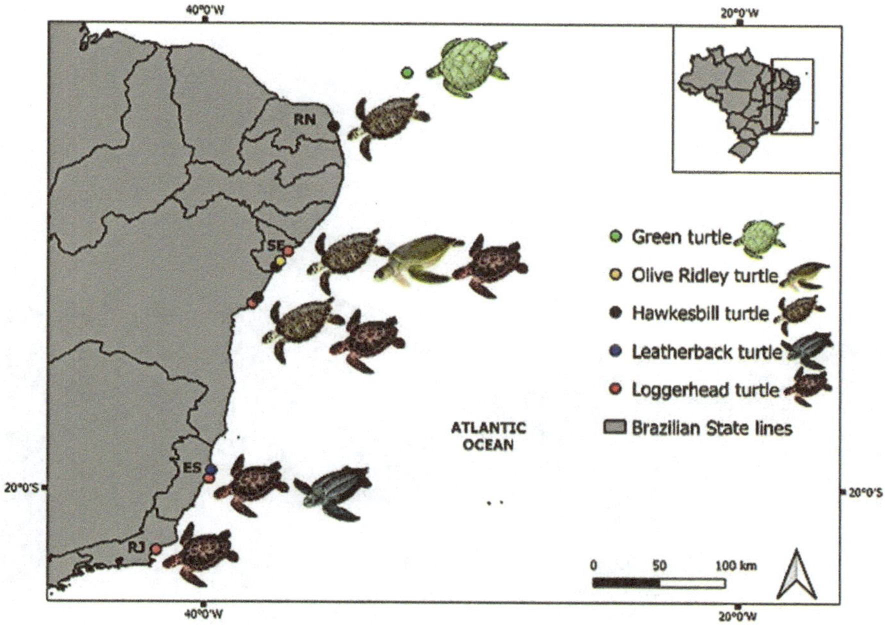

Fig. 5.6 Priority nesting sites of the five sea turtle species (Green turtle (*Chelonia mydas*), Olive Ridley turtle (*Lepidochelys olivacea*), Hawksbill turtle (*Eretmochelys imbricate*), Leatherback turtle (*Dermochelys coriacea*), and Loggerhead turtle (*Caretta caretta*)) on the Brazilian coast (states: Rio Grande do Norte (RN), Sergipe (SE), Espirito Santo (ES), and Rio de Janeiro (RJ))

areas between the states of Pará and Rio Grande do Sul and in international waters adjacent to Brazil's exclusive economic zone (De Castilhos et al. 2011).

The Leatherback turtle (*D. coriacea*) is a cosmopolitan species, usually living in the oceanic zone for most of its life. Its nesting area in Brazil is located on the northern coast of the state of Espírito Santo (Thomé et al. 2007; López-Mendilaharsu et al. 2009; De Padua Almeida et al. 2011) (Fig. 5.6).

The Hawksbill turtle (*E. imbricate*) is found globally, although more common in tropical waters. In Brazil, the priority breeding areas are located on the north coast of the states of Bahia and Sergipe and on the south coast of the state of Rio Grande do Norte (Fig. 5.6). The known feeding areas of this sea turtle are the oceanic islands of Fernando de Noronha and Atol das Rocas in northeast Brazil, and the Abrolhos archipelago (Marcovaldi et al. 2011).

Although all species of sea turtles are widely distributed along the Brazilian coast, green turtles are the best known among fishermen, because they are commonly found near the coast, where adults feed almost exclusively on seagrass and algae on rocky shores (Carvalho et al. 2016; Bahia and Bondioli 2010). Because of this behavior, they are also the species most caught incidentally in bottom trawl nets (Awabdi et al. 2018). Species such as the loggerhead turtle and leatherback turtle have more oceanic habits, as they feed on algae and animals in other habitats,

Fig. 5.7 Turtles nesting in Brazilian sandy beaches: (**a**) Female of Green turtle nesting and (**b**) returning to the sea, at Atol das Rocas/RN. (Photos: Robson Carvalho); (**c**) close-up of sea turtle eggs in a sandy beach nest and (**d**) researcher digging up sea turtle nest to count eggs in Sítio do Conde/BA. (Photos: Sarah da Silva Mendes)

making the species vulnerable to accidental capture in many kinds of fishing nets used by commercial fishermen (Carvalho et al. 2016).

Sea turtles are the only vertebrates that obligatorily nest and, consequently, depend on sandy beaches to complete their life cycle. Female sea turtles return to the same beach on which they emerged as hatchlings, to lay their eggs in the dry beach zones (Miller 1997) (Fig. 5.7c, d). Nest site selection is critical and can greatly impact offspring survival (Patrício et al. 2018). It is driven mainly by physical factors related to the probability of nest inundation (beach slope and distance from vegetation), suitability for excavation of the cavity and gas diffusion (grain size and compactness), and travel distance for females and hatchlings to reach the sea (beach width and slope) (Brown and McLachlan 2010).

At landscape scales, the presence of coastal lagoons and distance from human settlements have already been reported as important drivers of nesting beach selection by sea turtles (Kikukawa et al. 1999).

Table 5.1 The occurrence of cetaceans in Brazilian coastal waters (Monteiro-Filho et al. 2021; Wickert et al. 2016; Lodi and Borobia 2013; Siciliano et al. 2006)

	Taxon	Common name	Offshore	Coastal
Mysticeti	**Family Balaenidae**			
	Eubalaena australis	Southern right whale		X
	Family Balaenopteridae			
	Balaenoptera acutorostrata	Dwarf minke whale		X
	Balaenoptera edeni	Bryde's whale	X	X
	Megaptera novaeangliae	Humpback whale	X	X
Odontoceti	**Family Dephinidae**			
	Cephalorhynchus commersonii	Commerson's dolphin		X
	Delphinus delphis	Common dolphin	X	X
	Orcinus orca	Killer whale	X	X
	Sotalia fluviatilis[a]	Tucuxi		X
	Sotalia guianensis	Guiana dolphin, marine tucuxi		X
	Stenella frontalis	Atlantic spotted dolphin	X	X
	Steno bredanensis	Rough-toothed dolphin	X	X
	Tursiops truncatus	Common bottlenose dolphin	X	X
	Tursiops gephyreus	Lahille's bottlenose dolphin	X	
	Family Phocoenidae			
	Phocoena spinipinnis	Burmeister's porpoise		X
	Family Pontoporiidae			
	Pontoporia blainvillei	La Plata dolphin, Franciscana		X
	Family Iniidae[a]			
	Inia araguaiaensis[a]	Araguaia river dolphin		X
	Inia boliviensis[a]	Bolivian river dolphin		X
	Inia geoffrensis[a]	Amazon river dolphin, boto		X

[a]Freshwater occurrence

5.2.4 Cetaceans

Brazil is home to 49 of the 87 cetacean species in the world (Monteiro-Filho et al. 2021, Lodi and Borobia 2013), with around 18 inhabiting coastal areas (Table 5.1), such as estuaries, upwelling regions, and shallow waters near sandy beaches (Siciliano et al. 2006; Lodi and Borobia 2013) (Fig. 5.8a, b). Cetaceans exploit waters close to sandy beaches mainly for foraging and may develop specific behaviors and strategies depending on the shore characteristics (Santos et al. 2010). Although their presence in low depths comes with a stranding risk, food availability for piscivorous cetaceans is a highly attractive factor. Nevertheless, the association of cetaceans with sandy beaches is indeed more often made regarding stranding events (or beaching) (Fig. 5.8c), which is a conservation concern and a subject that has intrigued researchers as long as humans have kept records (Bradshaw et al. 2006). The causes of strandings are usually difficult to detect (Mann et al. 2010), but in many cases, they have been associated with anthropogenic impacts such as

Fig. 5.8 Dolphins found on the coast of Brazil: (**a**) *Sotalia guianensis* and (**b**) *Steno bredanensis*. (**c**) Decomposition of a Guiana dolphin (*Sotalia guianensis*) carcass stranded on a sand beach, Southeastern Brazil. (Photos: Bruna Pagliani). (**d**) Laguna Municipality, southern Brazil, where bottlenose dolphins hunt in cooperation with fishermen. (Photo: Camila Castroviejo)

bycatch, entanglement in fishing nets, and chemical and noise pollution (Gregory 2009; Baulch and Perry 2014; Unger et al. 2016). Cetacean carcasses often end up stranded on sandy beaches and can be a great source of information, contributing with the knowledge of basic data related to the biology, distribution, and occurrence mainly of rare species (Moura et al. 2010, 2016; Lemos et al. 2013), biology (Peltier et al. 2012), ecology, and physiological aspects (Jepson et al. 2003).

Additionally, the carcasses that are deposited along sandy beaches enter the trophic processes, where they are a potential food resource for thousands of endemic and transient scavengers (Schlacher et al. 2013).

Some species seem to take advantage of particular aspects of coastal areas and have developed specific behaviors associated with sandy beach habitats. Several foraging strategies have been reported, including intentional beaching[2] (Hoese 1971; Lopez and Lopez 1985; Guinet 1991; Peddemors and Thompson 1994), chasing, and enclosing prey (Peddemors and Thompson 1994; Santos et al. 2000; Sargeant et al. 2005). This last foraging behavior always includes a barrier such as sand bars, mud banks, mangrove islands, or seagrass beds in which the dolphin's group the fish up. Fishermen in Baía de São Marcos, state of Maranhão (North coast of Brazil), take advantage of such behavior. Despite not doing so intentionally, in

[2] Phenomenon in which dolphins strand themselves intentionally on a beach in order to catch prey that occur in the shallow parts of the beach.

these events, dolphins help the fishermen by scaring fish and pushing them close to the beaches where the fishermen put their fishing gear.

In Brazil, a famous case of associations between dolphins and humans occurs in Laguna, state of Santa Catarina, Southern Brazil (Fig. 5.8d). The register of cooperative fishing between Lahille bottlenose dolphins *Tursiops gephyreus* (previously recognized as the subspecies *T. truncatus gephyreus*) and fishermen dates back to the year 1850 and occurs primarily on the shores of the bay formed from the ocean to the lagoon (Pryor et al. 1990; Simões-Lopes et al. 1998; Peterson et al. 2008; Wickert et al. 2016).

The cooperation comprises fishermen positioned in line, parallel to the shore, in shallow waters waiting for the dolphins' sign. The group of dolphins swims through the bay searching for the school of mullets *Mugil cephalus*, enclosing the prey toward the fishermen's barriers. The fishermen recognize dolphin signs when the fish are grouped and immediately throw the fishing nets to trap the prey. At that moment, the dolphins are positioned at the bottom of the lagoon with their mouths opened, ready to capture the prey that escapes the trap.

Another reported case of sandy beach use by cetaceans is the Guiana dolphins *Sotalia guianensis*, in the Cananéia estuary, located in the state of São Paulo (Santos et al. 2000, 2010). The foraging and feeding behavior displayed close to sloping beaches, estuarine mudflats, or sandbanks is described as "beach hunting" (Sargeant et al. 2005). Usually, the dolphin emerges exposing its body, partially or totally, close to those barriers to catch the prey (Santos 2010). Interestingly, female-calf pairs composed the majority (92.4%) of beach hunters suggesting another possible example of cultural transmission (Santos 2010). Similarly, a female-calf association was observed in the intentional strandings behavior of killer whales *Orcinus orca* in Possession Island, Crozet Archipelago (Guinet 1991; Guinet and Bouvier 1995), and killer whales of Patagonia, Argentina (Lopez and Lopez 1985).

Similar behaviors associating dolphins and sandy beaches have also been documented in the complex Amazon River systems, where Amazon dolphins (*Inia* sp.) and Tucuxi botos (*Sotalia fluviatilis)* use many freshwater sandy beaches and mud banks to catch their prey, particularly in dry seasons (Boran et al. 2002). For more than four decades, those associations in sandy beaches have been known in different parts of the world, such as the Atlantic hump-backed dolphin *Souza teuszii* on the Mauritania coast (Busnel 1973) and the bottlenose dolphins *T. truncatus* in Georgia (Hoese 1971) and Florida (Torres and Read 2009) on the US coasts.

5.3 Main Threats

5.3.1 Urban Development

The increasing population densities on the Brazilian coast have led to significant levels of coastal urbanization. This comes with multiple stressors that act in the sandy beach environment linked to human activities and interventions, forming specific threats to vertebrates that frequent sandy beaches. Changes made to the physical environment related to urban development are numerous, such as landfills and silting for the construction of ports, industries, shrimp aquaculture, loss of sediment input from continental drainage, or the removal of neighboring vegetation such as mangroves (Chap. 9).

Although some studies have included the impact of anthropogenic factors at various spatial scales (e.g., nourishment, food depletion, and seascape fragmentation) on the diversity of surf zone fish (Costa et al. 2017b; Ortodossi et al. 2019; Wilber et al. 2003) and discussed the effects of urbanization (Shah Esmaeili et al. 2022), little is known on the degree of these impacts in driving surf zone fish communities in Brazil.

The increasing urbanization of sandy beaches and intense tourism in Brazil have been shown to reduce the richness, diversity, and abundance of fish in the surf zone of sandy beaches affected by pollution and increased tourism in the austral summer (Pereira et al. 2015; Franco et al. 2016; Costa et al. 2017b), overlapping with natural seasonal variations. Also, Lunardi et al. (2012) showed that migratory birds can avoid beaches closer to urban settlements along the northeastern coast of Brazil. Thus, beach quality becomes decisive for the energy intake and migration success of several threatened migratory shorebirds (Lunardi et al. 2012). Studies conducted in Australia also showed that increasing urbanization reduces the presence of large raptors that are known to scavenge on sandy beaches (Huijbers et al. 2013) while altering the habitat selection of many shorebirds (Meager et al. 2012) and contributing to the propagation of urban alien pigeons (*Columbia livia*).

Some species of migratory birds, such as the sandpiper (*Calidris fuscicollis*) and the red knot (*Calidris canutus*), have shown decreasing populations due to the decreased suitable habitats used during migration, including sandy beaches (Morrison et al. 2004; Anderson et al. 2019). The physical presence of humans on sandy beaches is known to scare away many bird species (Yasué 2006). Human settlements also coincide with the presence of domestic animals such as cats and dogs on beaches, as well as rats, all of which have been shown to prejudice the presence of some bird species by frightening and preying on birds (Banks and Bryant 2007; Beckerman et al. 2007). Additionally, vehicle traffic on the beach forms a threat to both birds and sea turtles (Costa et al. 2020) (Fig. 5.9).

5.3.2 Light and Sound Pollution

Artificial light, usually associated with man-made structures, causes stress or disorientation in several sandy beach vertebrates such as birds and sea turtles (de Freitas et al. 2017). For example, in coastal areas, these stressors leave sea turtle hatchlings disoriented, without being able to reach the sea (Mascarenhas et al. 2003), sometimes walking in circles for hours ("disorientation") or going directly toward the direction of light and not toward to sea ("misorientation") (Salmon 2003). Human disturbances at local scales, such as artificial light pollution and vehicle traffic,

Fig. 5.9 Sea turtle run over by a vehicle on a sandy beach. (Photo: Julyana Madureira)

disrupt nesting site selection (Sella and Fuentes 2019) and also threaten the offspring's survival by impacting both eggs and newborns (Fernandes et al. 2016).

In Brazil, more than half of the turtle nesting beaches were located in areas potentially exposed to light pollution and light levels appeared to have a significant effect on nest densities of hawksbills; however, other environmental factors are more significant in the choice of the nest site. Light pollution may affect more hatchlings in coastal waters; however, further studies on the impact of light pollution on sea turtle populations along the Brazilian coast should be carried out (Colman et al. 2020).

Human activity also generates anthropogenic noise, which mainly affects marine mammals, due to their accurate subaquatic hearing systems (Southall et al. 2008). Cetaceans use sound communication for many aspects of their life, such as locating prey, orientation, reproduction, defense, and other social skills (Tougaard et al. 2014). Anthropogenic noise can mask or avoid communication among animals, affecting social interaction, reproduction, as well as predator detection (Erbe et al. 2016). Maritime traffic and lack of regulated tourism imply serious risks to cetaceans since animals are subjected to stress and noise, which can leave them disorientated and, in many cases, can lead to injuries caused by boat engines (Camargo and Bellini 2007; Waerebeek et al. 2007; New et al. 2015). This issue, however, is still largely overlooked along the Brazilian coast, with very few studies relating the presence of cetaceans in uncommon areas to misorientation due to noise pollution (e.g., Neto et al. 2008).

5.3.3 Chemical Pollution

The lack of basic sanitation in coastal cities leads to the pollution of sandy beach habitats by domestic and industrial sewage. Trace elements have been detected in turtles (Barbieri 2009) and birds (Barbieri et al. 2010) along the Brazilian coast, since they, in general, are at the top of food chains in coastal areas, favoring the bioaccumulation and biomagnification of pollutants in their tissues and feathers (Burger and Gochfeld 2004; Bisi et al. 2012). Several studies indicate serious consequences in the life cycle of birds when their prey is contaminated by hydrocarbons or heavy metals (Bond et al. 2015; Burger 2002), including on Brazilian shores (e.g., Bisi et al. 2012). Due to their commercial value, bioaccumulation of trace metals in sandy beach fishes is not only of ecological interest but has also been studied on the Brazilian coast to monitor the risks of fishing and fish consumption by humans in coastal environments (Vieira et al. 2020).

Brazil is one of the countries most vulnerable to the ecological and socioeconomic impacts of pollution by crude oil in marine environments (Magris and Giarrizzo 2020), and sandy beaches are on the frontlines of possible disasters. A recent example of an oil spill reaching Brazilian beaches occurred in August 2019, when thousands of tons of crude oil residue from an unknown source began to wash up along an almost 3000-km stretch of shoreline and on >980 beaches (Escobar 2019; De Oliveira Soares et al. 2020). As expected, areas exposed to oil residues overlapped with the distribution range of various vertebrate species, including elasmobranchs, bony fishes, turtles, birds, and marine mammals; some of them were found covered in oil (Magris and Giarrizzo 2020; De Oliveira Soares et al. 2020).

Oil coverage and accumulation of toxic compounds disrupt physiological functioning, including nervous disorders, and increased susceptibility to predation (Johansen et al. 2017). Changes in the physicochemical mosaic of the water and sediments affect daily and seasonal activities, connectivity among coastal ecosystems, and, consequently, pivotal life cycle processes of vertebrates.

5.3.4 Solid Wastes Pollution

Marine debris discarded in rivers or left by beach users is known to cause harm to marine animals that mistake these items for food or nesting materials and get prejudiced by the consumption and exposure to adsorbed pollutants (Tavares et al. 2016). A well-known example of that is the ingestion of plastics by sea turtles (Mendes et al. 2015; Carvalho et al. 2015) and birds on Brazilian sandy beaches (Tavares et al. 2017). All turtles and 40% of seabirds found along the beaches in the state of Rio Grande do Sul, for example, ingested debris, with plastic being the main type of material ingested (Tourinho et al. 2009). Marine debris, especially plastic, and other anthropogenic sources of pollution are the principal threat associated with whale stranding (Baulch and Perry 2014).

Additionally, these larger plastics break down into microplastics, which have been shown to inhibit hatching, decrease growth rates, and alter the feeding preferences and innate behaviors of fish larvae (Foley et al. 2018). Also, predator-induced mortality rates could greatly increase, since individuals exposed to microplastics do not respond to olfactory threat cues (Lönnstedt and Eklöv 2016). Microplastic ingestion by surf zone fishes on the Northeast coast of Brazil has been reported regardless of their feeding habitats, although adverse effects at the individual level remain unknown (Dantas et al. 2020).

5.3.5 Harvest and Fishing

The main threat to sea turtles in Brazil until the 1980s was hunting and the consumption of eggs by humans. Nowadays, egg poaching has been reduced to very low levels (Thomé et al. 2007), but hunting persists in some regions of the Southeast, where turtles accidentally caught in nets are used by fishermen as food (Carvalho et al. 2015). The consequences of several human impacts, including poaching have led all five sea turtle species found on the Brazilian coast to be considered endangered.

Surf zone fishes have not been reported to be depleted by small-scale fishing typically practiced on beaches. Only one species that inhabits sandy beaches is currently known to be critically endangered (*Genidens planifrons*), whereas two are considered endangered (*Genidens barbus* and *Pogonias cromis*). On the Brazilian coast, hundreds of people participate in angling tournaments that remove tons of fish from the surf zone every year, but the long-term consequences of this activity are still unknown. Barrella et al. (2016) estimated catches by angling correspond to only 2% of commercial landings in Southeast Brazil. Traditional fishermen that continue to use beach seine nets often do it on smaller scales (Da Silva 2004). Some commercial species (e.g., mullets and blue fishes) inhabiting surf zones temporarily can become less frequent because of overfishing offshore, not necessarily as a result of fishing on the beach. Some species occurring in Brazilian surf zones, such as the Florida Pompano *Trachinotus carolinus,* have been classified as overfished elsewhere (Solomon and Tremain 2009).

5.3.6 Bycatch and Ghost Fishing

The incidental capture (i.e., bycatch) of non-target vertebrates and the subsequent disposal of these organisms constitute a major environmental problem that fishing activity has been dealing with (Rodrigues-Filho et al. 2015). Guimarães et al. (2018) reported a bycatch of five turtles/1000 h/unit effort by bottom trawlers along the Brazilian coast, mainly in the state of Rio de Janeiro. These authors found that turtles are caught mainly in shallow waters, where they are usually more abundant.

Shrimp trawling is used along almost the entire Brazilian coast (Branco et al. 2015; Junior et al. 2013) and has been estimated to produce more than one-third of the world's total bycatch because of its low selectivity (Alverson et al. 1994). Because shrimp trawling is conducted mainly in shallow waters, in some cases along surf zones, the bycatch composition consists mainly of juvenile fishes of demersal and benthopelagic species (Junior et al. 2013; Rodrigues-Filho et al. 2015).

Even though considered less invasive to surf zone fish, beach seine nets are known to capture threatened or endangered elasmobranchs (Yokota and Lessa 2006). The presence of both egg capsules and small individuals (e.g., *Carcharhinus porosus*) and large pregnant females of *Sympterygia* sp. in beach seine net hauls shows the use of surf zones as breeding areas as well as nurseries (ICMBio 2018). Some of these species have a high commercial value because of the quality of their fin meat (e.g., *Carcharhinus plumbeus*) or their ornamental/medicinal value (e.g., *Hippocampus reidi*).

Another significantly harmful phenomenon is "ghost fishing" when lost or abandoned fishing gear keeps capturing organisms (Matsuoka et al. 2005). The most recorded discharged fishing gears in Brazil are nets, ropes, and fishing lines found on beaches, submerged in coastal areas, and on oceanic islands (Link et al. 2019).

Fishing activities also significantly compromise the survival of cetaceans. Small cetaceans are often accidentally captured by fishing gear (Lewison et al. 2014; Quirós et al. 2018). The most critical example is the "Toninha" or "La Plata River dolphin," *Pontoporia blainvillei* (Fig. 5.10a), endemic to the Western South Atlantic and on the verge of extinction. It is one of the main species of small dolphins

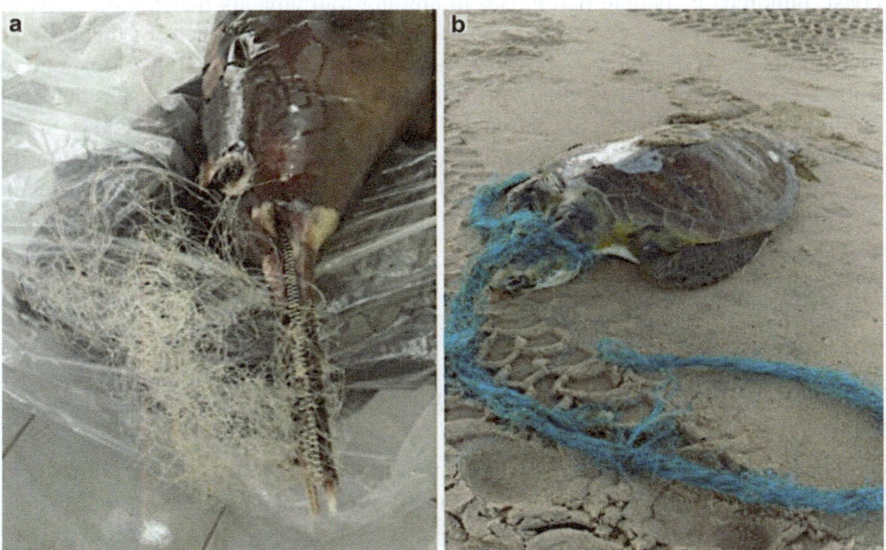

Fig. 5.10 (**a**) La Plata dolphin/toninha (*Pontoporia blainvillei*). (Photo: Bruna Pagliani). (**b**) Green sea turtle trapped in a fishing net. (Photo: Leonardo Costa)

recorded dead or seriously injured stranded on Brazilian sandy beaches. Being restricted to coastal waters, the La Plata dolphin faces a huge threat of incidental capture by fishing gear besides other environmental degradation (Rosas et al. 2002). Many gill nets are placed in shallow waters, up to five nautical miles away from the shore, and up to 20 m deep (Di Beneditto et al. 1998), and can accidentally trap dolphins. If not trapped, they tend to obtain food and damage and/or capture fish from the nets, consequently causing losses to the fishermen and giving them the image of "competitors" for the same resource.

5.3.7 Prey Depletion

On the Brazilian coast, benthic invertebrates, such as the mole crab *Emerita brasiliensis* and the trigonal clam *Tivella mactroides*, are important food sources for *Trachinotus* spp. and *Menticirrhus* spp., typical fishes inhabiting the surf zones (Rodrigues and Vieira 2010; Turra et al. 2015; Costa and Zalmon 2017). Polychaetes are also common food items for fishes on beaches (Amaral et al. 1994). Upper shore macroinvertebrates, such as sandhoppers, were also reported as important prey for surf zone fishes (Costa and Zalmon 2017). Similarly, the occurrence of shorebirds in beach habitats, such as the Royal tern (*Thalasseus maximus*) and the South American tern (*Sterna hirundinacea*), depends on the invertebrate density (Lunardi et al. 2012; Rangel et al. 2022).

However, due to the intense tourism and urban development, macroinvertebrates have been intensively disrupted on the Brazilian coast (see Chap. 9). Studies have shown that *Trachinotus* spp. are now feeding on litter-associated insects on urbanized sandy beaches, instead of preying on intertidal macroinvertebrates like in non-urbanized areas (Costa and Zalmon 2017; Júnior et al. 2020). In some cases, surf zone fish can avoid urbanized beaches during the high tourism season, in response to food depletion (Costa et al. 2017b). Consequently, predators of surf zone fish, such as seabirds, also avoid surf zones of disturbed beaches, and energy transfer between trophic levels tends to be less efficient leading to vertebrates chronically avoiding the beach ecosystem in disturbed areas (Costa et al. 2017a).

5.4 Conservation Challenges and Initiatives

To implement effective conservation actions, there is a need for solid baseline data, which for all the discussed vertebrate groups is either limited to certain species or specific areas along the coast (Amaral and Jablonski 2005). This is caused by both a lack of investment in scientific research and long-term monitoring programs. The main challenge in long-term monitoring is the sampling effort, which requires both heavy investment in time and resources and the implementation of standardization protocols at broad scales specifically for vertebrates. New techniques such as the use of Surf Baited Remote Underwater Videos (surf-BRUVs) (Vargas-Fonseca et al. 2016; Shah Esmaeili et al. 2021) or even implementing the use of eDNA (Jeunen et al. 2019) could be a possible low effort and low impact solution for the detection of aquatic vertebrates in sandy beaches.

The use of platforms such as e-bird (Sullivan et al. 2009), which uses the input of data provided by citizens, could also be performed for other vertebrate groups, increasing the number of reports on a large scale and helping the mapping of species distribution, including migrant ones. The study of coastal birds, for instance, could help in choosing beaches with fewer restrictions on human use, where activities such as tourism cause less damage to biodiversity, the ecosystem, and the services provided (McLachlan et al. 2013).

A possible solution for social-ecological conflicts on beaches is considering an ecosystem-based approach with an integration of physical, biological, and social factors (Chap. 10). For that, scientists from different areas need to collaborate with governance and the local population. Although this is very challenging, some examples of these efforts are productive in both generating baseline data as well as constructing conservation and management plans (e.g., such as the BIOTA-Araçá project, Amaral et al. 2018). The Ecosystem Approach to Fisheries (EAF) has been proposed as a holistic framework for managing fisheries. The purpose of this approach is to address the needs of societies and simultaneously promote ecosystem conservation in the long term. Co-management (Co-M) is a governance model that naturally accomplishes EAF goals, because its arrangement shares assignments between government, stakeholders, and, advisably, researchers for managing natural resources (Gianelli et al. 2015). Small-scale fisheries are predominant on Brazilian sandy beaches, and this chases the fundamental steps of EAF. However, in most cases, subsistence or commercial fishing is not included formally in management plans in Latin America (Gianelli et al. 2015). EAF/Co-M is undoubtedly a promising initiative for managing fishery on beaches, but further research is necessary on its performance at local and regional scales, particularly in areas where fishing occurs for commercial purposes.

Concerns are raised regarding the regulation of both commercial and recreational fishing. Notably, it is urgent to regulate fishing gear aiming to reduce loss and, consequently, bycatches and marine pollution. There is specific legislation to reduce incidental captures by certain fishing gear types, such as the use of turtle excluder devices (TEDs), but they are not specific for the predominant recreational angling or harvesting on sandy beaches vertebrates. There is a lack of surveillance of beach angling in Brazil (Barrella et al. 2016), and several angling tournaments (see Sect. 5.3.5) have applied the "catch-release" approach (promoted by IBAMA-Brazilian Institute for the Environment and Renewable Natural Resources) instead of donating the captured fishes to charities, as an attempt to reduce potential negative effects of angling on surf zone fish. This strategy is ecologically questionable, since the captured fishes can become vulnerable to predation and diseases after being disturbed and/or injured by fishing artifacts.

Implementation and inspection of regulations are most effective when there is support from local communities and the general public; therefore, investment in education and awareness is essential. Projects such as TAMAR (www.tamar.org.br) that work on the preservation of sea turtles, changed the history of sea turtle conservation in Brazil in the 1980s, by protecting the main breeding areas (Marcovaldi and Marcovaldi 1999). In the 1990s, the project had already worked with fishermen, environmental education, and ecotourism. The actions of TAMAR continue to this day, highlighting its enormous importance in the protection of sea turtles, mainly bringing more environmental awareness to the population and more action by government agencies in conservation efforts. Over the past few decades, several sea turtle conservation projects have added to their efforts to protect sea turtles in all regions of Brazil, making this charismatic species a great asset in passing the message on the importance of the conservation of sandy beaches as a whole (Table 5.2).

In conclusion, there are several challenges to achieving sandy beach conservation, and beach vertebrates could be excellent symbols for that. Several fishes, coastal birds, sea turtles, and marine mammals are charismatic and can contribute to attracting the conservation appeal of citizens. Marketing strategies based on public preferences must be applied to find potential flagship species, aiming to get more funding opportunities for both biodiversity popularization and environmental education initiatives, as well as to improve research and monitoring. As the above-mentioned vertebrates usually depend on several habitat features to select suitable beach habitats, they can also act as umbrella species, in which their protection will result in the conservation of the coexistent biota (Maslo et al. 2016). Therefore, it is time for networks and multidisciplinary efforts, especially if the main goal is the conservation of the ecosystem with one of the highest recreational values in the world.

5.5 Research Gaps and Avenues for Future Studies

For all vertebrate groups mentioned in this chapter, available information from different parts of the country is scarce. Since many vertebrates are migratory and not restricted to sandy beaches in one specific region, information about distribution ranges is essential. For example, little is known about the connectivity of coastal environments along the Brazilian coast. Nationwide or regional studies are rare but crucial when dealing with highly mobile species. Another problem is that most of the studies cover a short temporal scale (one season, < 2 years). This is likely related to the short duration of grants and research projects, on average lasting 2 years. This could be not sufficiently informative to detect impacts or to assess the recovery capacity from human disturbances that threaten sandy beach biodiversity in Brazil.

Although a few studies have focused on the variability of fish communities on sandy beaches, these studies were all highly concentrated in specific areas and so far, only one study comparing different regions is available (Araújo et al. 2018). One of the main problems is the labor intensity of surf zone fish sampling. Methodological constraints have been shown to compromise a full understanding of the surf zone functioning and biodiversity, given that generally only a single

Table 5.2 List of sandy beach vertebrates and their International Union for Conservation of Nature (IUCN) status, status in Portaria MMA N° 148, of June 7, 2022, and main threats they face, ranging from orange (mild) to dark red (heavy)

Group	Species	Status (IUCN)	Status (PMMA 2022)	Main treath					
				Urban development	Chemical pollution	Sound & light pollution	Harvesting and fishing	Bycatch and ghost fishing	Prey depletion
Fishes	*Genidens barbus*	Unknown	EN						
	Genidens planifrons	Unknown	CR						
	Carcharhinus obscurus	EN	EN						
	Carcharhinus plumbeus	EN	CR						
	Negaprion brevirostris	VU	EN						
	Gymnura altavela	EN	CR						
	Mustelus schmitti	CR	CR						
	Rhinoptera brasiliensis	CR	CR						
	Rioraja agassizii	VU	VU						
	Pseudobatos percellens	EN	VU						
	Zapteryx brevirostris	EN	VU						
	Pogonias cromis	LC	EN						
	Mycteroperca bonaci	NT	VU						
	Sphyrna lewini	CR	CR						
	Squatina guggenheim	EN	CR						
	Squatina occulta	CR	CR						
	Narcine brasiliensis	NT	VU						
Aves	*Charadrius wilsonia*	LC	VU						
	Sterna dougallii	LC	VU						
	Sterna hirundinacea	LC	VU						
	Thalasseus acuflavidus	Unknown	VU						
	Thalasseus maximus	LC	EN						
	Calidris canutus	NT	VU						
	Calidris pusilla	NT	EN						
	Calidris subruficollis	NT	VU						
	Limnodromus griseus	LC	EN						
	Numenius hudsonicus	Unknown	VU						
	Diomedea dabbenena	CR	CR						
	Diomedea epomophora	VU	VU						
	Diomedea exulans	VU	CR						
	Diomedea sanfordi	EN	EN						
	Thalassarche chlororhynchos	EN	EN						
	Procellaria aequinoctialis	VU	VU						
	Procellaria conspicillata	VU	VU						
	Pterodroma arminjoniana	VU	CR						
	Pterodroma deserta	VU	VU						
	Pterodroma incerta	EN	EN						
	Pterodroma madeira	EN	EN						
	Puffinus lherminieri	LC	CR						
Testudines	*Caretta caretta*	VU	VU						
	Eretmochelys imbricata	CR	EN						
	Lepidochelys olivacea	VU	VU						
	Dermochelys coriacea	VU	CR						
Marine mammals	*Eubalaena australis*	LC	EN						
	Sotalia guianensis	NT	VU						
	Tursiops gephyreus	Unknown	EN						
	Inia araguaiaensis	Unknown	VU						
	Inia geoffrensis	EN	EN						
	Pontoporia blainvillei	VU	CR						

sampling method (i.e., beach seine netting) is used. Additionally, few studies have focused on habitat requirements of specific fish species (e.g., Araújo et al. 2008a, b; Mattox et al. 2008; Andrades et al. 2012), particularly those with commercial interest (e.g., Mugilidae, Pomatomidae, Clupeidae, and Sciaenidae). As coastal fisheries are the source of more than 90% of the food provided by marine ecosystems (Garcia and de Leiva Moreno 2003) and have, in recent years, reached levels of approximately 130 million tons per year (Watson and Tidd 2018), it is important to study their direct and indirect impacts. Although some studies have focused on monitoring catch and bycatch numbers in Brazilian shallow waters (Silva Júnior et al. 2013; Rodrigues-Filho et al. 2015), there is still little information available on the species and the quantity of fish that end up as both targeted and non-targeted catches. Besides, mitigating actions for angling disturbances (e.g., catch and release approach) have not had their impacts and efficiency scientifically assessed in Brazil.

Studies on the population size of birds on sandy beaches are still scarce in Brazil (e.g., Lunardi et al. 2012; Barbieri and Hvenegaard 2008; Mancini et al. 2018). Most of the studies concerning habitat selection by birds are carried out on low-energy ecosystems like mangroves and coastal lagoons (Tavares et al. 2015), and thus, ocean beaches may be neglected as important feeding habitats. To argue about the importance of sandy beaches in that context (e.g., intertidal and surf zones), it is essential to determine how typical prey, usually invertebrates and fish, are important as a food resource for these bird species and the effects of human disturbances on both prey and predators (e.g., Costa et al. 2017a, b). It is not well-known if top-down or bottom-up processes concerning birds, fish, and invertebrates relationships are predominant on sandy beaches, including Brazilian ones (Dugan et al. 2003; Costa et al. 2017a, b). As seabirds and shorebirds require areas with low disturbance and a high abundance of food, they can be important conservation shortcuts that deserve more attention in ecological assessments on sandy beaches.

Studies on the genetic information of beach organisms are important for the strategies of conservation and management of vertebrates, since the knowledge they bring can help to understand biological factors, such as interconnectivity between nesting and foraging metapopulations (Jordão et al. 2015), and other factors related to conservation, such as the impacts of incidental catches by fishing nets (Proietti et al. 2014a). In Brazil, genetic studies are still lacking, especially those seeking to understand the frequency and importance of hybrids for the conservation of species, since hybrids may have a high frequency in some regions (Soares et al. 2017). According to Proietti et al. (2014b), understanding how hybridization affects the distribution and ecology of sea turtles in Brazil is a complex task that is nevertheless fundamental when defining conservation strategies.

Considering strandings on sandy beaches and the use of shallow waters associated with that environment, the main gaps concerning cetacean conservation are related to the lack of basic information on the biology and ecology of species, mainly the ones classified as Data Deficient by the IUCN. Basic information (such as population growth, for example) is necessary for better understanding the dynamics of populations and their interaction with environmental factors, as well as making predictions about future changes in that population. Besides the aforementioned information, studies on how anthropogenic stressors affect marine mammal species should be the principal research direction for setting specific conservation and management targets and maintaining the biodiversity of the group. Especially for coastal areas, bycatches and fishing gear entanglement are threats to small cetaceans and must be prioritized in management actions. More recently, as a result of globalization and maritime transportation in the past decades, another gap consists of how noise pollution affects cetacean populations around the world. This information can directly or indirectly assist in decision-making and monitoring species to prevent (or minimize) stranding situations.

Overall, drivers of community composition or focal species of sandy beach vertebrates have not been revealed so far. Very few studies have also discussed the connectivity with the open sea and terrestrial environments (Liebowitz et al. 2016), regarding the exportation of biomass for example. Lastly, many studies focus on a small number of species within a certain phylum. The interaction between different phyla within the sandy beach ecosystem (e.g., trophic connections), including all the ones discussed in other chapters of this book, is still very little reported, including effects caused by urbanization (Costa et al. 2017a, b).

5.6 Final Remarks

This chapter emphasizes the diversity and importance of vertebrate species found on Brazilian sandy shores, including fishes, birds, turtles, and cetaceans. Unfortunately, these species are severely threatened by several human activities such as fishing, coastal urbanization, and discharge of pollutants, and many of them face risk of extinction. Solid baseline data is still necessary to implement effective conservation actions for all the discussed vertebrates. Studies comprising broad and integrative perspectives, including the propagation of impacts through the whole coastal food-web, may provide subsidies for the implementation of public policies for the conservation and sustainable management of coastal zones, including sandy beaches and their interfaces. Despite the research gaps, vertebrates are very often present on beaches, allowing scientific studies that support planning for sustainable ecotourism. Notoriously, vertebrates raise more public engagement for conservation than resident fauna mainly composed of inconspicuous crustaceans, mollusks, polychaetes, and insects. We need to better explore this potential of using vertebrates in outreach and providing a more biocentric view of sandy beaches.

Acknowledgments This work was supported by the Fundação de Amparo à Pesquisa do Estado de São Paulo (FAPESP) through scholarship grants to YSE (grant 2018/0955-5), and through the financial support linked to the FAPESP BIOTA project Program (grant 2018/19776-2) and the FAPESP Global Climate Change Research Program thematic project (grant 2015/03804-9). This project also counted on the financial support of Fundação Grupo Boticário (grant 1133_20182). This study was additionally financed in part by the Coordenação de Aperfeiçoamento de Pessoal de Nível Superior – Brasil (CAPES) – Finance Code 001 through a scholarship grant to YSE. LLC is supported by Fundação de Amparo à Pesquisa do Estado do Rio de Janeiro – FAPERJ (grant E-26/200.620/2022 and E-26/210.384/2022). RHC is supported by the Coordenação de Aperfeiçoamento de Pessoal de Nível Superior (CAPES 88881.133275/2016-01).

References

Alverson DL, Freeberg MH, Murawski SA et al (1994) A global assessment of fisheries bycatch and discards. Food & Agriculture Organization, Rome

Amaral ACZ, Jablonski S (2005) Conservation of marine and coastal biodiversity in Brazil. Conserv Biol 19:625–631

Amaral ACZ, Nonato EF, Petti MV (1994) Contribution of the polychaetous annelids to the diet of some Brazilian fishes. Mém Mus Natl Hist Nat 162:331–337

Amaral ACZ, Corte GN, Denadai MR, Colling LA, Borzone C, Veloso V, ... & Almeida TCMD (2016). Brazilian sandy beaches: characteristics, ecosystem services, impacts, knowledge and priorities. Brazilian Journal of Oceanography, 64, 5–16.

Amaral ACZ, Ciotti AM, Fonseca G (2018) Biodiversity and functioning of a subtropical coastal ecosystem: subsidies for integrated management. Ocean Coast Manag 164:1–3

Anderson AM, Friis C, Gratto-Trevor CL et al (2019) Consistent declines in wing lengths of Calidridine sandpipers suggest a rapid morphometric response to environmental change. PLoS One 14(4):e0213930

Andrades R, Bolzan MS, Contaifer LS (2012) Scientific note evidence of sandy beaches as growth grounds for commercial fish in the south-western Atlantic. Pan-Am J Aquat Sci 7(2):107–110

Andrades R, Gomes MP, Pereira-Filho GH et al (2014) The influence of allochthonous macroalgae on the fish communities of tropical sandy beaches. Estuar Coast Shelf Sci 144:75–81

Araújo FG, Silva MA, Azevedo MCC et al (2008a) Spawning season, recruitment and early life distribution of *Anchoa tricolor* (Spix and Agassiz, 1829) in a tropical bay in southeastern Brazil. Braz J Biol 68(4):823–829

Araújo FG, Silva MA, Santos JN et al (2008b) Habitat selection by anchovies (Clupeiformes: Engraulidae) in a tropical bay at southeastern Brazil. Neotrop Ichthyol 6(4):583–590

Araújo FG, Rodrigues FL, Teixeira-Neves TP et al (2018) Regional patterns in species richness and taxonomic diversity of the nearshore fish community in the Brazilian coast. Estuar Coast Shelf Sci 208:9–22

Awabdi DR, Tavares DC, Bondioli ACV et al (2018) Influences of conservation action on attitudes and knowledge of fishermen towards sea turtles along the southeastern Brazil. Mar Policy 95:57–68

Azevedo MCC, de Sousa Gomes-Gonçalves R, Mattos TM et al (2017) Taxonomic and functional distinctness of the fish assemblages in three coastal environments (bays, coastal lagoons and oceanic beaches) in southeastern Brazil. Mar Environ Res 129:180–188

Bahia NCF, Bondioli ACV (2010) Interação das tartarugas marinhas com a pesca artesanal de cerco-fixo em Cananéia, litoral sul de São Paulo. Biotemas 23(3):203–213

Baldassin P, Santos RA, Da Cunha JMM et al (2010) Cephalopods in the diet of Magellanic penguins *Spheniscus magellanicus* found on the coast of Brazil. Mar Ornithol 38:55–57

Banks PB, Bryant JV (2007) Four-legged friend or foe? Dog walking displaces native birds from natural areas. Biol Lett 3(6):611–613

Barbieri E (2009) Concentration of heavy metals in tissues of green turtles (Chelonia mydas) sampled in the Cananéia estuary, Brazil. Braz J Oceanogr 57:243–248

Barbieri E, Hvenegaard GT (2008) Seasonal occurrence and abundance of shorebirds at Atalaia Nova Beach in Sergipe State, Brazil. Waterbirds 31(4):636–644

Barbieri E, Passos EDA, Filippini A, dos Santos IS, Garcia CAB (2010) Assessment of trace metal concentration in feathers of seabird (Larus dominicanus) sampled in the Florianópolis, SC, Brazilian coast. Environ Monit Assess 169(1):631–638

Barrella W, Ramires M, Rotundo MM et al (2016) Biological and socio-economic aspects of recreational fisheries and their implications for the management of coastal urban areas of southeastern Brazil. Fish Manag Ecol 23(3–4):303–314

Baulch S, Perry C (2014) Evaluating the impacts of marine debris on cetaceans. Mar Pollut Bull 80:210–221

Beckerman AP, Boots M, Gaston KJ (2007) Urban bird declines and the fear of cats. Anim Conserv 10(3):320–325
Bellini C, Santos AJ, Grossman A et al (2013) Green turtle (*Chelonia mydas*) nesting on Atol das Rocas, north-eastern Brazil, 1990–2008. J Mar Biol Assoc UK 93(4):1117–1132
Bisi TL, Lepoint G, de Freitas Azevedo A, Dorneles PR, Flach L, Das K, Malm O, Lailson-Brito J (2012) Trophic relationships and mercury biomagnification in Brazilian tropical coastal food webs. Ecol Indic 18:291–302
Bond AL, Hobson KA, Branfireun BA (2015) Rapidly increasing methyl mercury in endangered ivory gull (*Pagophila iocone*) feathers over a 130-year record. Proc Royal Soc B Biol Sci 282(1805):20150032
Boran JR, Evans PGH, Rosen MJ (2002) Behavioural ecology of cetaceans. In: Evans PGH, Raga JA (eds) Marine mammals. Springer, Boston, pp 197–242
Borland HP, Schlacher TA, Gilby BL et al (2017) Habitat type and beach exposure shape fish assemblages in the surf zones of ocean beaches. Mar Ecol Prog Ser 570:203–211
Bornatowski H, Heithaus MR, Abilhoa V et al (2012) Feeding of the Brazilian sharpnose shark Rhizoprionodon lalandii (Müller & Henle, 1839) from southern Brazil. J Appl Ichthyol 28(4):623–627
Bradshaw CJA, Evans K, Mark A (2006) Mass cetacean strandings: a plea for empiricism. Conserv Biol 20(2):584–586
Branco JO, Freitas Júnior F, Christoffersen ML (2015) Bycatch fauna of seabob shrimp trawl fisheries from Santa Catarina State, southern Brazil. Biota Neotrop 15(2):e20140143
Brown AC, McLachlan A (2010) The ecology of sandy shores. Elsevier.
Burger J (2002) Food chain differences affect heavy metals in bird eggs in Barnegat Bay, New Jersey. Environ Res 90(1):33–39
Burger J, Gochfeld M (2004) Marine birds as sentinels of environmental pollution. EcoHealth 1(3):263–274
Busnel RG (1973) Symbiotic relationship between man and dolphins. Trans NY Acad Sci 35(2):112–131
Camargo FS, Bellini C (2007) Report on the collision between a spinner dolphin and a boat in the Fernando de Noronha Archipelago, Western Equatorial Atlantic, Brazil. Biota Neotrop 7(1):bn00807012007
Carvalho RH, Lacerda PD, da Silva Mendes S et al (2015) Marine debris ingestion by sea turtles (Testudines) on the Brazilian coast: an underestimated threat? Mar Pollut Bull 101(2):746–749
Carvalho RH, Mamede N, et al (2016) Attitudes towards conservation and fishing interaction with sea turtles in the southeast coast of Brazil. Ocean & Coastal Management 127:55-62
Clark BM (1997) Variation in surf-zone fish community structure across a wave-exposure gradient. Estuar Coast Shelf Sci 44(6):659–674
Colman LP, Lara PH, Bennie J et al (2020) Assessing coastal artificial light and potential exposure of wildlife at a national scale: the case of marine turtles in Brazil. Biodivers Conserv 29:1135–1152
Connell JH (1978) Diversity in tropical rain forests and coral reefs. Science 199(4335):1302–1310
Conservação da Biodiversidade ICMBio Instituto Chico Mendes (2019) Relatório anual de rotas e áreas de concentração de aves migratórias no Brasil (in Portuguese). URL https://www.gov.br/icmbio/pt-br/centrais-de-conteudo/publicacoes/relatorios/relatorio_de_rotas_e_areas_de_concentracao_de_aves_migratorias_brasil_3edicao_2019.pdf
Costa LL, Zalmon IR (2017) Surf zone fish diet as an indicator of environmental and anthropogenic influences. J Sea Res 128:61–75
Costa LL, Tavares DC, Suciu MC et al (2017a) Human-induced changes in the trophic functioning of sandy beaches. Ecol Indic 82:304–315
Costa LL, Landmann JG, Gaelzer R et al (2017b) Does human pressure affect the community structure of surf zone fish in sandy beaches? Cont Shelf Res 132:1–10

Costa LL, Secco H, Arueira VF et al (2020) Mortality of the Atlantic ghost crab Ocypode quadrata (Fabricius, 1787) due to vehicle traffic on sandy beaches: a road ecology approach. J Environ Manag 260:110168
Da Silva PP (2004) From common property to co-management: lessons from Brazil's first maritime extractive reserve. Mar Policy 28(5):419–428
Dantas NCFM, Silva Júnior CABD, Feitosa CV et al (2016a) Seasonal influence of drifting seaweeds on the structure of fish assemblages on the eastern equatorial Brazilian coast. Braz J Oceanogr 64(4):365–374
Dantas NCFM, Silva Júnior CAB, Lippi DL et al (2016b) Diel variations and ecological aspects in fish assemblages of a sandy beach in the semi-arid region of northeast Brazil. Braz Arch Biol Technol 59:e16160076
Dantas NC, Duarte OS, Ferreira WC et al (2020) Plastic intake does not depend on fish eating habits: identification of microplastics in the stomach contents of fish on an urban beach in Brazil. Mar Pollut Bull 153:110959
De Carvalho-Souza GF, Miranda DDA, Pataro L (2016) Hazards in hanging gardens: a report on failures of recognition by green turtles and their conservation implications. Mar Pollut Bull 105(1):98–101
De Castilhos JC, Coelho CA, Argolo JF et al (2011) Avaliação do estado de conservação da tartaruga marinha Lepidochelys iocone (Eschscholtz, 1829) no Brasil. Biodivers Bras 1:28–36
De Freitas JR, Bennie J, Mantovani W et al (2017) Exposure of tropical ecosystems to artificial light at night: Brazil as a case study. PLoS One 12(2):e0171655
De Oliveira Soares M, Teixeira CEP, Bezerra LEA et al (2020) Oil spill in South Atlantic (Brazil): environmental and governmental disaster. Mar Policy 115:103879
De Padua Almeida A, Moreira LM, Bruno SC et al (2011) Green turtle nesting on Trindade Island, Brazil: abundance, trends, and biometrics. Endanger Species Res 14(3):193–201
Del Favero JM, Dias JF (2013) Spatio-temporal variation in surf zone fish communities at Ilha do Cardoso State Park, São Paulo, Brazil. Lat Am J Aquat Res 41(2):239–253
Del Favero JM, Dias JF (2015) Juvenile fish use of the shallow zone of beaches of the Cananéia-Iguape coastal system, southeastern Brazil. Braz J Oceanogr 63(2):103–114
Di Beneditto AP, Ramos R, Lima NRW (1998) Fishing activity in Northern Rio de Janeiro State (Brazil) and its relation with small cetaceans. Braz Arch Biol Technol 41(3):296–302
Dugan JE, Hubbard DM, McCrary MD et al (2003) The response of macrofauna communities and shorebirds to macrophyte wrack subsidies on exposed sandy beaches of southern California. Estuar Coast Shelf Sci 58:25–40
Erbe C, Reichmuth C, Cunningham K et al (2016) Communication masking in marine mammals: a review and research strategy. Mar Pollut Bull 103:15–38
Escalante R (1970) Aves marinas del Rio de la Plata y aguas vecinas del Oceano Atlantico. Barreiro e Ramos, Montevideo, 199 p
Escobar H (2019) Mystery oil spill threatens marine sanctuary in Brazil. Science 366:672
Félix FC, Spach HL, Moro OS et al (2007) Ichthyofauna composition across a wave: energy gradient on southern Brazil beaches. Braz J Oceanogr 55(4):281–292
Fernandes MLB, Silva LCC, Moura GJB (2016) Influência dos impactos ambientais na escolha da praia de desova da espécie Eretmochelys imbricata. Biota Amazônia 6(4):44–48
Foley CJ, Feiner ZS, Malinich TD, Höök TO (2018) A meta-analysis of the effects of exposure to microplastics on fish and aquatic invertebrates. Sci Total Environ 631:550–559
Franco ACS, Ramos Chaves MCN et al (2016) Responses of fish assemblages of sandy beaches to different anthropogenic and hydrodynamic influences. J Fish Biol 89(1):921–938
Gaelzer LR, Zalmon IR (2008a) Diel variation of fish community in sandy beaches of southeastern Brazil. Braz J Oceanogr 56(1):23–39
Gaelzer LR, Zalmon IR (2008b) Tidal influence on surf zone ichthyofauna structure at three sandy beaches, southeastern Brazil. Braz J Oceanogr 56(3):165–177
Garcia SM, de Leiva Moreno I (2003) Global overview of marine fisheries. In: Sinclair M, Valdimarsson G (eds) Responsible fisheries in the marine ecosystem. FAO, Rome, pp 1–24

Gianelli I, Martínez G, Defeo O (2015) An ecosystem approach to small-scale co-managed fisheries: the yellow clam fishery in Uruguay. Mar Policy 62:196–202

Gondolo GF, Mattox GMT, Cunningham PTM (2011) Ecological aspects of the surf-zone ichthyofauna of Itamambuca Beach, Ubatuba, SP. Biota Neotrop 11(2):183–192

Gregory MR (2009) Environmental implications of plastic debris in marine settings—entanglement, ingestion, smothering, hangers-on, hitch-hiking and alien invasions. Philos Trans R Soc Lond B Biol Sci 364:2013–2025

Guimarães SM, Tavares DC, Monteiro-Neto C (2018) Incidental capture of sea turtles by industrial bottom trawl fishery in the Tropical South-western Atlantic. J Mar Biol Assoc UK 98(6):1525–1531

Guinet C (1991) Intentional stranding apprenticeship and social play in killer whales (*Orcinus orca*). Can J Zool 69(11):2712–2716

Guinet C, Bouvier J (1995) Development of intentional stranding hunting techniques in killer whale (*Orcinus orca*) calves at Crozet Archipelago. Can J Zool 73:27–33

Hoese HD (1971) Dolphin feeding out of water in a salt marsh. Journal of Mammalogy, 52(1), 222–223

Hubbard DM, Dugan JE (2003) Shorebird use of an exposed sandy beach in southern California. Estuarine, Coastal and Shelf Science, 58, 41–54

Huijbers CM, Schlacher TA, Schoeman DS, Weston MA, Connolly RM (2013) Urbanisation alters processing of marine carrion on sandy beaches. Landscape and Urban Planning, 119, 1–8

Instituto Chico Mendes de Conservação da Biodiversidade. (2018). Livro Vermelho da Fauna Brasileira. Ameaçada de Extinção. Brasília: ICMBio. 4162p

Jepson PD, Arbelo M, Deaville R, Patterson IAP, Castro P, Baker JR, ... & Fernández A (2003) Gas-bubble lesions in stranded cetaceans. Nature, 425(6958), 575–576

Jeunen GJ, Knapp M, Spencer HG, Taylor HR, Lamare MD, Stat M, ... & Gemmell NJ (2019) Species-level biodiversity assessment using marine environmental DNA metabarcoding requires protocol optimization and standardization. Ecology and evolution, 9(3), 1323–1335

Johansen JL, Esbaugh AJ (2017) Sustained impairment of respiratory function and swim performance following acute oil exposure in a coastal marine fish. Aquatic Toxicology, 187, 82–89

Júnior TV, Mancini BF, Knoeller JDSM (2020) Feeding habits of the barred grunt (Conodon nobilis) (Haemulidae: Perciformes) in the surf zone of Praia Grande, São Paulo, Brazil. Unisanta BioSci 9(3):194–204

Jordão JC, Bondioli ACV, Guebert FM, Thoisy BD, & Toledo LFDA (2015) Green turtle (Chelonia mydas) genetic diversity at Paranaguá Estuarine Complex feeding grounds in Brazil. Genetics and Molecular Biology, 38, 346–352

Kikukawa A, Kamezaki N, Ota H (1999) Factors affecting nesting beach selection by loggerhead turtles (Caretta caretta): a multiple regression approach. Journal of Zoology, 249(4), 447–454

Lemos LS, de Moura JF, Hauser-Davis RA, de Campos RC, Siciliano S (2013) Small cetaceans found stranded or accidentally captured in southeastern Brazil: Bioindicators of essential and non-essential trace elements in the environment. Ecotoxicology and environmental safety, 97, 166–175

Lercari D, Bergamino L, Defeo O (2010) Trophic models in sandy beaches with contrasting morphodynamics: comparing ecosystem structure and biomass flow. Ecological Modelling, 221(23), 2751–2759

Lewison RL, Crowder LB, Wallace BP, Moore JE, Cox T, Zydelis R, ... & Safina C (2014) Global patterns of marine mammal, seabird, and sea turtle bycatch reveal taxa-specific and cumulative megafauna hotspots. Proceedings of the National Academy of Sciences, 111(14), 5271–5276

Liebowitz DM, Nielsen KJ, Dugan JE, Morgan SG, Malone DP, Largier JL, ... Carr MH (2016) Ecosystem connectivity and trophic subsidies of sandy beaches. Ecosphere, 7(10), e01503

Link J, Segal B, Casarini LM (2019) Abandoned, lost or otherwise discarded fishing gear in Brazil: A review. Perspectives in Ecology and Conservation, 17(1), 1–8

Lodi L, Borobia M (2013) Anomalous colouration in an Atlantic spotted dolphin (Stenella frontalis) from southeastern Brazil. Brazilian Journal of Aquatic Science and Technology, 17(2), NB1-3

Lönnstedt O M, Eklöv P (2016) Environmentally relevant concentrations of microplastic particles influence larval fish ecology. Science, 352(6290), 1213–1216
Lopez JC, Lopez D (1985) Killer whales (Orcinus orca) of Patagonia, and their behavior of intentional stranding while hunting nearshore. Journal of Mammalogy, 66(1), 181–183
López-Mendilaharsu M, Rocha CF, Miller P, Domingo A, Prosdocimi L (2009) Insights on leatherback turtle movements and high use areas in the Southwest Atlantic Ocean. Journal of Experimental Marine Biology and Ecology, 378(1–2), 31–39
Lunardi VO, Macedo RH, Granadeiro JP, Palmeirim JM (2012) Migratory flows and foraging habitat selection by shorebirds along the northeastern coast of Brazil: the case of Baía de Todos os Santos. Estuar Coast Shelf Sci 96:179–187
Magris RA, Giarrizzo T (2020) Mysterious oil spill in the Atlantic Ocean threatens marine biodiversity and local people in Brazil. Mar Pollut Bull 153:110961
Mann D, Hill-Cook M, Manire C, Greenhow D, Montie E, Powell J, ... & Hoetjes, P (2010) Hearing loss in stranded odontocete dolphins and whales. PLoS One, 5(11), e13824
Mancini PL, Matinata BS, Fischer LG (2017) Aves da Baía do Araçá e arredores
Mancini PL, Reis-Neto AS, Fischer LG, Silveira LF, Schaeffer-Novelli Y (2018) Differences in diversity and habitat use of avifauna in distinct mangrove areas in São Sebastião, São Paulo, Brazil. Ocean & coastal management, 164, 79–91
Marcovaldi MÂ, Dei Marcovaldi GG (1999) Marine turtles of Brazil: the history and structure of Projeto TAMAR-IBAMA. Biological conservation, 91(1), 35–41
Marcovaldi MÂ, Lopez GG, Soares L, Belini C, dos Santos AS, Lopez M (2011) Avaliação do estado de conservação da tartaruga marinha Eretmochelys imbricata (Linnaeus, 1766) no Brasil. Biodiversidade Brasileira-BioBrasil, (1)
Maslo B, Leu K, Faillace C, Weston MA, Pover T, Schlacher TA (2016) Selecting umbrella species for conservation: A test of habitat models and niche overlap for beach-nesting birds. Biological Conservation, 203, 233–242
Mascarenhas R, Zeppelini D, Moreira VS (2003) Observations on sea turtles in the State of Paraíba, Brazil. Marine Turtle Newsletter, 101(1), 16–18
Matsuoka T, Nakashima T, Nagasawa N (2005) A review of ghost fishing: scientific approaches to evaluation and solutions. Fisheries Science, 71, 691–702
McLachlan A, Defeo O, Jaramillo E, & Short AD (2013) Sandy beach conservation and recreation: guidelines for optimising management strategies for multi-purpose use. Ocean & coastal management, 71, 256–268
Meager JJ, Schlacher TA, Nielsen T (2012) Humans alter habitat selection of birds on ocean-exposed sandy beaches. Divers Distrib 18(3):294–306
Mendes SS, Carvalho RH, Faria AF et al (2015) Marine debris ingestion by Chelonia mydas (Testudines: Cheloniidae) on the Brazilian coast. Mar Pollut Bull 92(1–2):8–10
Miller JD (1997) Reproduction in sea turtles. In: Lutz PL, Musick JA (eds) The biology of sea turtles. CRC Press, Boca Raton, pp 151–181
Monteiro-Filho ELA, Oliveira LV, Monteiro KDA et al (2021) Guia Ilustrado de Mamíferos Marinhos, 2a edn. Instituto de Pesquisas Cananéia, Cananéia.
Morrison RIG, Ross KR, Niles LJ (2004) Declines in wintering populations of Red Knots in southern South America. Condor 106(1):60–70
Mattox GMT, Gondolo GF, Cunningham PTM (2008) Atherinella blackburni (Schultz, 1949) at Itamambuca Beach, Ubatuba, SP: ecological characterization and distribution on the Brazilian coast (Teleostei: Atheriniformes: Atherinopsidae). Brazilian Journal of Biology, 68, 307–313
Moura JF, Di Dario BPS, Lima LM et al (2010) A stranded pygmy killer whale on the coast of Rio de Janeiro State, Brazil. Mar Biodivers Rec 3:1–4
Moura JF, Acevedo-Trejos E, Tavares DC et al (2016) Stranding events of Kogia whales along the Brazilian coast. PLoS One 11(1):e0146108
Naves LC (1999) Ecologia alimentar do talha-mar Rhynchops nigra (Aves: Rhynchopidae) na desembocadura da Lagoa dos Patos. Dissertation. Fundação Universidae do Rio Grande

Neto ES, Rossi-Santos MR, Baracho CG, Cipolotti SR, Sampaio CL, Velozo RS, Souto LR (2008) A case study of a lone humpback whale calf (Megaptera novaeangliae) inside Baía de Todos os Santos, Bahia State, north-eastern Brazil, with implications for rescue procedures. Mar Biodivers Rec 1:E97

New LF, Hall AJ, Harcourt R et al (2015) The modelling and assessment of whale-watching impacts. Ocean Coast Manag 115:10–16

Olds AD, Vargas-Fonseca E, Connolly RM et al (2017) The ecology of fish in the surf zones of ocean beaches: a global review. Fish Fish 19:78–89

Olds AD, Nagelkerken I, Huijbers CM et al (2018) Connectivity in coastal seascapes. In: Pittman SJ (ed) Seascape ecology. Wiley-Blackwell, Oxford, pp 261–292pp

Oliveira RE, Pessanha AL (2014) Fish assemblages along a morphodynamic continuum on three tropical beaches. Neotrop Ichthyol 12(1):165–175

Ortodossi NL, Gilby BL, Schlacher TA et al (2019) Effects of seascape connectivity on reserve performance along exposed coastlines. Conserv Biol 33(3):580–589

Patrício AR, Varela MR, Barbosa C et al (2018) Nest site selection repeatability of green turtles, *Chelonia mydas*, and consequences for offspring. Anim Behav 139:91–102

Peddemors VM, Thompson G (1994) Beaching behaviour during shallow water feeding by humpback dolphins *Sousa plumbea*. Aquat Mamm 20(2):65–67

Peltier H, Dabinb W, Danielc P et al (2012) The significance of stranding data as indicators of cetacean populations at sea: modelling the drift of cetacean carcasses. Ecol Indic 18:278–290

Pereira HH, Neves LM, da Costa MR et al (2015) Fish assemblage structure on sandy beaches with different anthropogenic influences and proximity of spawning grounds. Mar Ecol 36(1):16–27

Peterson D, Hanazaki N, Simões-Lopes PC (2008) Natural resource appropriation in cooperative artisanal fishing between fishermen and dolphins (*Tursiops truncatus*) in Laguna, Brazil. Ocean Coast Manag 51:469–475

Plotkin PT (2010) Nomadic behaviour of the highly migratory olive ridley sea *turtle Lepidochelys olivacea* in the eastern tropical Pacific Ocean. Endanger Species Res 13(1):33–40

Proietti MC, Reisser J, Marins LF et al (2014a) Hawksbill× loggerhead sea turtle hybrids at Bahia, Brazil: where do their offspring go? PeerJ 2:e255

Proietti MC, Reisser J, Marins LF et al (2014b) Genetic structure and natal origins of immature hawksbill turtles (Eretmochelys imbricata) in Brazilian waters. PLoS One 9(2):e88746

Pryor K, Lindbergh J, Lindbergh S et al (1990) A dolphin-human fishing cooperative in Brazil. Mar Mamm Sci 6(1):77–82

Quirós YB, Hartwick M, Rotstein DS et al (2018) Discrimination between bycatch and other causes of cetacean and pinniped stranding. Dis Aquat Org 127:83–95

Rangel DF, Da Silva EFN, Costa LL (2022) Occurrence and behaviour of shorebirds depend on food availability and distance of beaches from urban settlements. Acta Ornithol 56(2):217–226

Rodrigues FL, Vieira JP (2010) Feeding strategy of *Menticirrhus americanus* and *Menticirrhus littoralis* (Perciformes: Sciaenidae) juveniles in a sandy beach surf zone of southern Brazil. Fortschr Zool 27(6):873–880

Rodrigues-Filho JL, Branco JO, Monteiro HS et al (2015) Seasonality of ichthyofauna bycatch in shrimp trawls from different depth strata in the southern Brazilian coast. J Coast Res 31(2):378–389

Rosas FCW, Monteiro-Filho ELA, Oliveira MR (2002) Incidental catches of franciscana (*Pontoporia blainvillei*) on the southern coast of São Paulo state and the coast of Paraná state, Brazil. Latin Am J Aquat Mamm 1:161–168

Salmon M (2003) Artificial night lighting and sea turtles. Biologist 50(4):163–168

Santos MCO (2010) Guiana dolphins (*Sotalia guianensis*) displaying beach hunting behavior in the Cananéia estuary, Brazil: social context and conservation issues. Braz J Oceanogr 58(2):143–152

Santos MCO, Rosso S, Siciliano S et al (2000) Behavioral observation of the marine tucuxi dolphins (*Sotalia fluviatilis*), in São Paulo estuarine waters, southeastern Brazil. Aquat Mamm 26(3):260–267

Santos MCO, Siciliano S, Vicente AFC et al (2010) Cetacean records along São Paulo state coast, southeastern Brazil. Braz J Oceanogr 58(2):123–142
Santos AJB, Soares LS, Marcovaldi MA, et al. (2011) Avaliação do Estado de Conservação da Tartaruga Marinha *Caretta caretta* Linnaeus, 1758 no Brasil. Biodiversidade Brasileira (1):3–11
Sargeant BL, Mann J, Berggren P et al (2005) Specialization and development of beach hunting, a rare foraging behavior, by wild bottlenose dolphins (*Tursiops sp.*). Can J Zool 83(11):1400–1410
Schlacher TA, Strydom S, Connolly RM et al (2013) Multiple scavengers respond rapidly to pulsed carrion resources at the land-ocean interface. Acta Oecol 48:7–12
Selig ER, Turner WR, Troëng S et al (2014) Global priorities for marine biodiversity conservation. PLoS One 9(1):e82898
Sella KAN, Fuentes MMPB (2019) Exposure of marine turtle nesting grounds to coastal modifications: implications for management. Ocean Coast Manag 169:182–190
Shah Esmaeili Y, Corte GN, Checon HH et al (2021) Comprehensive assessment of shallow surf zone fish biodiversity requires a combination of sampling methods. Mar Ecol Prog Ser 667:131–144
Shah Esmaeili Y, Corte GN, Checon HH et al (2022) Revealing the drivers of taxonomic and functional diversity of nearshore fish assemblages: implications for conservation priorities. Divers Distrib 28(8):1597–1609
Siciliano S, Moreno IB, Silva ED et al (2006) Baleias, botos e golfinhos na Bacia de Campos. ENSP/FIOCRUZ, Rio de Janeiro, 100p
Sick H (1997) Ornitologia Brasileira, 3rd edn. Nova Fronteira, Rio de Janeiro, 912 p
Silva Júnior CABD, Araújo MED, Feitosa CV (2013) Sustainability of capture of fish bycatch in the prawn trawling in northeastern Brazil. Neotrop Ichthyol 11(1):133–142
Simões-Lopes PC, Fabián ME, Menegueti JO (1998) Dolphin interactions with the mullet artisanal fishing on southern Brazil: a qualitative and quantitative approach. Rev Bras Zool 15(3):709–726
Soares LS, Bolten AB, Wayne ML et al (2017) Comparison of reproductive output of hybrid sea turtles and parental species. Mar Biol 164(1):9
Soeth M, Spach HL, Ribeiro GC et al (2014) Fish temporal variation in different ontogenetic stages in a sheltered beach of the Baía Norte, southern Brazil. Neotropical Biol Conserv 9(1):27–41
Solomon JJ, Tremain DM (2009) Recruitment timing and spatial patterns of estuarine use by young-of-the-year Florida pompano, *Trachinotus carolinus*, in northeastern Florida. Bull Mar Sci 85(2):133–148
Southall BL, Bowles AE, Ellison WT et al (2008) Marine mammal noise-exposure criteria: initial scientific recommendations. Bioacoustics 17(1–3):273–275
Sullivan BL, Wood CJ, Iliff MJ et al (2009) eBird: a citizen-based bird observation network in the biological sciences. Biol Conserv 142:2282–2292
Tavares DC, Guadagnin DL, de Moura JF et al (2015) Environmental and anthropogenic factors structuring waterbird habitats of tropical coastal lagoons: implications for management. Biol Conserv 186:12–21
Tavares DC, Costa LL, Rangel DF et al (2016) Nests of the brown booby (Sula leucogaster) as a potential indicator of tropical ocean pollution by marine debris. Ecol Indic 70:10–14
Tavares DC, de Moura JF, Merico A et al (2017) Incidence of marine debris in seabirds feeding at different water depths. Mar Pollut Bull 119(2):68–73
Thomé JC, Baptistotte C, Moreira LMP et al (2007) Nesting biology and conservation of the leatherback sea turtle (Dermochelys coriacea) in the state of Espírito Santo, Brazil, 1988–1989 to 2003–2004. Chelonian Conserv Biol 6(1):15–27
Torres LG, Read AJ (2009) Where to catch a fish? The influence of foraging tactics on the ecology of bottlenose dolphins (*Tursiops truncatus*) in Florida Bay, Florida. Mar Mamm Sci 25:797–815
Tougaard J, Wright AJ, Madsen PT (2014) Cetacean noise criteria revisited in the light of proposed exposure limits for harbour porpoises. Mar Pollut Bull 90(1–2):196–208

Tourinho PS, Ivar do Sul JA, Fillmann G (2009) Is marine debris ingestion still a problem for the coastal marine biota of southern Brazil? Mar Pollut Bull 60:396–401

Turra A, Fernandez WS, Bessa E et al (2015) Multi-species generalist predation on the stochastic harvested clam *Tivela mactroides* (Mollusca, Bivalvia). Estuar Coast Shelf Sci 166:115–123

Unger B, Rebolledo ELB, Deaville R et al (2016) Large amounts of marine debris found in sperm whales stranded along the North Sea coast in early 2016. Mar Pollut Bull 112:134–141

Vargas-Fonseca E, Olds AD et al (2016) Combined effects of urbanization and connectivity on iconic coastal fishes. Divers Distrib 22(12):1328–1341

Vieira TC, Rodrigues APDC, Amaral PM, de Oliveira DF, Gonçalves RA, Rodrigues e Silva C, Vasques RO, Malm O, Silva-Filho EV, Godoy JM d O, Machado W, Filippo A, Bidone ED (2020) Evaluation of the bioaccumulation kinetics of toxic metals in fish (A. brasiliensis) and its application on monitoring of coastal ecosystems. Mar Pollut Bull 151:110830

Vooren CM, Brusque LF (1999) As aves do ambiente costeiro do Brasil: biodiversidade e conservação. Programa Nacional da Diversidade Biológica—PRONABIO

Vooren CM, Chiaradia A (1990) Seasonal abundance and behaviour of coastal birds on Cassino Beach, Brazil. Ornitol Neotrop 1:9–24

Waerebeek KV, Baker AN, Félix F et al (2007) Vessel collisions with small cetaceans worldwide and with large whales in the Southern Hemisphere, an initial assessment. Latin Am J Aquat Mamm 6(1):43–69

Wallace BP, DiMatteo AD, Bolten AB et al (2011) Global conservation priorities for marine turtles. PLoS One 6(9):e24510

Watson RA, Tidd A (2018) Mapping nearly a century and a half of global marine fishing: 1869–2015. Mar Policy 93:171–177

Wickert JC, Eye SMV, Oliveira LR et al (2016) Revalidation of *Tursiops gephyreus* Lahille, 1908 (Cetartiodactyla: Delphinidae) from the southwester Atlantic Ocean. J Mammal 97(6):1728–1737

Wilber DH, Clarke DG, Ray GL et al (2003) Response of surf zone fish to beach nourishment operations on the northern coast of New Jersey, USA. Mar Ecol Prog Ser 250:231–246

Yasué M (2006) Environmental factors and spatial scale influence shorebirds' responses to human disturbance. Biol Conserv 8:2–9

Yokota L, Lessa RP (2006) A nursery area for sharks and rays in northeastern Brazil. Environ Biol Fish 75(3):349–360

Yorio P, Bertellotti M, Gandini P et al (1998) Kelp gulls *Larus dominicanus* breeding on the Argentine coast: population status and relationship with coastal management and conservation. Mar Ornithol 26(1):11–18

Zacharias MA, Roff JC (2001) Use of focal species in marine conservation and management: a review and critique. Aquat Conserv 11(1):59–76

Chapter 6
Population Biology

Marcelo Petracco, Guilherme Nascimento Corte, Daiane Aviz, Rayane Romão Saad Abude, Matheus Augusto, Carlos Henrique Soares Caetano, Ricardo Silva Cardoso, and Tatiana Medeiros Barbosa Cabrini

6.1 Introduction

Population biology can be simply defined as the study of a group of individuals of the same species with a high probability of interacting with each other (Hastings 1996). Nevertheless, this term has been used both in the sense of applying mathematical models to describe and predict population dynamics (e.g., Hastings 1996; Wilson and Bossert 1971) or in a wider context, usually as an interdisciplinary field that combines concepts from ecology, evolution, systematics, genetics, and mathematics (e.g., Neal 2019). Here, we use population biology as the investigation of population parameters, such as abundance, growth, production, and reproduction, and how environmental features and biological interactions influence them.

Marcelo Petracco, Guilherme Nascimento Corte and Tatiana Medeiros Barbosa Cabrini contributed equally with all other contributors.

M. Petracco · D. Aviz
Laboratório de Pesquisa em Monitoramento Ambiental Marinho (LAPMAR), Laboratório de Oceanografia Biológica (LOB), Faculdade de Oceanografia, Instituto de Geociências, Universidade Federal do Pará (UFPA), Belem, Para, Brazil

G. N. Corte
College of Science and Mathematics, University of Virgin Islands, Saint Thomas, USVI, USA

R. R. S. Abude · M. Augusto · R. S. Cardoso · T. M. B. Cabrini (✉)
Laboratório de Ecologia Marinha (ECOMAR), Departamento de Ecologia e Recursos Marinhos, Instituto de Biociências, Universidade Federal do Estado do Rio de Janeiro (UNIRIO), Rio de Janeiro, RJ, Brazil
e-mail: tatiana.cabrini@unirio.br

C. H. S. Caetano
Laboratório de Zoologia Marinha (ZOOMAR), Departamento de Zoologia, Instituto de Biociências, Universidade Federal do Estado do Rio de Janeiro (UNIRIO), Rio de Janeiro, RJ, Brazil

A. C. Z. Amaral et al. (eds.), *Brazilian Sandy Beaches*, Brazilian Marine Biodiversity, https://doi.org/10.1007/978-3-031-30746-1_6

Population biology studies are essential as they provide robust information for understanding how populations change in space and time, thereby assisting in the management of stocks of commercially and ecologically important species (Dolbeth et al. 2012). They have also been recognized as a powerful tool in ecosystem conservation programs with applications in energy – biomass flow studies, trophic ecology, and assessment of environmental stress (Dolbeth et al. 2012).

In this chapter, we present the main population parameters estimated for sandy beach macrobenthic species, the most well-known component of Brazilian sandy beach biodiversity, and how they are obtained. Then, we review the population studies on Brazilian sandy shores emphasizing the most studied species and sites investigated. Finally, we highlight gaps and avenues that should be explored in future studies and provide some final remarks on the population biology of Brazilian sandy beach species.

6.2 Population Parameters of Sandy Beach Species

6.2.1 Abundance

Abundance is perhaps the most important parameter for investigating populations' dynamics in natural ecosystems (Yin and He 2014). Knowledge about abundance variations is critical to understanding underlying processes influencing the distribution of organisms in space and time (Defeo and Rueda 2002), as well as to inform decision-making in biological management and conservation, for example, assessing the extinction risk of endangered and/or over-exploited species. Furthermore, knowledge about species abundance is essential to estimate other population parameters such as recruitment, growth, biomass, production, and turnover rate.

Given its importance, abundance estimates should be reliable to avoid erroneous interpretations of populations' conditions. However, most sandy beach populations show aggregated patterns of distribution related to environment gradients (e.g., zonation, Chap. 4), which may lead to under or overestimates of their abundance. Thus, to better quantify sandy beach populations, researchers should sample the whole area where the species is likely to be found, attempting to include sites with high and low abundances.

For species inhabiting different zones of the across-shore gradient (e.g., supralittoral, intertidal, and/or subtidal), this is done usually following two sampling designs: (1) species-driven and (2) environmentally driven (Defeo and Rueda 2002). In both designs, sampling points are systematically placed along transects perpendicularly fixed to the shoreline, and samples are collected along these transects. However, while the length of the transects is determined by species' across-shore distribution in the former (at least two or more samples with zero abundance are sampled in both the up shore and down shore directions before terminating the transect), it is defined by the environmental references in the second (e.g., tidal marks, swash levels, or drift lines in the second) (see Defeo and Rueda (2002) and Schlacher et al. (2008) for a revision on sandy beach sampling issues). In Brazil, most population biology studies on sandy shores used the environmentally driven/fixed transects approach to investigate population parameters of local species (e.g., Cardoso and Veloso 2003; Petracco et al. 2003; Turra et al. 2014). However, preliminary surveys are usually conducted to better establish the distribution range of the studied species (e.g., Corte et al. 2020; Denadai et al. 2004; Petracco et al. 2014).

After sampling, sandy beach researchers usually present quantitative estimates of abundance as mean densities per square meter, where the number of individuals is averaged by the sampled area. It is also common to provide abundance estimates as the number of individuals per running meter or strip transect (IST: ind/m) (Defeo and Rueda 2002), an approach that considers the abundance of individuals along the whole across-shore gradient. The IST can be obtained by multiplying the mean density (ind/m^2) of all n samples pertaining to the transect multiplied by the corresponding width (across-shore component) of the transect (Defeo and Rueda 2002). In Brazil, approximately 75% of the studies presenting the abundance of sandy beach populations used the number of individuals per square meter, while the remaining 25% estimated the abundance as individuals per running meter or strip transect (Table 6.1). Only a limited number of studies (e.g., Corte et al. 2015; dos Anjos 2016) used both approaches, which may facilitate comparisons with studies performed in different locations on the Brazilian coast and worldwide.

6.2.2 *Recruitment*

In general terms, recruitment refers to the addition of new individuals to a population and is strongly linked to changes in the abundance and size structure of a population. Recruitment of marine organisms, including sandy beach species, involves four phases: (1) development (including dispersal as a planktonic form), (2) testing of habitat for suitability, (3) settlement, and (4) survival until the organisms are counted by an observer (Keough and Downes 1982). It is important to note that the organism needs to go through these four phases to be considered as part of the population (recruit) (Keough and Downes 1982).

Several studies on the population biology of Brazilian sandy beach species have used the software FAO-ICLARM Stock Assessment Tools (FISAT II – version 1.1.0) (Gayanilo et al. 2005) to estimate recruitment rates of macrobenthic organisms. The FISAT II provides multiple analyses on population parameters of fish and invertebrate species and is a valuable tool for sandy beach researchers. However, from our experience, it does not provide accurate recruitment estimates, especially for populations with continuous recruitment – a common pattern for Brazilian sandy beach species (Table 6.1).

Therefore, probably the most effective way to estimate recruitment events for sandy beach populations is simply grouping sampled individuals into size classes, plotting periodical length-frequency distributions (i.e., the number of individuals in each size class), and estimating recruitment patterns by observing peaks in the

Table 6.1 Summary of population biology studies of macrofaunal species on Brazilian sandy beaches

Authors	State	Species	Zone	Dens m^2	Reprod	Recruit	WP	L_∞	C	K	LE	Z	P	B	P/B	N	T	Method
Annelida																		
Souza and Borzone (2007)	PR	*Euzonus furciferus*	I	681.8	–	Wint	0.10	2.32	1.00	0.76	1.80	3.20	0.47	0.22	2.13	1	1	ELEF
Omena and Amaral (2000)	SP	*Laeonereis acuta*	I	200–3500	Cont	Cont	0.20	1.54	1.00	2.20	0.10	–	–	–	–	1	1	ELEF
Santos et al. (2003)	PE	*Laeonereis acuta*		–	Cont	–	–	–	–	–	–	–	–	–	–	–	–	–
Santos (1994)	RS	*Scolelepis gaucha*	I	920–1106250[a]	–	–	–	0.89	0.85	3.40	–	10.08	–	–	12.92	1	1	ELEF
Leão et al. (2012)	RJ	*Scolelepis goodbodyi*	I	9811	Cont	Cont	–	1.23	–	1.20	2.50	0.73	–	–	–	1	1	–
MacCord and Amaral (2007)	SP	*Scolelepis goodbodyi*	I	–	Cont	Cont	–	–	–	–	–	–	–	–	–	1	1	–
Shimizu (1997)	SP	*Scolelepis squamata*	I	–	–	–	0.55	0.62	1.00	2.55	0.10	–	–	–	–	–	–	–
Souza and Borzone (2000)	PR	*Scolelepis squamata*	I	20,277	Cont	–	0.70	0.70	1.00	2.60	0.70	2.38	0.57	0.21	2.70	1	1	–
Otegui et al. (2012)	SC	*Thoracophelia furcifera*	I	1630.6	Wint-Spr	Aut	0.97	3.60	0.30	0.63	2.60	3.80	–	–	–	1	1	ELEF
Crustacea																		
Cardoso and Veloso (1996)	RJ	*Atlantorchestoidea brasiliensis*	I	80	Cont	Cont	0.20	13.40	0.90	1.36	1.20	2.92	0.29	0.13	2.20	1	2	ELEF
Pennafirme and Soares-Gomes (2022)	RJ	*Atlantorchestoidea brasiliensis*	I	–	Cont	Cont	0.40	12.60	–	2.15	1.00	–	–	–	–	1	2	BERT
Manoel (2008)	RJ	*Austinixa patagoniensis*	I	–	–	–	0.21	12.18	0.24	3.14	0.64	4.50	–	–	–	–	–	–
Shimizu (1997)	SP	*Callichirus major*	I & S	–	–	–	0.75	15.37	0.52	0.90	4.00	–	–	–	–	–	–	–
Souza et al. (1998)	PR	*Callichirus major*	I	–	–	–	0.55	15.00	0.80	0.90	3.00	0.95	–	–	–	–	–	–
Alves-Júnior et al. (2014)	PE	*Callichirus major*	I	–	Sum	Cont	–	–	–	–	–	–	–	–	–	1	1	–

Alves-Júnior et al. (2018)	PE	*Callichirus major*	I & S	–	–	Sum	–	–	–	–	–	–	–	–	–	1	1	–
Botter-Carvalho et al. (2002)	PE	*Callichirus major*	I	1.6	–	–	–	–	–	–	–	–	–	–	–	1	1	–
Mcallan (2006)	RJ	*Callichirus major*	I & S	–	–	–	0.63	17.35	0.52	2.50	0.95	1.62	–	–	–	–	–	–
Botter-Carvalho et al. (2007)	PE	*Callichirus major*	I	–	Sum-Aut	Aut-Wint	0.61	14.30	0.96	0.86	3.44	3.92	–	–	–	1	1	ELEF
Simão and Soares-Gomes (2017)	RJ	*Callichirus major*	I & S	42.16	Sum	Wint	0.80	15.00	0.90	0.85	3.16	0.92	3.9–12.7	4.3–7.2	0.9–1.9	1	2.5	ELEF
Rosa et al. (2018)	SE	*Callichirus major*	I & S	–	Sum	–	0.60	15.50	0.20	0.78	3.45	1.40	–	–	–	–	–	–
Laurino et al. (2020)	SP	*Callichirus major*	I & S	0–28	–	–	–	–	–	–	–	–	–	–	–	2	2	–
Moschetto et al. (2020)	SP	*Callichirus major*	I & S	–	–	–	–	–	–	–	–	–	–	–	–	–	–	–
Costa et al. (2022)	RJ	*Callichirus major*	I & S	0.09	–	–	–	–	–	–	–	–	–	–	–	2	–	–
Baptista-Metri et al. (2005)	PR	*Callinectes danae*	I	–	Cont	–	–	–	–	–	–	–	–	–	–	–	–	–
Danin et al. (2020)	PA	*Clibanarius symmetricus*	I & S	11.05	Sum	Cont	–	12.60	0.00	0.48	1.55	2.50	0.35	0.15	2.40	1	1	–
Nakagaki and Pinheiro (1999)	RJ	*Emerita brasiliensis*	I	–	Wint	Sum	–	–	–	–	–	–	–	–	–	–	–	–
Cardoso et al. (2003)	RJ	*Emerita brasiliensis*	I	–	–	–	0.40	17.00	1.00	1.90	1.40	7.37	–	–	–	2	1	ELEF
Cardoso et al. (2003)	RJ	*Emerita brasiliensis*	I	–	–	–	0.70	16.50	1.00	2.10	0.90	9.92	–	–	–	2	1	ELEF
Petracco et al. (2003)	RJ	*Emerita brasiliensis*	I & S	790	–	–	0.90	24.50	0.30	2.20	0.75	8.20	46.88	4.91	9.55	1	2	ELEF

(continued)

Table 6.1 (continued)

Authors	State	Species	Zone	Dens m^2	Reprod	Recruit	WP	L_∞	C	K	LE	Z	P	B	P/B	N	T	Method
Petracco et al. (2003)	RJ	*Emerita brasiliensis*	I & S	1889	–	–	0.30	25.30	1.00	2.10	0.75	10.30	156.07	23.09	6.76	1	2	ELEF
Veloso and Sallorenzo (2010)	RJ	*Emerita brasiliensis*	I	3478.7	–	Spr-Sum	0.55	22.86	0.45	1.53	–	–	96.80	41.74	2.31	2	1	MPA
Veloso and Sallorenzo (2010)	RJ	*Emerita brasiliensis*	I	2039.5	–	Spr-Sum	0.45	23.04	0.72	2.26	–	–	193.70	33.26	5.82	2	1	MPA
Petracco et al. (2017)	RJ	*Emerita brasiliensis*	I	2039	–	–	–	–	–	1.61	1.63	–	96.80	41.71	2.31	1	–	–
Petracco et al. (2017)	RJ	*Emerita brasiliensis*	I	850	–	–	–	–	–	1.95	1.02	–	106.81	27.44	3.89	1	–	–
Petracco et al. (2017)	RJ	*Emerita brasiliensis*	I	1139	–	–	–	–	–	2.09	0.69	–	95.45	14.00	6.81	1	–	–
Petracco et al. (2017)	RJ	*Emerita brasiliensis*	I	3478	–	–	–	–	–	2.10	1.83	–	193.74	33.26	5.82	1	–	–
Petracco et al. (2017)	RJ	*Emerita brasiliensis*	I	3275	–	–	–	–	–	2.25	0.74	–	217.86	39.00	5.59	1	–	–
Petracco et al. (2017)	RS	*Emerita brasiliensis*	I	2300	–	–	–	–	–	–	1.33	–	238.50	32.24	7.30	1	–	–
Petracco et al. (2010)	SP	*Excirolana armata*	I	6025.5[a]	Cont	Spr	0.54	8.03	0.50	1.71	0.80	5.50	17.25	5.50	3.14	1	2	MPA
Silva (2013)	RJ	*Excirolana armata*	I	22.1	Cont	Cont	0.95	8.53	0.57	1.28	2.88	2.15	0.03	0.01	3.00	3	2	–
Silva (2013)	RJ	*Excirolana armata*	I	62.34	Cont	Cont	0.48	9.20	0.33	1.24	1.11	4.47	0.12	0.05	2.40	3	2	–
Fonseca et al. (2000)	RJ	*Excirolana braziliensis*	I	312.8	–	–	0.72	9.50	0.90	0.70	2.20	2.93	–	–	–	1	2	ELEF
Caetano et al. (2006)	RJ	*Excirolana braziliensis*	I	216.2–17195.8[a]	Cont	Cont	0.59	8.34	0.81	1.14	1.90	2.20	0.13	0.08	1.62	2	2	MPA
Caetano et al. (2006)	RJ	*Excirolana braziliensis*	I	26.5–1264.2[a]	Cont	Cont	0.57	10.41	0.71	0.94	1.80	2.82	0.56	0.25	2.24	2	2	MPA

Silva (2013)	RJ	*Excirolana braziliensis*	I	55.87	Cont	Cont	0.81	9.90	0.31	1.14	1.23	3.81	0.08	0.03	2.67	3	2	–
Calil and Borzone (2008)	PR	*Metamysidopsis neritica*	I	–	Cont	Cont	–	–	–	–	–	–	–	–	–	1	1	–
Leite et al. (2003)	SP	*Monokalliapseudes schubarti*	I	–	Cont	Cont	–	10.70	–	3.00	–	–	–	–	–	1	1	–
Pennafirme and Soares-Gomes (2017)	RJ	*Monokalliapseudes schubarti*	I	–	–	Cont	0.46	0.12	–	2.91	0.80	1.00	10.60	1.95	5.44	1	1.5	Gulland
Alberto and Fontoura (1999)	RS	*Ocypode quadrata*		0.12–0.7	–	–	–	–	–	–	3.00	–	–	–	–	1	3	–
Negreiros-Fransozo et al. (2002)	SP	*Ocypode quadrata*	B & I	–	Spr-Sum	Sum	–	–	–	–	–	–	–	–	–	–	–	–
Turra et al. (2005)	SP	*Ocypode quadrata*	B & I	0–2.85	–	–	–	–	–	–	–	–	–	–	–	–	–	–
Neves and Bemvenuti (2006)	RS	*Ocypode quadrata*	B & I	0.4–5	–	–	–	–	–	–	–	–	–	–	–	3	–	–
Neves and Bemvenuti (2006)	RS	*Ocypode quadrata*	B & I	0.4–1.5	–	–	–	–	–	–	–	–	–	–	–	3	–	–
Neves and Bemvenuti (2006)	RS	*Ocypode quadrata*	B & I	0.3–1	–	–	–	–	–	–	–	–	–	–	–	3	–	–
Araújo et al. (2008)	ES	*Ocypode quadrata*	B & I	0.33	–	–	–	–	–	–	–	–	–	–	–	3	–	–
Araújo et al. (2008)	ES	*Ocypode quadrata*	B & I	0.08	–	–	–	–	–	–	–	–	–	–	–	3	–	–
Araújo et al. (2008)	ES	*Ocypode quadrata*	B & I	0.16	–	–	–	–	–	–	–	–	–	–	–	3	–	–
Rosa and Borzone (2008)	PR	*Ocypode quadrata*	B & I	0–0.4	–	–	–	–	–	–	–	–	–	–	–	13	–	–
Souza et al. (2008)	PE	*Ocypode quadrata*	B & I	0–1	–	–	–	–	–	–	–	–	–	–	–	12	–	–

(continued)

Table 6.1 (continued)

Authors	State	Species	Zone	Dens m^2	Reprod	Recruit	WP	L_∞	C	K	LE	Z	P	B	P/B	N	T	Method
Magalhães et al. (2009)	BA	*Ocypode quadrata*	B & I	–	–	–	–	–	–	–	–	–	–	–	–	9	–	–
Maia-Carneiro et al. (2013)	RJ	*Ocypode quadrata*	B & I	0.67	–	–	–	–	–	–	–	–	–	–	–	–	–	–
Pombo and Turra (2013)	SP	*Ocypode quadrata*	B & I	–	–	–	–	46.40	0.30	0.90	–	–	–	–	–	3	1	MPA
Pombo and Turra (2013)	SP	*Ocypode quadrata*	B & I	–	–	–	–	57.95	0.40	0.90	–	–	–	–	–	3	1	MPA
Pombo and Turra (2013)	SP	*Ocypode quadrata*	B & I	–	–	–	–	53.01	0.40	1.11	–	–	–	–	–	3	1	MPA
Oliveira et al. (2016)	RJ	*Ocypode quadrata*	B & I	0.05–0.7	–	–	–	–	–	–	–	–	–	–	–	–	–	–
Oliveira and Yokoyama (2021)	SP	*Ocypode quadrata*	B & I	0–0.07	–	–	–	–	–	–	–	–	–	–	–	1	1.5	–
Souza et al. (2021)	PA	*Ocypode quadrata*	B & I	0.6[a]	–	–	–	53.36	0.00	0.76	–	–	–	–	–	1	1	ELEF
Zavarize (2009)	RJ	*Pagurus criniticornis*	I	–	–	–	0.05	4.96	0.40	0.62	2.35	3.37	–	–	–	–	–	–
Alves and Pezzuto (1998)	RS	*Pinnixa patagoniensis*	I	–	Spr-Sum	Aut	0.40	–	0.42	1.40	2.04	3.60	–	–	–	–	1.5	ELEF
Pezzuto (1998)	RS	*Sergio mirim*	I	–	Spr-Sum	Spr	0.20	29.02	0.50	0.71	4.14	1.54	–	–	–	1	3.5	ELEF
Ramos (2014)	PE	*Talorchestia tucurauna*	I	866.3	Cont	Cont	0.07	11.12	0.01	2.12	1.10	4.36	–	–	–	2	1	MPA
Ramos (2014)	PE	*Talorchestia tucurauna*	I	53.2	Cont	Cont	–	–	–	–	–	–	–	–	–	2	1	–
Mollusca																		
Fontoura-Silva et al. (2016)	RJ	*Agathistoma viridulum*	I	–	–	–	0.30	28.84	0.25	0.83	3.03	5.36	–	–	–	1	1	MPA
Fontoura-Silva et al. (2016)	RJ	*Agathistoma viridulum*	I	–	–	–	0.30	27.44	0.30	0.84	–	–	–	–	–	–	–	TAG
Narchi (1976)	SP	*Anomalocardia brasiliana*	I	–	Cont	–	–	–	–	–	–	–	–	–	–	–	–	–

Salvador (2001)	SP	*Anomalocardia brasiliana*	I	–	–	–	0.50	35.50	0.80	1.30	2.00	1.41	–	–	–	–	–	–
Rocha-Barreira and Araújo (2005)	CE	*Anomalocardia brasiliana*	I	–	Cont	–	–	–	–	–	–	–	–	–	–	–	–	–
Boehs et al. (2008)	PR	*Anomalocardia brasiliana*	I	94–489	–	–	–	–	–	–	–	–	–	–	–	–	–	–
Luz and Boehs (2011)	BA	*Anomalocardia brasiliana*	I	–	–	–	–	–	–	–	–	–	–	–	–	–	–	–
Mattos and Cardoso (2012)	RJ	*Anomalocardia brasiliana*	I & S	289.4	–	–	0.41	37.62	0.85	0.71	2.64	3.17	11.60	9.14	1.27	1	2	ELEF
Rodrigues et al. (2013)	RN	*Anomalocardia brasiliana*	I	1215	Cont	Cont	–	28.70	–	0.61	–	–	–	–	–	2	3	–
Rodrigues et al. (2013)	RN	*Anomalocardia brasiliana*	I	773	Cont	Cont	–	29.87	–	0.48	–	–	–	–	–	2	3	–
Corte et al. (2014)	SP	*Anomalocardia brasiliana*	I	–	Cont	–	–	–	–	–	–	–	–	–	–	–	–	–
Corte et al. (2014)	SP	*Anomalocardia brasiliana*	I	–	Continuous	–	–	–	–	–	–	–	–	–	–	–	–	–
Corte et al. (2015)	SP	*Anomalocardia brasiliana*	I	–	Sum-Aut	–	–	–	–	–	–	–	–	–	–	–	–	–
Corte et al. (2015)	SP	*Anomalocardia brasiliana*	I	–	Cont	–	–	–	–	–	–	–	–	–	–	–	–	–
Corte et al. (2015)	SP	*Anomalocardia brasiliana*	I	52.92	–	Cont	–	42.47	0.34	0.53	3.34	1.54	2.86	3.60	0.79	1	1	MPA
Corte et al. (2017)	SP	*Anomalocardia brasiliana*	I	51.8	–	Aut	–	46.03	0.29	0.37	3.41	–	2.89	4.75	0.61	1	1	MPA
Silva-Cavalcanti et al. (2018)	PE	*Anomalocardia brasiliana*	I	319–496	–	–	–	–	–	–	–	–	–	–	–	–	–	–

(continued)

Table 6.1 (continued)

Authors	State	Species	Zone	Dens m^2	Reprod	Recruit	WP	L_∞	C	K	LE	Z	P	B	P/B	N	T	Method
Cardoso and Cabrini (2016)	RJ	*Cerithium atratum*	I	86.7–366.7	–	–	0.89	36.30	0.70	0.58	4.26	4.20	4.56	3.90	1.17	1	1	–
Denadai et al. (2004)	SP	*Cerithium atratum*	I	79.7	–	Wint	0.60	33.84	0.75	1.30	1.60	2.67	1.15	0.29	3.90	1	1	–
Salvador (2001)	SP	*Corbula cubaniana*	I	–	–	–	–	12.00	–	0.60	1.50	1.88	–	–	–	–	–	–
Mattos and Cardoso (2012)	RJ	*Diplodonta punctata*	I & S	77.9	–	–	0.20	18.48	0.39	1.10	1.63	3.53	0.79	0.43	1.82	1	2	ELEF
Cardoso and Veloso (2003)	RJ	*Donax hanleyanus*	I	–	–	–	0.40	26.40	0.80	0.80	1.60	1.70	2.20	1.50	1.47	1	2	ELEF
Dos Anjos (2016)	SP	*Donax hanleyanus*	I	27.6	–	Cont	–	32.25	0.00	0.80	3.20	1.59	13.44	10.77	1.25	1	1	ELEF
Rocha-Barreira et al. (2002)	CE	*Donax striatus*	I	–	–	Cont	0.66	25.10	1.00	1.16	1.00	–	–	–	–	1	1	ELEF
Yokoyama (2010)	SP	*Nassarius vibex*	I	–	–	–	0.59	18.51	0.42	0.58	–	1.01	–	–	–	–	–	MPA
Cardoso and Cabrini (2016)	RJ	*Nassarius vibex*	I	8.8–55.9	–	–	0.10	19.10	0.76	0.94	2.50	4.10	0.67	0.58	1.16	1	1	–
Rocha-Barreira (2010)	RS	*Olivancillaria vesica*	I	–	Sum	–	–	–	–	–	–	–	–	–	–	–	–	–
Caetano et al. (2003)	RJ	*Olivancillaria vesica*	I	0.15	–	–	0.30	59.00	0.60	0.70	4.28	3.12	0.19	0.19	1.00	1	2	ELEF
Araújo and Rocha-Barreira (2012)	CE	*Olivella minuta*	I	78.5	–	–	0.30	13.50	0.20	0.70	4.00	4.70	4.13	2.21	1.87	1	2	ELEF
Petracco et al. (2014)	SP	*Olivella minuta*	I	3295[a]	–	Spr-Sum	0.40	11.16	0.35	1.25	1.50	3.55	12.12	6.35	1.91	2	1	MPA
Petracco et al. (2014)	SP	*Olivella minuta*	I	120[a]	–	Sum	0.40	11.43	0.57	1.45	1.82	3.55	0.82	0.77	1.06	2	1	MPA
Corte et al. (2020)	SP	*Olivella minuta*	I & S	224.4	–	–	–	–	–	–	–	–	–	–	–	2	0.5	–
Corte et al. (2022)	SP	*Olivella minuta*	I & S	23.7	–	–	–	–	–	–	–	–	–	–	–	2	0.5	–

Authors	State	Species	Zone	Dens m^2	Reprod	Recruit	WP	L_∞	C	K	LE	Z	P	B	P/B	N	T	Method
Abrahão et al. (2010)	SP	*Tagelus plebeius*	I	17.1	–	Spr-Sum	0.97	67.01	0.43	1.73	1.70	3.12	1.53	1.12	1.37	1	1	–
Farias and Rocha-Barreira (2017)	CE	*Tagelus plebeius*	S	–	Cont	–	–	–	–	–	–	–	–	–	–	1	1.2	–
Silva et al. (2015)	SP	*Tagelus plebeius*	I	–	Cont	–	0.94	74.14	0.47	0.52	2.58	2.16	–	–	–	–	–	E&M
Cardoso et al. (2015)	RJ	*Tellina lineata*	I	5.2–123.7	–	–	1.00	32.33	0.41	0.54	3.15	1.96	1.39	2.29	1.65	1	1	–
Cardoso et al. (2015)	RJ	*Tellina versicolor*	I	1.7–36.7	–	–	0.70	35.49	0.23	0.83	2.86	6.37	0.79	0.58	1.36	1	1	–
Turra et al. (2014)	SP	*Tivela mactroides*	I	867[a]	–	–	–	40.75	0.00	1.00	2.38	2.81	–	–	1.02	–	–	ELEF
Turra et al. (2015)	SP	*Tivela mactroides*	I	125000[a]	–	–	–	38.60	0.60	0.40	3.30	2.45	–	–	1.45	–	–	ELEF
Echinodermata																		
Garcia and Borzone (2015)	PR	*Lytechinus variegatus*	S	–	Sum-Aut	–	–	–	–	–	–	–	–	–	–	1	1	–
Tavares and Borzone (2006)	PR	*Mellita quinquiesperforata*	S	–	Spr-Sum	–	–	–	–	–	–	–	–	–	–	–	–	–
Yokoyama et al. (2008)	SP	*Ophionereis reticulata*	I	–	–	Spr-Sum	–	–	–	–	–	–	–	–	–	1	1	–
Yokoyama and Amaral (2011)	SP	*Ophionereis reticulata*	I	0.46–9.46	–	Cont	0.90	11.47	0.97	0.42	6.00	0.49	–	–	–	1	1	MPA

Dens m^2 number of individuals per square meter (a denote individuals per linear meter), *Reprod* period of more intense spawning, *Recruit* period of more intensive recruitment, *WP* winter point, *L∞* theoretical maximum length, *C* seasonal growth oscillation, *K* growth rate, *LE* life expectancy, *Z* mortality rate, *P* production (g AFDW/m^2/year), *B* biomass (g AFDW/m^2), *P/B* turnover rate (year), *N* number of populations investigated, *T* duration of the study (year), *Method* method used to estimate growth parameter

Zone: B backshore, *I* intertidal, *S* sublitoral, *Method: ELEF* ELEFAN routine (FISAT software), *MPA* modal progression analysis, *BERT* BERTHAL® software, *Gulland* Gulland & Holt graphic method, *TAG* tag and recapture, *E&M* ELEFAN and MPA

[a]Number of individuals per running meter or strip transect (IST)

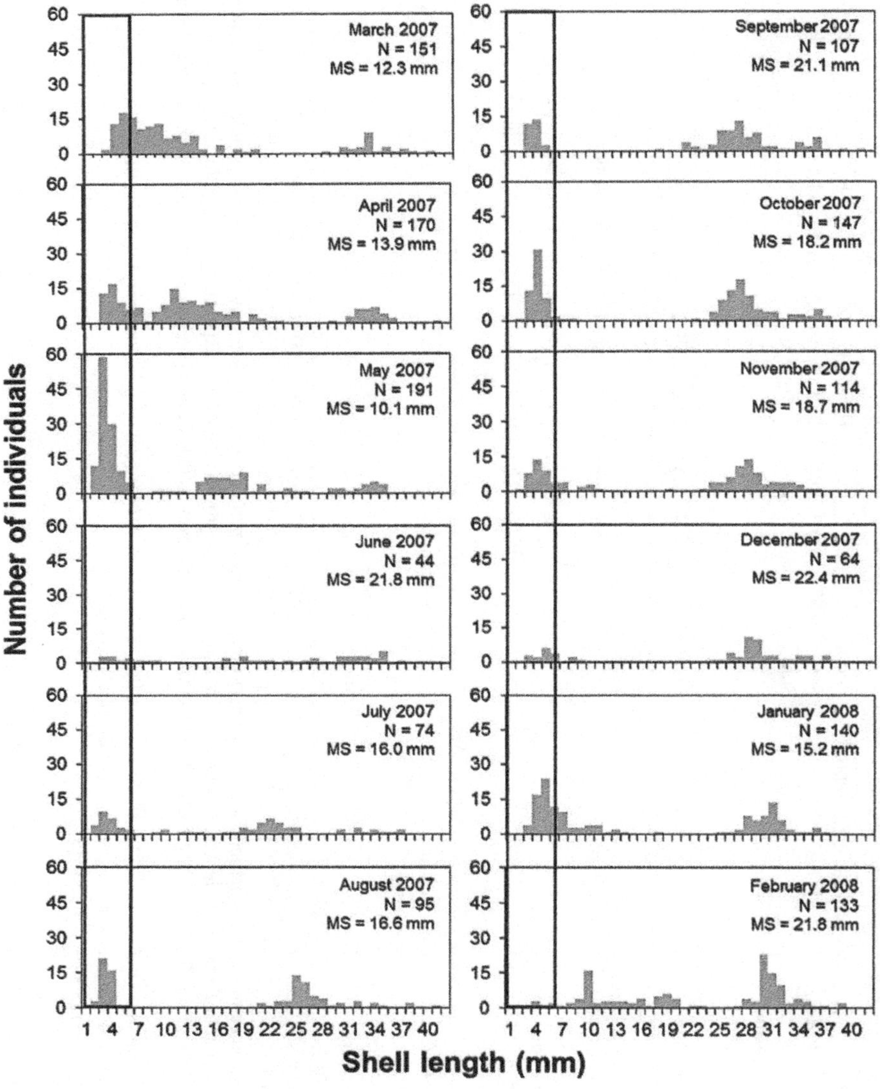

Fig. 6.1 Length-frequency distributions (LFD) of *Anomalocardia brasiliana* (*flexuosa*) on the North coast of São Paulo state. N: number of individuals; MS: mean size of individuals

frequency of juveniles in the length-frequency distributions. Figure 6.1 shows the length-frequency distributions of one population of the species *Anomalocardia brasiliana* (*flexuosa*) inhabiting the North coast of São Paulo investigated by Corte et al. (2015). The rectangles on the left side of each plot highlight the presence of young individuals (<5 mm length) throughout the year but with higher intensity in May 2007. This pattern, however, is not demonstrated by the recruitment estimates obtained with the FISAT II software (Fig. 6.2).

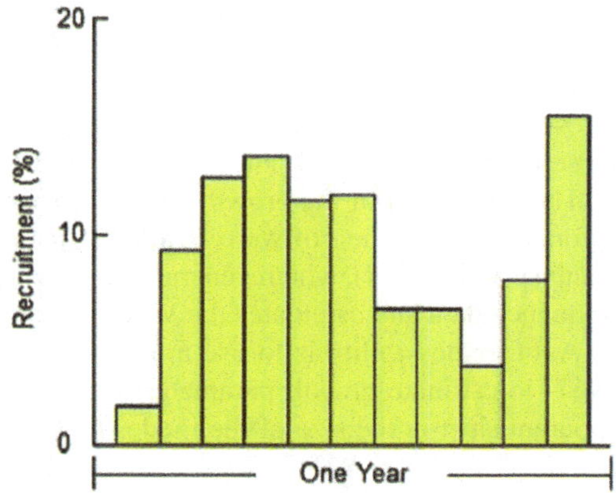

Fig. 6.2 Recruitment estimates for *Anomalocardia brasiliana* (*flexuosa*) on the North coast of São Paulo state obtained with the software FAO-ICLARM Stock Assessment Tools (FISAT II – version 1.1.0). (The data was the same used in Fig. 6.1)

6.2.3 Growth

Growth is generally defined as the increase in size and/or mass of an organism with time. It is the main characteristic of most organisms and the essential information to estimate population mortality or production (Brey 2001). Although growth is an individual feature (i.e., each individual of a population exhibits its pattern of growth during its life), it is expected to be similar among all specimens within a population. Thus, growth may be estimated as growth function parameters describing the growth of an "average individual" in the population (Brey 2001).

Growth parameters of sandy beach species are usually estimated using the generalized von Bertalanffy growth function (VBGF).

$$L_t = L_\infty \left\{ 1 - \exp^{\left[-K(t-t_0) + \left(\frac{CK}{2\pi}\right) \mathrm{sinsin}\left(2\pi(t-t_s)\right) - \left(\frac{CK}{2\pi}\right) \mathrm{sinsin}\left(2\pi(t_0-t_s)\right) \right]} \right\}$$

Where *Lt* is the length (size) at time *t*; L_∞ is the theoretical maximum length that one individual of the studied population would reach; *K* is the curvature parameter (a measure of the rate at which maximum size is reached); *C* is the parameter reflecting the amplitude of seasonal growth oscillation (varies between 0 and 1, with 0 indicating no oscillation and 1 interrupted growth rate at a given season); *t*0 is the time at length 0, and ts is used to estimate the "winter point," that is, the time where growth is slowest (WP = *ts* + 0.5).

To estimate the growth parameters, it is necessary to know how the population changes in individual size over time (i.e., obtain an age-length key) and use these values as input for the VBGF. This can be done directly, for example, using mark-recapture (tagging) experiments, or indirectly by analyzing the length-frequency distribution (LFD) of the population at different times (Fig. 6.2). The mark-recapture approach involves sampling and measuring the same individual in different times and can provide detailed and high-quality information on individual growth.

However, the costs of its implementation on large scales and/or in natural ecosystems can be prohibitive as only a very small fraction of organisms are recaptured (Fontoura-da-Silva et al. 2016). Therefore, the LFD approach has been widely used to estimate the growth parameters of marine invertebrates and fishes, including those found on Brazilian sandy beaches (Table 6.1).

The estimation of the growth parameters through the LFD analysis can be done automatically in the software FISAT II using the Electronic Length-Frequency Analysis (ELEFAN), which generates a growth curve from the best fit to the length-frequency data and calculates the VBGF parameters (Fig. 6.3).

Another possibility is to use modal progression analysis (MPA) (Bhattacharya 1967) to estimate growth parameters. Briefly, this method infers growth from the apparent shift in the size of the modes (cohorts) over time (i.e., in a time series of length-frequency samples). This is done in three steps: (1) decomposition of LFD to identify modes/cohorts (i.e., separation of normally distributed components of size-frequencies – Bhattacharya's method and NORMSEP in FISAT II) (Fig. 6.4), (2) "linking" of the modes perceived to belong to the same cohorts, and (3) using the growth increments and size-at-age data to estimate growth adopting a theoretical growth function (such as VBGF) (Gayanilo et al. 2005; Gomez and Defeo 1999) (Fig. 6.4). While using the ELEFAN routine is more straightforward, MPA provides the statistical significance of each parameter and may, therefore, provide more reliable estimates. Furthermore, the MPA is likely a more appropriate method to estimate the growth parameters of sandy beach invertebrates due to their relatively lower mobility and large variations in recruitment (Chatzinikolaou and Richardson 2008).

In Brazil, the LFD analysis is by far the most used method to assess the growth of sandy beach species. Most studies done thus far applied the ELEFAN routine to obtain growth parameters; however, the use of MPA has increased over the last decade (Table 6.1). On the coast of the state of São Paulo, Silva et al. (2015) used the ELEFAN routine and MPA to estimate the growth parameters of the razor clam *Tagelus plebeius*. The authors found similar values when using both methods, thereby strengthening their findings.

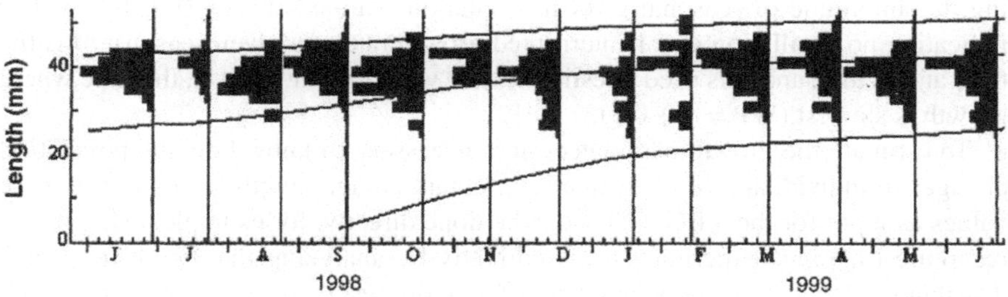

Fig. 6.3 Seasonal growth curve fitted for *Olivancillaria vesica* on a sandy beach in Rio de Janeiro State (SE Brazil). ELEFAN I routine

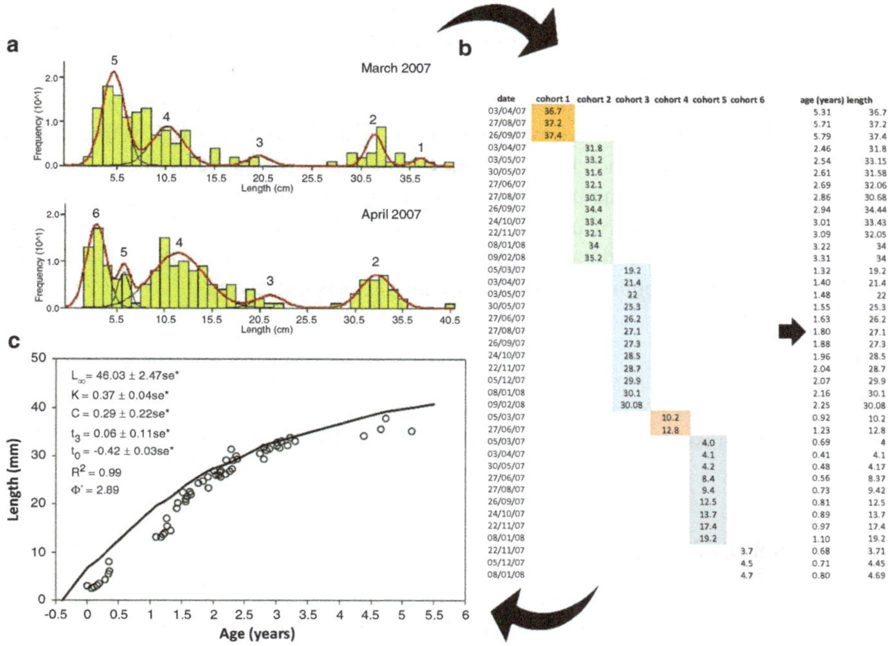

date	cohort 1	cohort 2	cohort 3	cohort 4	cohort 5	cohort 6	age (years)	length
03/04/07	36.7						5.31	36.7
27/08/07	37.2						5.71	37.2
26/09/07	37.4						5.79	37.4
03/04/07		31.8					2.46	31.8
03/05/07		33.2					2.54	33.15
30/05/07		31.6					2.61	31.58
27/06/07		32.1					2.69	32.06
27/08/07		30.7					2.86	30.68
26/09/07		34.4					2.94	34.44
24/10/07		33.4					3.01	33.43
22/11/07		32.1					3.09	32.05
08/01/08		34					3.22	34
09/02/08		35.2					3.31	34
05/03/07			19.2				1.32	19.2
03/04/07			21.4				1.40	21.4
03/05/07			22				1.48	22
30/05/07			25.3				1.55	25.3
27/06/07			26.2				1.63	26.2
27/08/07			27.1				1.80	27.1
26/09/07			27.3				1.88	27.3
24/10/07			28.5				1.96	28.5
22/11/07			28.7				2.04	28.7
05/12/07			29.9				2.07	29.9
08/01/08			30.1				2.16	30.1
09/02/08			30.08				2.25	30.08
05/03/07				10.2			0.92	10.2
27/06/07				12.8			1.23	12.8
05/03/07					4.0		0.69	4
03/04/07					4.1		0.41	4.1
30/05/07					4.2		0.48	4.17
27/06/07					8.4		0.56	8.37
27/08/07					9.4		0.73	9.42
26/09/07					12.5		0.81	12.5
24/10/07					13.7		0.89	13.7
22/11/07					17.4		0.97	17.4
08/01/08					19.2		1.10	19.2
22/11/07						3.7	0.68	3.71
05/12/07						4.5	0.71	4.45
08/01/08						4.7	0.80	4.69

Fig. 6.4 Growth parameters estimation through modal progression analysis (MPA). First (**a**) normal distributions of length frequency data are separated using Bhattacharya's method (FiSAT II). Then, (**b**) modes perceived to belong to the same cohorts are linked to building an age-length key which is (**c**) used to estimate growth by adopting a theoretical growth function (such as VBGF). Numbers above each normal distribution correspond to different cohorts (the cohort 1 (older) was not identified in April 2007). The mean size of each cohort is provided in the analyses and used in step 2 of the Modal Progression Analysis

Along the Brazilian coast, only Fontoura-da-Silva et al. (2016) used mark-recapture methods to estimate the growth parameters of sandy beach populations. They assessed the growth of the sea snail *Agathistoma viridulum* over 2 years in the state of Rio de Janeiro and found significant differences between this method and LFD analysis in the first year, likely due to the low number of recaptured individuals. Yet, both methods estimated similar parameters during the second year, when a large number of individuals were recaptured and the sample size was adequate.

6.2.4 Biomass

Biomass, frequently denominated as the standing crop, is a measurement of how much living tissue mass of a population is present at one instant in time (or averaged over several periods) per unit area (e.g., g/m^2). Since the mass of sandy beach species may vary significantly depending on the amount of water in their body and/or the inclusion of inorganic parts such as shells and exoskeleton, the biomass of sandy beach populations is generally expressed in ash-free dry mass (AFDM). This is achieved by first drying individuals at a moderate temperature (~60 °C) until they reach a constant weight (dry mass). Then, these individuals are ignited in a muffle furnace (~500 °C for 5 h) to estimate the ash (inorganic) content. Finally, the AFDM is obtained by the difference between the dry mass and the ash content.

To correctly estimate the biomass of a sandy beach population, it is necessary to obtain the average mass of all size classes. Then each value is multiplied by the number of individuals in the corresponding size class and added together to obtain the biomass of the population. Mathematically, this is expressed as

$$\underline{B} = \Sigma N_i M_i \Delta t_i$$

where N_i is the average number (or density) of individuals in size class i, M_i is the average body mass of size class i, and Δt_i is the period evaluated.

6.2.5 Secondary Production

Secondary production is generally defined as the formation of heterotrophic biomass through time, which is potentially available as food for the next trophic level (Waters 1977). It integrates measures of abundance/density and individual growth and can be viewed as a measure of the fitness of the population (Dolbeth et al. 2012). For a single animal, secondary production represents the formation of biomass through its growth (mass/ind/time). When scaled to the population level, secondary production expresses the amount of biomass produced per unit of area over time (e.g., g/m^2/year) (Benke and Huryn 2017).

In Brazil, most studies used the mass-specific growth rate method (MSGRM) to estimate the secondary production of sandy beach populations (Petracco et al. 2012). This method is usually based on the von Bertalanffy growth function (VBGF see Sect. 6.2.3) and is adequate for populations with continuous recruitment and/or whose cohorts cannot be distinguished – a pattern commonly found for Brazilian sandy beach species (Table 6.1). In the MSGR method, a length-frequency distribution is estimated from samples taken at intervals over time (e.g., Fig. 6.2), and the production is obtained from the mass-size relationship according to the equation:

$$P = \Sigma N_i M_i G_i \Delta t_i$$

where N_i, M_i, and G_i are, respectively, the mean annual density (e.g., ind/m^2), the mean individual body mass (e.g., g AFDM/m^2), and the mass-specific growth rate in class i. Δt_i is the period evaluated (generally years, but it can be expressed in other time units) (Crisp 1984; Brey 2001). G_i is determined by:

$$G_i = bK\left[\left(\frac{L\infty}{L_i}\right) - 1\right]$$

where b is the exponent (b) of the power mass-size (length) relationship ($M = aL^b$), K, L_∞, are VBGF parameters, and L_i is the mean size in class i.

Despite its importance for the ecological assessment of population and ecosystem status, the number of studies evaluating general patterns of secondary production in Brazilian sandy beaches is still very incipient (Petracco et al. 2017, 2019). Almost all studies on the secondary production of sandy beach fauna have been published since the beginning of this century, and the number of species investigated is considerably low (Table 6.1).

All species investigated in secondary production studies along the Brazilian coast belong to the four main macrobenthic groups: mollusks (11 species; 6 bivalves and five gastropods), crustaceans (6 species; 3 decapods, 2 isopods, and 1 amphipod), annelids (5 species), and echinoderms (1 species). As observed worldwide, bivalves are the group with more populations studied, probably because they are frequently harvested and used as food or income sources (Petracco et al. 2013). In Brazil, we can find information on the secondary production of the clams *Donax hanleyanus*, *Anomalocardia brasiliana*, *Diplodonta punctata*, *Eurytellina lineata*, *Tivela mactroides*, and *Mesodesma mactroides*. Among gastropods, we have estimates for the species *Hastula cinerea*, *Olivella minuta*, *Olivancillaria vesica*, *Nassarius vibex*, and *Cerithium atratum*.

Although crustaceans are the most investigated group on Brazilian sandy beaches, secondary production estimates are available for only a few species, such as the cirolanid isopods *Excirolana braziliensis* and *E. armata*, the talititrid *Pseudorchestoidea* (*Atlantorchestoidea*) *brasiliensis*, the mole crab *Emerita brasiliensis* (with estimates mainly from Rio de Janeiro), and the thalassinid decapod *Callichirus major*, which has been studied in beaches from the states of Rio de Janeiro, São Paulo, and Paraná.

6.2.6 *Turnover Rate (P/B Ratio)*

The turnover rate, also known as the production-to-biomass ratio (P/B ratio) or productivity rate, is a measure of the amount of new biomass produced by the current standing crop of the population. It can be understood as a weighted mean value of biomass growth rates of all individuals in the population and, therefore, is a powerful parameter that describes the dynamic relationship between individual growth, biomass, and production (Benke and Huryn 2010). The turnover rate is related to the instantaneous mortality rate (Z) (Sect. 6.2.8) and other relevant life-history traits such as body mass and, particularly, life span (Cusson and Bourget 2005; Petracco et al. 2013). Usually, the smaller the body mass and the shorter the lifespan of the population, the higher the turnover rate.

6.2.7 Longevity (Life Expectancy)

Longevity is defined as the maximum age range that the population reaches. Estimation of the longevity of sandy beach populations can be performed using the inverse of VBGF, assuming a maximum shell length of 99% of the asymptotic length (Sparre and Venema 1992)

$$t_{\max} = \frac{\left[\ln\ln L_{99\%} - \left(L_{\infty} - L_{99\%}\right)\right]}{k}$$

Nevertheless, there are different methods used to estimate the longevity of sandy beach populations, such as (1) age at observed maximum length (Cardoso and Veloso 1996), (2) the relation L_t/L_∞ equal to 0.95 (L_t are 95% of the L_∞), and (3) L_t as the length representing 99% of the population (L99%) (Cardoso and Veloso 1996).

Studies to estimate the longevity of the Brazilian sandy beach macrofauna were performed in most of these ways. For example, Cardoso and Veloso (2003) estimated life span based on the L95%; Petracco et al. (2010) used the length representing the 99th percentile of the population (L99%); whereas Cardoso and Cabrini (2016) applied an inverse von Bertalanffy growth equation considering the observed maximum length.

6.2.8 Mortality

The mortality rate is equivalent to the number of individuals removed from the population by death for a unit of time in relation to population size. It regulates the size and structure of a population and is usually related to the monotonous decrease in the number of individuals in an age class with time, a process caused by predation, parasitism, and diseases (Brey 2001).

As observed for longevity, there are several methods used to estimate the mortality rate of sandy beach species, most of them implemented in the FISAT II software (Gayanilo et al. 2005). Total mortality (*Z*) of sandy beach populations is usually estimated from the slope of the descending straight segment of age-structured catch curves as:

$$\log_e\left(N_i\right) = a + bt_i$$

where N is the number of individuals of age t, a and b are estimated through regression analysis, and b, with the sign changed, is an estimate of Z (Pauly et al. 1995).

When no information regarding the age of individuals is available, the total mortality (Z) can be obtained from length-converted catch curves following the equation:

$$\ln\left(\frac{N}{t_i}\right) = a + bt_i$$

where N, a, and b are estimated as above, and t_i is the time needed for the individual of a given length class i to grow to the next class. t_i can be estimated as:

$$t_i = \frac{-1}{k}\ln\ln\left[\left(\frac{L_\infty - L_{2i}}{L_\infty - L_{1i}}\right)\right] + t_0$$

where $L1_i$ and $L2_i$ are the lower and upper limits, respectively, of length class i (Pauly et al. 1995).

Mortality estimations can also describe the actual loss of specimens due to mortality at a certain time t, a parameter known as the instantaneous mortality rate. This parameter is obtained from the Malthusian (i.e., exponential) model of population growth, but with signs changed (Brey 2001):

$$Z = \frac{-\ln\ln\left(\frac{N_t}{N_0}\right)}{t}$$

where N_t is the number of individuals at time t and N_0 is the initial number of individuals.

Finally, it is possible to obtain estimates of the number of individuals that die during a year, the annual mortality rates (A), through the formula:

$$A = 1 - e^{-Z}$$

where Z corresponds to the instantaneous mortality rate of the population (Ricker 1977).

Since mortality is usually linked to the transference of energy from one trophic level to the next (i.e., prey-predator), it is related to the turnover rates of a population (P/B ratio, Sect. 6.2.6). Allen (1971) and Brey (1999) showed that the mortality rate Z is linearly related to the P/B ratio of the population provided that (i) the population is in a steady state, (ii) growth can be described by the von Bertalanffy growth model (VBGF), and (iii) mortality can be described by the single negative exponential mortality model which assumes that the instantaneous rate of mortality Z is constant over the whole lifespan of the cohort (i.e., the number of deaths in a small interval of time is proportional to the number of animals present at that time) and is estimated as:

$$N_t = N_0 * e^{-Z*t}$$

Several studies were performed on Brazilian sandy beaches to assess the mortality rate of macrofauna and used different methods. For instance, in Rio de Janeiro, Southeast Brazil, Mattos, and Cardoso (2012) estimated the annual finite mortality rates (*A*) for the bivalves *Anomalocardia brasiliana* and *Diplodonta punctata* (0.96 year^{-1} and 0.97 year^{-1}, respectively), whereas Caetano et al. (2003) evaluated the instantaneous mortality (*Z*) of *Olivancillaria vesica* over 2 years (3.12 year^{-1} in the first year and 2.56 year^{-1} in the second year).

6.2.9 Reproductive Cycle

The reproductive strategies of a species play a major role in its population dynamics and perpetuity. Knowledge of the reproductive cycles is fundamental to determining population parameters such as reproductive output and timing of spawning, starting points to better understand the recruitment and growth patterns of populations (Corte 2015). Additionally, studies of reproductive cycles support strategies of closed fishing seasons over the year during reproductive activity peaks (Denadai et al. 2015) and are an essential tool for the correct management of harvesting practices and development of mariculture of exploited species (Morsan and Kroeck 2005).

The reproductive cycle of marine invertebrates includes the entire chain of events from the proliferation and differentiation of germ cells (gametogenesis) to spawning and subsequent recession and/or recovery of the gonad. Its duration varies among species and populations and may be limited to days or weeks or even lasts the entire year (continuous reproduction) (MacCord and Amaral 2007; Corte et al. 2014).

The production of gametes, especially eggs, is energetically expensive and strongly influenced by endogenous characteristics (i.e., genetic and physiological features), environmental variables (e.g., food availability, temperature, and salinity), and biological interactions (e.g., parasitism) (Corte 2015; Llodra 2002). Therefore, differences in reproductive cycles can be found even between closely located populations.

Because of the variety of reproductive strategies and morphological forms of sandy beach macrofauna, a wide range of methodologies has been developed to characterize their reproduction. For example, the analyses of the reproductive cycle of polychaetes, mollusks, and echinoderms from Brazilian sandy shores usually demand time-consuming histological analysis of their gonads (e.g., Tavares and Borzone 2006; MacCord and Amaral 2007; Yokoyama and Amaral 2011; Corte et al. 2014; Farias and Rocha-Barreira 2017) (Fig. 6.5). On the other hand, many species of crustaceans can have their reproductive cycle more easily assessed as they carry larger eggs that can be detached from their bodies (e.g., Fonseca et al. 2000; Pennafirme and Soares-Gomes 2022) (Fig. 6.6).

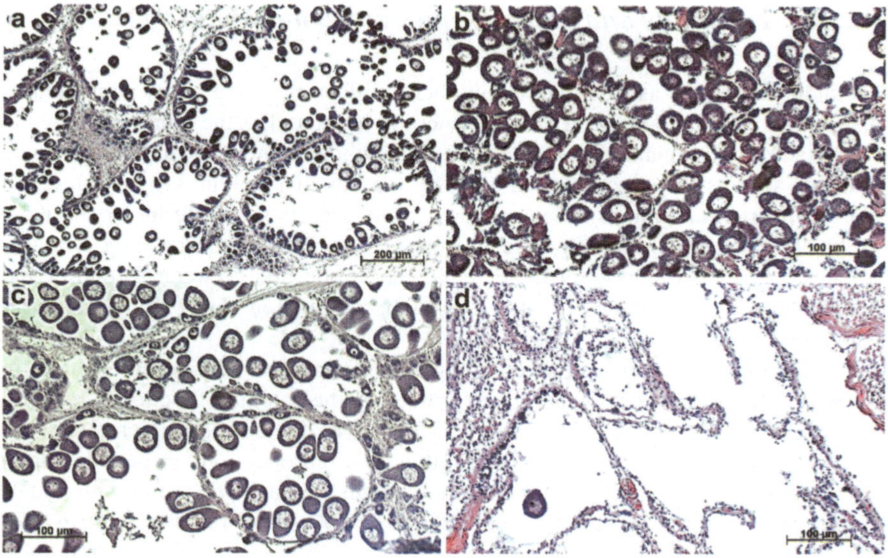

Fig. 6.5 Histological analyses of the gametogenesis in females of *Anomalocardia brasiliana* (*flexuosa*): (**a**) gametogenesis stage, (**b**) mature, (**c**) spawning, and (**d**) spent/recovering

Fig. 6.6 *Atlantorchestoidea brasiliensis:* (**a**) male, (**b**) female, and (**c**) ovigerous female. Scale bar: 5 mm. (Reprinted from Pennafirme and Soares-Gomes (2022)). Population dynamics and reproductive biology of the sandhopper *Atlantorchestoidea brasiliensis* (Amphipoda: Talitridae) on a sandy beach in the Southwestern Atlantic Coast. (Licensed by Taylor & Francis and Copyright Clearance Center)

6.2.10 Fecundity and Mean Length at Sexual Maturity (L_{50})

Fecundity is defined as the potential number of offspring produced by a female in a determined time (Llodra 2002) and is generally measured as the number of eggs in sandy beach individuals. When considering the whole population, fecundity can be understood as its reproductive potential. This reproductive output is highly plastic in marine invertebrates and is influenced by the bioenergetics and life-history strategy of the organism (Llodra 2002). Since the production of eggs requires large amounts of energy, there are important relationships and trade-offs between fecundity and other population attributes, such as growth, production, and sexual maturity (Llodra 2002).

The relationship between the total length and fecundity of ovigerous females in sandy beach populations is generally expressed by the potential function $F = a * L^b$, where F is the number of eggs/embryos by ovigerous females, L is the total length, and a and b are constants. The mean length at first reproduction, or mean length at sexual maturity (L_{50}), may be defined as the length at which 50% of all individuals are sexually mature (e.g., as the length at which 50% of all females in a stock of crustaceans are ovigerous or the length at which 50% of all female mollusks have gonads in an advanced stage of development).

Petracco et al. (2010) analyzed the reproductive biology of the cirolanid isopod *Excirolana armata* on an exposed sandy beach in Southeastern Brazil and recorded that the length-fecundity relationship did not differ significantly between years. The individual fecundity ranged from 2 to 18, with a mean (±SD) value of 10 ± 3.12 eggs/embryos per female. The mean length at sexual maturity (L_{50}) was 4.66 mm.

6.3 Studies on Brazilian Sandy Beaches

Most likely, studies on the population biology of species inhabiting Brazilian sandy beaches began in the 1970s, with Walter Narchi's investigation of the gametogenic cycle of the clam *Anomalocardia brasiliana* (*flexuosa*) on the coast of São Paulo (Narchi 1976) as one of the first published records. However, it was only in the 1990s that the number of publications on this topic increased consistently (Fig. 6.7).

Sandy beach populations have been investigated in the four Brazilian coastal geographical regions; however, the number of publications from each region is very unbalanced. Most studies were performed in the southeast region, mainly in the Rio de Janeiro and São Paulo states, where the number of experts on the topic is higher (Table 6.1). On the other hand, a very limited number of studies assessed the population biology of species from the North region of Brazil.

Approximately 40 species had their population parameters studied on Brazilian sandy beaches (Table 6.1). Crustacea is the most investigated group, followed by mollusks, annelids, and echinoderms. The crabs *Ocypode quadrata*, *Emerita brasiliensis*, and the clam *Anomalocardia brasiliana* (*flexuosa*) are the species with the higher number of publications on population biology parameters, tailed by the ghost

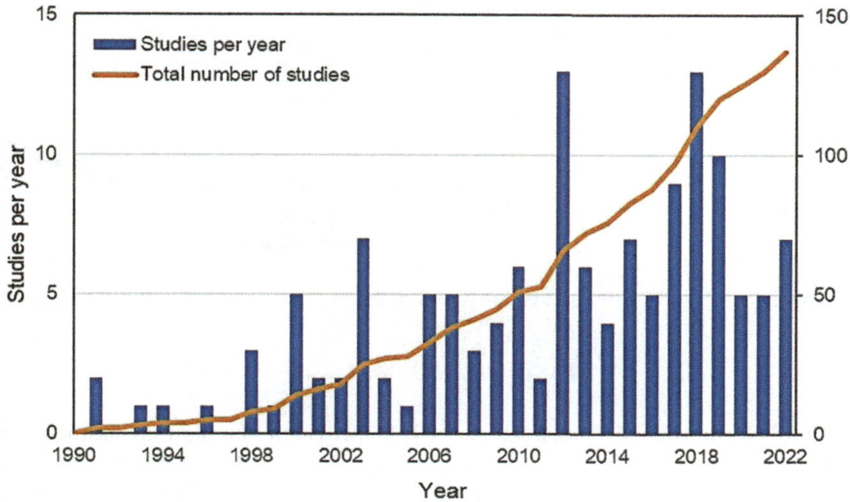

Fig. 6.7 Number of studies carried out on population biology of Brazilian sandy beach species. Only studies listed on the ISI Web of Science® were considered. Keywords: "beach*" AND "Brazil*" AND ("population dynamic*" OR "population biology"). (The search was performed on 08 August, 2022)

shrimp *Callichirus major*, the gastropod *Olivella minuta*, and the isopod *Excirolana braziliensis*. Among annelids, species belonging to the genus *Scolelepis* are the most studied. Only three species of echinoderms (i.e., *Ophionereis reticulata*, *Lytechinus variegatus*, and *Mellita quinquiesperforata*) had their population parameters investigated along the Brazilian coast. In the next pages, we discuss the population biology of the most studied species on Brazilian sandy shores.

6.3.1 Ocypode quadrata

Ghost crabs of the genus *Ocypode* are one of the most conspicuous organisms on sandy beaches around the globe (Pombo et al. 2017). In Brazil, the ghost crab *Ocypode quadrata* (Fig. 6.8), popularly known as "Maria-farinha," can be found along the whole coast, mainly in the supralittoral and upper intertidal zones of sandy beaches (Turra et al. 2005; Rosa and Borzone 2008). *Ocypode quadrata* is strongly affected by coastal urbanization (Blankensteyn 2006; Magalhães et al. 2009; Neves and Bemvenuti 2006) and is considered an efficient indicator of human impacts on sandy beaches (Barboza et al. 2021; Costa and Zalmon 2019). Due to its wide distribution, easy detection (most studies estimate population parameters of *O. quadrata* through burrows measurements), and use as a biological indicator, the population biology of *O. quadrata* has been investigated along the whole coast of Brazil.

Fig. 6.8 The ghost crab *Ocypode quadrata*. (Photo: Alvaro E. Migotto. Source: cifonauta.cebimar.usp.br)

In the North region, Souza et al. (2021) provided the first assessment of *O. quadrata* population parameters on the Amazon coast. By conducting monthly sampling throughout 1 year, they found that *O. quadrata* is strongly influenced by the two hydrological seasons on the Amazon coast, with higher abundance values during the rainy season. Also, burrows were deeper under drier conditions, suggesting that the spatial distribution of *O. quadrata* is influenced by sediment moisture. Growth parameters in the Amazon coast ($L_{\infty} = 53.36$ mm, $K = 0.76$) were similar to those found in Southeast Brazil (e.g., de Oliveira et al. 2016; Pombo and Turra 2017), indicating that local environmental characteristics may be more important on the population biology of *O. quadrata* than latitudinal effects.

In Northeast Brazil, Souza et al. (2008) examined the abundance, distribution, and morphometry of *O. quadrata* burrows on 12 sandy beaches with different degrees of anthropic impact in the Pernambuco state. The authors reinforced the bioindicator potential of *O. quadrata*, given that higher densities of burrows were recorded at non-urban beaches. Moreover, burrow diameter increased landward and was positively correlated with organic matter content. The potential of *O. quadrata* as an indicator of human action impacts on northeastern Brazilian beaches was also demonstrated by Magalhães et al. (2009), who reported a lower number of burrows on urban beaches on the coast of Bahia state.

The population biology of *O. quadrata* was also studied in all three coastal states of the Southeast region: Espírito Santo, Rio de Janeiro, and São Paulo. Araujo et al. (2008) examined the density and spatial distribution of *O. quadrata* on three sandy beaches of Espírito Santo and found a greater abundance of small burrows near the waterline, whereas larger burrows were mostly registered on the upper beach. As observed in the Northeast region, human activities such as the number of people and the traffic of vehicles negatively influenced the density and spatial distribution of *O. quadrata* in Espírito Santo.

In Rio de Janeiro, Maia-Carneiro et al. (2013) evaluated the spatial distribution of *O. quadrata* in an insular environment (Ilha Grande) and found that larger burrows were located near the vegetation while smaller-sized crabs occupied areas near the sea. According to the authors, this contrasting distribution is likely related to intraspecific competition, with larger-sized individuals excluding smaller ones from areas more protected against tides and flooding and with higher availability of food resources. Lower capacity of excavation in small individuals, which would need to stay close to the water to avoid the risk of desiccation due to their high surface/volume ratio, was also suggested as a reason for the differences in the spatial distribution of small and large individuals. In Guanabara Bay, close to the city of Rio de Janeiro, Oliveira et al. (2016) registered higher burrow density and an overall higher number of adults than juveniles in the dry season (winter and autumn). Maximum size and growth rate were estimated as $L_\infty = 58$ mm and $K = 0.730$, respectively. On the North coast of Rio de Janeiro state, Costa and Zalmon (2019) sampled six beaches with different degrees of urbanization during high and low tourism seasons and found that burrow density and size of *O. quadrata* were mainly related to urbanization and temperature. Furthermore, the number of nocturnal foraging individuals was a relevant metric for beach impact evaluation.

On the north coast of São Paulo state, Negreiros-Fransozo et al. (2002) studied the reproductive cycle and recruitment of *O. quadrata* and found reduced reproduction during winter. Recruitment of young individuals, however, was detected throughout the year (although higher during summer). Turra et al. (2005) assessed the spatial distribution of *O. quadrata* in five low-energy tide-dominated sandy beaches and registered larger burrows toward the upper zones of the beaches. Lower values of density were found in areas with very fine sand and most used by beachgoers. Pombo and Turra (2013) estimated the burrow occupation rate by ghost crabs on nine beaches and found that all burrows less than 10 cm deep were unoccupied, illustrating the possible degree of the unreliability of abundance estimates based on indirect measures. Pombo and Turra (2017) sampled three populations of *O. quadrata* on sandy beaches with contrasting morphodynamics (dissipative, intermediate, and reflective) and found that maximum size (L_∞) and the growth rate increased from dissipative to reflective conditions. Pombo et al. (2018) assessed the activity of *O. quadrata* over daily cycles and registered peaks of activity during both day and night, mostly influenced by tidal rhythms, while Gomes et al. (2019) analyzed the natural diet of *O. quadrata* and found a large number of food items in their stomach, confirming the generalist behavior of this species. Still, on the North coast of São Paulo, Pombo and Turra (2017) sampled nine pristine areas with distinct morphodynamics and wave exposure during 1 year and found that crab density and mean size were higher in sites with smaller grain sizes and steeper slopes. Moreover, they found that larger crabs were more present on upper zones of the beaches and that storms were an important factor in regulating ghost crab populations, causing individuals to contract their distribution landward and increasing the burrow occupation rate. The response of *O. quadrata* to storms was also studied on the central coast of São Paulo by Oliveira and Yokoyama (2021), which reported lower density and distribution of burrows after storms.

In South Brazil, Rosa and Borzone (2008) examined the spatial distribution of *O. quadrata* on 13 sandy beaches of the Paranaguá Bay Estuarine Complex, Paraná. Overall, burrow densities were similar to those observed in oceanic beaches, ranging from 0 to 5.7 burrows/m^2. Yet, no burrow was recorded on the four innermost beaches, indicating that low salinity may prevent the occurrence of ghost crabs. Also, they found strong seasonal variability in burrow densities, with lower densities during the winter. In the Rio Grande do Sul, Neves and Bemvenuti (2006) compared the density and across-shore distribution of ghost crab *O. quadrata* burrows at three beaches with different degrees of anthropic impact and found lower density at beaches with higher anthropic impact. Burrows along the most urbanized beach were also concentrated in areas protected from traffic.

6.3.2 Emerita brasiliensis

The mole crab *Emerita brasiliensis* (Fig. 6.9) is commonly found along the Atlantic coast of South America, from Rio de Janeiro (Brazil) to Uruguay, where it inhabits the intertidal zone of reflective, intermediate, and dissipative beaches (Defeo and Cardoso 2002). *Emerita brasiliensis,* popularly known in Brazil as "tatuí" or "tatuíra," is a filter-feeding species that may reach high values of biomass and production, being an important component of the trophic web of sandy beaches ecosystems (Cardoso et al. 2003; Petracco et al. 2003, 2012, 2013).

Studies performed on the Brazilian coast have shown that the population biology of *E. brasiliensis* is influenced by beach morphodynamic features and latitudinal patterns. Cardoso et al. (2003) tested the influence of beach morphodynamics on the population biology of *E. brasiliensis* on two sandy beaches with different morphodynamic characteristics (one intermediate and one reflective) on the Rio de Janeiro coast and found higher mortality and lower recruitment and growth rates in the more reflective beach. Similarly, Petracco et al. (2017) showed that secondary production and turnover rates (P/B) of *E. brasiliensis* are negatively related to grain size (mm) and positively related to beach width. Furthermore, higher recruitment and growth rates (K) were found in more dissipative conditions.

Latitudinal variation in population parameters of *E. brasiliensis* was assessed by Defeo and Cardoso (2002). By examining the population biology of *E. brasiliensis*

Fig. 6.9 *Emerita brasiliensis*. (Photo: Cláudio Dias Timm. Source: inaturalist.ca)

along the entire range of its distribution, they found clear geographical patterns such as a shift from continuous to seasonal reproduction and recruitment from subtropical to temperate sandy beaches. Moreover, life span was lower and mortality higher on more tropical sandy beaches, whereas individual biomass increased from subtropical to temperate beaches.

6.3.3 Anomalocardia Brasiliana *(*flexuosa*)*

The venerid clam *Anomalocardia brasiliana* (Fig. 6.10), also known as "vôngole" or "berbigão," occurs along the whole Brazilian coast and inhabits a wide range of beach morphodynamic states, from intermediate beaches to tidal flats (Chap. 1). This species lives shallowly burrowed in the intertidal region and has great economic importance in Brazil (Corte et al. 2015; Pezzuto et al. 2010). Due to its widespread occurrence and economic importance, studies focusing on the population biology of *A. brasiliana* have been conducted in three of the four Brazilian coastal geographical regions.

On the coast of Rio Grande do Norte, Northeast Brazil, Rodrigues et al. (2013) found that *A. brasiliana* may reach densities up to 1800 ind/m^2 and show continuous reproduction and recruitment. Similar results were recorded by Rocha-Barreira and Araújo (2005), at Ceará, and Luz and Boehs, at Bahia, who also registered a continuous reproductive cycle in *A. brasiliana* populations. In Pernambuco, Silva-Cavalcanti et al. (2018) found more intense recruitment during summer and a lower density of individuals in the rainy season (0–1170 ind/m^2) when compared to the dry season (0–5798 ind/m^2).

In Rio de Janeiro, Southeast Brazil, Mattos and Cardoso (2012) compared the population parameters of *A. brasiliana* with another suspension-feeding bivalve, the ungulinid *Diplodonta punctata*. The authors found that *A. brasiliana* showed higher abundance, growth rate, and production, whereas mortality and turnover rates were higher for *D. punctata*. Moreover, *A. brasiliana* was predominantly found on the

Fig. 6.10 *Anomalocardia brasiliana (flexuosa)*. (Photo: Alvaro E. Migotto. Source: cifonauta.cebimar.usp.br)

upper levels of the intertidal zone, while *D. punctata* was restricted to the lower levels, indicating that the two species show different abilities to exploit resources. On the north coast of São Paulo, the population biology of *A. brasiliana* was investigated by Corte et al. (2015) and Corte et al. (2014, 2015, 2017) on sandy beaches with different morphodynamics (one intermediate beach and one tidal flat). Significant differences were recorded between populations and mainly attributed to beach morphodynamics. The population from the tidal flat reached a higher size (L_∞) and total biomass; however, recruitment, growth rate (K), production, turnover rate (P/B), and reproductive effort were higher in the population from the intermediate beach.

The population biology of *A. brasiliana* was also investigated in a tidal flat on the coast of Paraná state, South Brazil. Boehs et al. (2008) found higher abundance and recruitment of *A. brasiliana* on the higher levels of the flat. The population density was lower during summer, likely due to higher predation rates and pluviosity.

The comparison of population parameters of *A. brasiliana* along the Brazilian coast suggests a latitudinal pattern for its reproductive cycle, with resting stages occurring only at high-latitude sites (Corte 2015). Nevertheless, other population parameters, such as abundance, productivity, and growth rate, are likely more influenced by local characteristics, such as sediment type and salinity, than large-scale geographical patterns.

6.3.4 Callichirus major

The ghost shrimp *Callichirus major* (Family Callianassidae, Fig. 6.11), popularly known as "corrupto" in Brazil, is a solitary burrowing shrimp distributed from North Carolina, USA, to Santa Catarina, southern Brazil (Botter-Carvalho et al. 2007). It is generally found below the mean water level of sheltered and dissipative sandy beaches, where it builds deep burrows in the sediment.

Callichirus major has great ecological importance in sandy beach ecosystems as it increases the cycling of organic matter and nutrients through bioturbation and favors the establishment of other species in its galleries, such as bivalves, polychaetes, and other crustaceans (Souza and Borzone 2003). Along the Brazilian coast, *C. major* is commonly used as bait for fishing (Botter-Carvalho et al. 2007), with its catch being estimated as 10% of the population stock (Souza and Borzone 2003).

Average densities of *C. major* along the Brazilian coast range from <0.1 (Rio de Janeiro: Costa et al. 2022) to >42 burrows/m^2 (Rio de Janeiro: Simão and Soares-Gomes 2017). Seasonal variations in burrow density have been recorded in Northeast (Rosa et al. 2018) and South Brazil (Souza et al. 1998; Souza and Borzone 2003), with no clear latitudinal pattern. On the Rio de Janeiro coast, Southeast Brazil, Costa et al. (2022) sampled 11 beach sites and found higher abundance and body size in sites with smaller sediment particle size and well-sorted sediments. The body size of *C. major* was larger in areas close to major rivers, probably due to greater

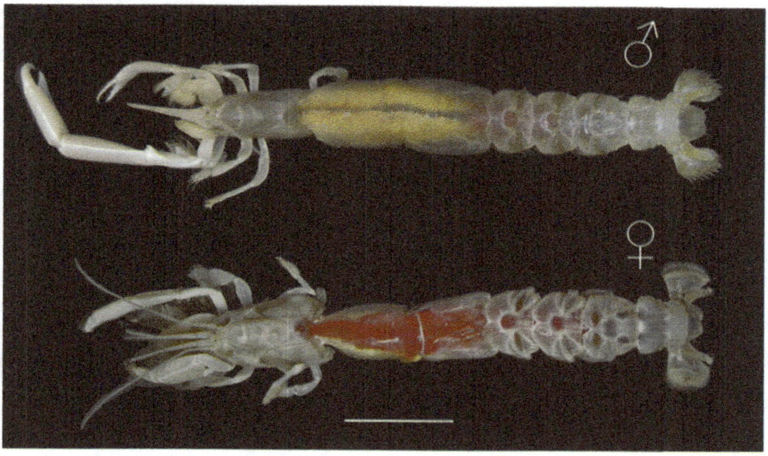

Fig. 6.11 *Callichirus major*. (Reprinted from Scientia Marina, 83, 2019, "Relative growth, sexual maturity and handedness in the ghost shrimp *Callichirus major* (Decapoda: Callianassidae) from the southwestern Atlantic," Copyright (c) 2011 Consejo Superior de Investigaciones Científicas (CSIC). Licensed under a Creative Commons Attribution 4.0 International (CC BY 4.0) License. Scale bar = 1 cm)

trophic subsidies. Similarly, Laurino et al. (2020) recorded a higher burrow abundance of *C. major* in beaches with very fine sand and low variability of tide amplitude, suggesting that tidal regime and sediment type are stronger drivers of *C. major* density than temperature and precipitation variations. Besides environmental characteristics, the intensity of exploitation is also a major factor driving changes in the abundance of *C. major*, leading to a decline in the species' population in many sites along the Brazilian coast (Moschetto et al. 2020).

While beach morphodynamics features seem to be the main environmental driver of changes in abundance and size of *C. major*, variations in temperature and rainfall appear as important factors controlling its reproductive biology. Studies performed in the Northeast (e.g., Botter-Carvalho et al. 2007; Alves-Júnior et al. 2014; Rosa et al. 2018), Southeast (Peiró et al. 2014; Moschetto et al. 2020), and South (Souza et al. 1998) regions of Brazil have consistently reported more intense reproductive periods of *C. major* (i.e., frequency of ovigerous females) during summer.

6.3.5 Olivella minuta

Olivella minuta (Fig. 6.12) is one of the most abundant sandy beach gastropods along the western Atlantic, inhabiting the swash zone and subtidal of dissipative and intermediate beaches from Texas, USA, to Santa Catarina, Southern Brazil (Checon et al. 2021). This snail remains semi-buried, moving by superficial excavation, and

Fig. 6.12 *Olivella minuta.* (Photo: Alvaro E. Migotto. Source: cifonauta.cebimar.usp.br)

feeds on a large variety of food items, including algae, small mollusks, crustaceans, and polychaetes (Checon et al. 2021; Corte et al. 2019b).

Araújo and Rocha-Barreira (2012) studied the population biology of *O. minuta* for 2 years in Ceará, Northeastern Brazil, and found mean densities of almost 80 ind/m^2. Total maximum length ($L_\infty \approx 13$ mm), growth rate ($k = 0.7$), longevity ($t$max ≈ 4 years), and turnover rate (P/B ≈ 1.7) were similar for both years. Growth was constant over the period, with no significant seasonal oscillation. The population biology of *O. minuta* was also assessed over 2 years on two sandy beaches on the coast of São Paulo, Southeast Brazil, by Petracco et al. (2014). Total maximum length ($L_\infty \approx 11$ mm) and longevity ($t_{max} < 2$ years) were similar at both beaches but smaller than registered in Ceará by Araújo and Rocha-Barreira (2012). Conversely, the growth rate was higher in São Paulo ($k > 1$). The abundance of individuals and turnover rates differed significantly between the two areas analyzed in São Paulo (Pernambuco Beach: 3295.3 IST and P/B = 1.91; Barequeçaba; 120 IST and P/B = 1.06), likely due to higher environmental stability at the more sheltered environment (Pernambuco beach). The size structure of both populations was also different, with a larger number of young individuals being recorded at Pernambuco beach throughout the year. Similar results were recorded by Corte et al. (2020), who also investigated the across-shore distribution of *O. minuta* on Pernambuco and Barequeçaba beaches. As registered by Petracco et al. (2014), Corte et al. (2020) found a higher abundance and smaller size of *O. minuta* individuals in the more sheltered area. Moreover, by sampling the intertidal and subtidal zones of both areas, the authors found that individuals were more homogeneously distributed in the across-shore gradient of Pernambuco beach, with a similar number of individuals inhabiting both intertidal and subtidal zones. On the other hand, most *O. minuta* individuals from Barequeçaba beach were found in the subtidal zone (especially young ones).

The generalist feeding behavior of *O. minuta* was demonstrated by Corte et al. (2019a, b) and Checon et al. (2020), which found >40 food items in its stomach. Moreover, through field and laboratory experiments, Checon et al. (2020, 2021) showed that the tidal cycle and light availability affect the feeding activity of

O. minuta. During conditions of light availability, *O. minuta* is more active during low tides (emerged). However, a higher foraging rate was registered when individuals were immersed in lightless conditions. According to the authors, enhanced activity during dark immersed periods is linked to higher food resources brought by the tide and avoidance of visual predators such as fish and birds.

Despite its wide distribution and high abundance on Brazilian sandy beaches, the reproductive biology of *O. minuta* remains largely overlooked. Species belonging to the genus *Olivella* usually encase their eggs in dome-shaped capsules and lay their egg capsules in hard substrates found in sandy beach environments, such as small stones or fragments of bivalve shells (Araújo and Rocha-Barreira 2012; Matthews-Cascon et al. 2011). Recently, however, Corte et al. (2019a, b) reported a massive event of *O. minuta* egg-capsule deposition on the shells of living, buried specimens of the trigonal clam *Tivela mactroides*, emphasizing the importance of biological interactions in sandy beach ecosystems.

6.4 Research Gaps and Avenues for Future Studies

Brazilian sandy beaches host a large number of species that play a key role in ecosystem functioning and the support of human populations. Despite the increasing number of studies investigating the population biology of species inhabiting Brazilian sandy shores (Fig. 6.7), the available information on this topic is still limited and precludes the correct understanding of Brazilian beach biodiversity. For example, we gathered data on population attributes of less than 40 species inhabiting the Brazilian coast; however, Corte et al. (2022) registered more than 170 macrobenthic species on only 30 sandy beaches of the North coast of São Paulo. Moreover, of the approximately 120 studies compiled in Table 6.1, 64 (53%) are about crustaceans' species, while only nine studies (7%) investigate the population biology of annelids, the taxa with the larger number of species in sandy beach ecosystems (Chap. 4). Given the large number and variety of beaches resulting from the large latitudinal extent and the influence of various tidal and climatic patterns along the Brazilian coast (Chap. 1), it is reasonable to infer that only a small fraction of the populations inhabiting the Brazilian sandy shores have been studied. This reduced amount of information gets even smaller when we consider that most studies focused on the population biology of macrobenthic invertebrates. Information on population parameters of important ecological and economic groups such as meiofaunal organisms (Chap. 3) and vertebrates (Chap. 5) is absent for almost all species.

While an increasing number of studies have been assessing how anthropic impacts such as vehicle traffic and tourism affects the abundance of Brazilian sandy beach populations (mostly crustaceans), the number of investigations on the reproductive biology and secondary production of Brazilian species is still low. Since this knowledge is essential to support the correct management of harvesting practices and the development of mariculture of exploited species, future studies must investigate these parameters in a larger pool of species and beach environments.

Studies on population biology have been conducted on sandy beaches in the four coastal geographical regions of Brazil; however, the number of publications relating to each of the regions is very unbalanced. The vast majority of studies were conducted in the southeastern and southern regions, while very few investigations were carried out in the northern region. When compared to the other coastal regions of Brazil, performing research on sandy beaches of the northern coast is considerably more challenging due to their difficult accessibility (most beaches are accessed only by boat), high intertidal amplitude – which results in beaches with extremely wide intertidal zones (often >500 m), and relatively lower number of universities and research centers. Nevertheless, the northern region presents environmental characteristics that differentiate it from the rest of the country, such as the influence of a macrotidal regime, a low degree of human influence on sandy beaches, and the massive fluvial discharge of freshwater from the Amazon River plume. This environmental uniqueness of the northern Brazilian coast highlights the need for ecological investigations on the population of local sandy beaches, a need further intensified by the high erosive rates which threaten the northern beaches of Brazil (~65% of its coastline retreating – the higher rate along the whole Brazilian coast).

A better comprehension of the population biology of a large number of species and in a wider latitudinal range is important to better know the dynamics of Brazilian sandy beaches biodiversity and support the management of harvesting practices. It is also very important to assess the potential effects of climate change. With global temperatures rising, a poleward shift in the distribution of many organisms is expected and indeed being observed already (Defeo et al. 2021). For example, these climate-related stressors have already caused mass mortalities events and shifts in the distribution of sandy beach species in South Brazil, such as the yellow clam *Mesodesma mactroides* (Defeo et al. 2021), and are likely to affect other species along the Brazilian coast. Hindering this situation, the temperature of the southwestern Atlantic Ocean is rising at several times the average global rate, shifting the tropical Brazilian Current poleward and increasing the speed and frequency of onshore winds and storm surges (Franco et al. 2020; Gianelli et al. 2021).

Recent investigations suggest that coastal biodiversity is reorganizing under the influence of climate and anthropic-driven pressures. However, the lack of long-term data precludes the assessment of changes in the population parameters of Brazilian sandy beaches. All population studies on Brazilian beaches had a duration of 3 years or less (most with 1 year), highlighting the sparse temporal replication of studies conducted on Brazilian beaches. Long-term data series are therefore necessary to understand the breadth of the potential climate- and anthropic-driven changes in Brazilian sandy beach populations. Moreover, important processes related to sandy beach populations, such as recruitment, zonation, and intra- and inter-specific interactions, may change at intervals of years or decades, and only from long-term monitoring can we better understand them (Turra et al. 2013; Amaral et al. 2016).

Besides expanding the latitudinal and temporal reach of studies conducted on Brazilian sandy shores, future population studies should also attempt to sample different zones in the across-shore gradient. Most population studies done on Brazilian sandy beaches have concentrated their sampling on the upper zones (i.e., supralittoral and intertidal), while the sublittoral is often overlooked. Recent investigations, however, have shown that young individuals of "intertidal species," such as the clam *Tivela mactroides* and the sea snail *Olivella minuta*, may be more present in the subtidal (Turra et al. 2014; Corte et al. 2019a). By restricting population assessments to only one zone of the beach ecosystem, measurements of abundance and size structure of sandy beach populations are likely to be biased, thereby compromising the correct estimation of recruitment, growth, and production rates.

Finally, the use of recent technologies can be a valuable addition to better characterize and understand the dynamics of sandy beach populations. Recently, Shah Esmaeili et al. (2021) used surf zone Baited Remote Underwater Video Stations (surf-BRUVS) to sample the biodiversity of shallow surf zone fish communities, a poorly investigated component of Brazilian biodiversity. Molecular techniques, such as environmental DNA (eDNA) and analysis of mitochondrial genes, may also help us better understand the connectivity among sandy beach populations and unveil their phylogeographical patterns (e.g., Mattos et al. 2019).

6.5 Final Remarks

Population biology can be viewed as a record of major events relating to its growth, reproduction, and survival. Understanding how and why these parameters change over time is of utmost importance to preserve the functioning of natural ecosystems and manage the stocks of commercially exploited species. Over the past few decades, Brazilian researchers have published a significant amount of information on population attributes of local sandy beach species; however, most of this knowledge is limited to a small number of species. There is also a significant time and geographical constraint, with most studies performed for 1 year or less in the Southeast and South region of Brazil. On the other hand, populations inhabiting the unique sandy beaches of the North region remain largely unexplored. Since life histories vary tremendously from one species to the next and are strongly influenced by environmental features, future studies should investigate a large set of species on different regions of the Brazilian coast. Furthermore, it is also essential to establish long-term monitoring studies so that variations in population parameters of Brazilian sandy beaches related to anthropic actions and climate change can be perceived and predicted. Only through a more comprehensive effort along the country's coast can we better understand the natural dynamic of Brazilian sandy beach populations and manage the stock of ecological and economically important species.

Acknowledgments This work was supported by the Fundação de Amparo à Pesquisa do Estado de São Paulo (FAPESP) (GNC, 2017/17071-9) and Conselho Nacional de Desenvolvimento Científico e Tecnológico (CNPq) (GNC, 165320/2020-6).

References

Abrahão JR, Cardoso RS, Yokoyama LQ, Amaral ACZ (2010) Population biology and secondary production of the stout razor clam Tagelus plebeius (Bivalvia, Solecurtidae) on a sandflat in southeastern Brazil. Zoologia 27:54–64

Alberto RMF, Fontoura NF (1999) Distribuição e estrutura etária de *Ocypode quadrata* (Fabricius, 1787) (Crustacea, Decapoda, Ocypodidae) em praia arenosa do litoral sul do Brasil. Rev Bras Biol 59(1):95–108

Allen KR (1971) Relation between production and biomass. J Fish Res Board Can 28(10):1573–1581

Alves-Júnior FA, Araújo MSLC, Coelho PA (2014) Population biology of Callichirus major (Say, 1818) (Crustacea: Callianassidae) at Piedade Beach, Brazil. Revista de Gestão Costeira Integrada 14(1):109–118. https://doi.org/10.5894/rgci435

Alves-Júnior FA, Santana JL, Figueiredo AMF, Araújo, MSLC (2018) Population and reproductive biology of *Callichirus major* (Say, 1818) (Decapoda: Axiidae: Callianassidae) in an urban beach, Northeastern Brazil. PANAMJAS 13(2):166–178

Amaral ACZ, Corte GN, Filho JSR et al (2016) Brazilian sandy beaches: characteristics, ecosystem services, impacts, knowledge and priorities. Braz J Oceanogr 74(Special Issue 2):5–16

Araújo PH, Rocha-Barreira C (2012) Population dynamic and secondary production of *Olivella minuta* (Gastropoda: Olividae) on sandy beach in northeastern Brazil environmental influences on seagrass ecosystems of the semiarid coast of Brazil. Amici Molluscarum 20(1):7–15

Araujo CCV, Rosa DM, Fernandes JM (2008) Densidade e distribuição espacial do caranguejo *Ocypode quadrata* (Fabricius, 1787) (Crustacea, Ocypodidae) em três praias arenosas do Espírito Santo, Brasil. Biotemas 21(4):73–80

Baptista-Metri C, Pinheiro MAA, Blankensteyn A et al (2005) Biologia populacional e reprodutiva de *Callinectes danae* Smith (Crustacea, Portunidae), no Balneário Shangri-lá, Pontal do Paraná, Paraná, Brasil. Rev Bras Zool 22(2):446–453

Barboza CAM, Mattos G, Soares-Gomes A et al (2021) Low densities of the ghost crab *Ocypode quadrata* related to large scale human modification of sandy shores. Front Mar Sci 8:589542

Benke AC, Huryn AD (2010) Benthic invertebrate production—facilitating answers to ecological riddles in freshwater ecosystems. J North Am Benthol Soc 29(1):264–285

Benke AC, Huryn AD (2017) Secondary production and quantitative food webs. In: Hauer F, Lamberti G (eds) Methods in stream ecology, 3rd edn. Academic Press, London, pp 235–254

Bhattacharya CG (1967) A simple method of resolution of a distribution into Gaussian components. Biometrics 23:115–135

Blankensteyn A (2006) O uso do caranguejo maria-farinha *Ocypode quadrata* (Fabricius) (Crustacea, Ocypodidae) como indicador de impactos antropogênicos em praias arenosas da Ilha de Santa Catarina, Santa Catarina, Brasil. Rev Bras Zool 23:870–876

Boehs G, Absher TM, Cruz-Kaled AC (2008) Ecologia populacional de *Anomalocardia brasiliana* (Gmelin, 1791) (Bivalvia, Veneridae) na Baía de Paranaguá, Paraná, Brasil. Bol Inst Pesca 34(2):259–270

Botter-Carvalho ML, Santos PJP, Carvalho PVVC (2002) Spatial distribution of Callichirus major (Say, 1818) (Decapoda, Callianassidae) on a sandy beach, Piedade, Pernambuco, Brazil. Nauplius 10(2):97–109

Botter-Carvalho ML, Santos PJP, Carvalho PVVC (2007) Population dynamics of *Callichirus major* (Say, 1818) (Crustacea, Thalassinidea) on a beach in northeastern Brazil. Estuar Coast Shelf Sci 71(3–4):508–516

Brey T (1999) A collection of empirical relations for use in ecological modelling. Available at: http://hdl.handle.net/1834/25708

Brey T (2001) Population dynamics in benthic invertebrates. A virtual handbook. Version 01.2. Available at: http://www.thomas-brey.de/science/virtualhandbook

Caetano CHS, Veloso VG, Cardoso RS (2003) Population biology and secondary production of *Olivancillaria vesica vesica* (Gmelin, 1791) (Gastropoda: Olividae) on a sandy beach in southeastern Brazil. J Molluscan Stud 69:67–73

Caetano CHS, Cardoso RS, Veloso VG et al (2006) Population biology and secondary production of *Excirolana braziliensis* (Isopoda: Cirolanidae) in two sandy beaches of southeastern Brazil. J Coast Res 22(4):825–835

Calil P, Borzone CA (2008) Population structure and reproductive biology of *Metamysidopsis neritica* (Crustacea: Mysidacea) in a sand beach in south Brazil. Rev Bras Zool 25(3):403–412

Cardoso RS, Cabrini TMB (2016) Population dynamics and secondary production of gastropods on a sheltered beach in south-eastern Brazil: a comparison between an herbivore and a scavenger. Mar Freshw Res 68(1):87–94

Cardoso RS, Veloso VG (1996) Population biology and secondary production of the sandhopper *Pseudorchestoidea brasiliensis* (Amphipoda: Talitridae) at Prainha Beach, Brazil. Mar Ecol Prog Ser 142:111–119

Cardoso RS, Veloso VG (2003) Population dynamics and secondary production of the wedge clam *Donax hanleyanus* (Bivalvia: Donacidae) on a high-energy, subtropical beach of Brazil. Mar Biol 142:153–162

Cardoso RS, Veloso VG, Caetano CHS (2003) Life history of *Emerita Brasiliensis* (Decapoda: Hippidae) on two beaches with different morphodynamic characteristics. J Coast Res 35:392–401

Cardoso RS, Galhardo LB, Cabrini TMB (2015) Population ecology and secondary production of congeneric bivalves on a sheltered beach in southeastern Brazil. J Shellfish Res 34(3):931–938

Chatzinikolaou E, Richardson CA (2008) Population dynamics and growth of *Nassarius reticulatus* (Gastropoda: Nassariidae) in Rhosneigr (Anglesey, UK). Mar Biol 153(4):605–619

Checon HH, Silva MO, Corte GN et al (2020) Full stomachs at empty tides: tidal cycle affects feeding activity and diet of the sandy beach gastropod *Olivella minuta*. J Molluscan Stud 86(3):219–227

Checon HH, Silva MO, Corte GN et al (2021) Night underwater rides: the activity of a sandy beach gastropod is affected by interactive effects of light availability and water level. Mar Biol Res 17(5–6):523–528

Corte GN (2015) Reproductive cycle and parasitism in the clam *Anomalocardia brasiliana* (Bivalvia: Veneridae). Invertebr Reprod Dev 59(2):66–80

Corte GN, Yokoyama LQ, Amaral ACZ (2014) An attempt to extend the habitat harshness hypothesis to tidal flats: a case study of *Anomalocardia brasiliana* (Bivalvia: Veneridae) reproductive biology. Estuar Coast Shelf Sci 150:136–141

Corte GN, Yokoyama LQ, Coleman RA et al (2015) Population dynamics of the harvested clam *Anomalocardia brasiliana* (Bivalvia: Veneridae) in Cidade Beach, south-east Brazil. J Mar Biol Assoc UK 95(6):1183–1191

Corte GN, Coleman RA, Amaral ACZ (2017) Environmental influence on population dynamics of the bivalve *Anomalocardia brasiliana*. Estuar Coast Shelf Sci 187:241–248

Corte GN, Yokoyama LQ, Checon HH et al (2019a) Spatial and temporal variation in the diet of the sandy beach gastropod *Olivella minuta*. Invertebr Biol 138(4):e12269

Corte GN, Yokoyama LQ, Denadai MR et al (2019b) Egg-capsule deposition of the marine gastropod *Olivella minuta*: the importance of an interspecific relationship with the soft-bottom bivalve Tivela mactroides. J Molluscan Stud 85(1):126–132

Corte GN, Yokoyama LQ, Tardelli DT et al (2020) Spatial patterns of the gastropod *Olivella minuta* reveal the importance of tide-dominated beaches and the subtidal zone for sandy beach populations. Reg Stud Mar Sci 39:101454

Corte GN, Checon HH, Shah EY et al (2022) Evaluation of the effects of urbanization and environmental features on sandy beach macrobenthos highlights the importance of submerged zones. Mar Pollut Bull 182:113962

Costa LL, Zalmon IR (2019) Multiple metrics of the ghost crab *Ocypode quadrata* (Fabricius, 1787) for impact assessments on sandy beaches. Estuar Coast Shelf Sci 218:237–245

Costa LL, Constantino WD, de Aquino Ferreira K et al (2022) Sandy beach ecology in the Anthropocene: ghost shrimp (Crustacea: Callianassidae) as a study model. Estuar Coast Shelf Sci 275:107999

Crisp DJ (1984) Energy flow measurements. In: Holme NA, McIntyre AD (eds) Methods for the study of marine benthos, 2nd edn. Blackwell Scientific Publications, Norfolk, pp 284–372

Cusson M, Bourget E (2005) Global patterns of macroinvertebrate production in marine benthic habitats *. Mar Ecol Prog Ser 297:1–14

Danin APF, Pombo M, Martinelli-Lemos JM et al (2020) Population ecology of the hermit crab *Clibanarius symmetricus* (Anomura: Diogenidae) on an exposed beach of the Brazilian Amazon coast. Reg Stud Mar Sci 33:100944

Defeo O, Cardoso R (2002) Macroecology of population dynamics and life history traits of the mole crab *Emerita brasiliensis* in Atlantic sandy beaches of South America. Mar Ecol Prog Ser 239:169–179

Defeo O, Rueda M (2002) Spatial structure, sampling design and abundance estimates in sandy beach macroinfauna: some warnings and new perspectives. Mar Biol 140(6):1215–1225

Defeo O, McLachlan A, Armitage D et al (2021) Sandy beach social–ecological systems at risk: regime shifts, collapses, and governance challenges. Front Ecol Environ 19(10):564–573

Denadai MR, Amaral ACZ, Turra A (2004) Biology of a tropical intertidal population of *Cerithium atratum* (Born, 1778) (Mollusca, Gastropoda). J Nat Hist 38(13):1695–1710

Denadai MR, le Sueur-Maluf L, Marques CG et al (2015) Reproductive cycle of the trigonal clam *Tivela mactroides* (Bivalvia, Veneridae) in Caraguatatuba Bay, southeastern Brazil. Mar Biol Res 11(8):847–858

Dolbeth M, Cusson M, Sousa R et al (2012) Secondary production as a tool for better understanding of aquatic ecosystems. Can J Fish Aquat Sci 69(7):1230–1253

dos Anjos DWC (2016) Dinâmica populacional e produção secundária de *Donax hanleyanus* (Philippi 1842) (Bivalvia, Donacidae) em uma praia exposta refletiva no litoral sudeste do Brasil. Undergraduate Thesis. Universidade Federal do Pará

Farias MF, Rocha-Barreira CDA (2017) Ciclo reproductivo de *Tagelus plebeius* (bivalvia) en el estuario del río ceará, nordeste do Brasil. Bol Inst Pesca 43(4):578–592

Fonseca DB, Veloso VG, Cardoso RS (2000) Growth, mortality, and reproduction of *Excirolana braziliensis* Richardson, 1912 (Isopoda,Cirolanidae) on the Prainha Beach, Rio De Janeiro, Brazil. Crustaceana 73(5):535–545

Fontoura-da-Silva V, Cardoso RS, Caetano CHS (2016) Mark-recapture versus length-frequency based methods: evaluation using a marine gastropod as a model. J Exp Mar Biol Ecol 474:171–179

Franco BC, Defeo O, Piola AR et al (2020) Climate change impacts on the atmospheric circulation, ocean, and fisheries in the southwest South Atlantic Ocean: a review. Clim Chang 162(4):2359–2377

Garcia AYT, Borzone CA (2015) The reproductive cycle of the sea urchin *Lytechinus variegatus* (Echinodermata: Echinoidea) in southern Brazil. Rev Biol Trop 63:243–250

Gayanilo FC, Sparre P, Pauly D (2005) FAO-ICLARM stock assessment tools II (FISAT user guide). Worldfish Center, Rome

Gianelli I, Ortega L, Pittman J et al (2021) Harnessing scientific and local knowledge to face climate change in small-scale fisheries. Glob Environ Change 68:102253

Gomes TT, Gheler-Costa C, Rinaldi CA et al (2019) Natural diet of *Ocypode quadrata* (Fabricius, 1787) (Crustacea, Decapoda, Brachyura) from the Northern Coast of São Paulo, Brazil. Pap Avulsos Zool 59:e20195957

Gomez J, Defeo O (1999) Life history of the sandhopper *Pseudorchestoidea brasiliensis* (Amphipoda) in sandy beaches with contrasting morphodynamics. Mar Ecol Prog Ser 182:209–220

Hastings A (1996) Population biology: concepts and models, 1st edn. Springer, New York

Keough MJ, Downes BJ (1982) Recruitment of marine invertebrates: the role of active larval choices and early mortality. Oecologia 54:348–352

Laurino IRA, Buchmann FS, Hernáez P (2020) Spatial-temporal distribution of the burrowing shrimp *Callichirus major* (Say, 1818) (Decapoda, Callichiridae) in preserved populations of southeastern Brazil. Thalassas 36(2):333–342

Leão LSD, Soares-Gomes A, Costa T et al (2012) Population dynamics and reproductive strategy of *Scolelepis goodbodyi* (Polychaeta: Spionidae) in a subtropical Atlantic beach. Fortschr Zool 29(3):195–202

Leite FP, Turra A, Souza EC (2003) Population biology and distribution of the tanaid *Kalliapseudes schubarti* Mañé-Garzon, 1949, in an intertidal flat in southeastern Brazil. Braz J Biol 63(3):469–479

Llodra ER (2002) Fecundity and life-history strategies in marine invertebrates. Adv Mar Sci 43:88–154

Luz JR, Boehs G (2011) Reproductive cycle of *Anomalocardia brasiliana* (Mollusca: Bivalvia: Veneridae) in the estuary of the Cachoeira River, Ilhéus, Bahia. Braz J Biol 71(3):679–686

MacCord FS, Amaral ACZ (2007) The reproductive cycle of *Scolelepis goodbodyi* (Polychaeta, Spionidae). Mar Biol 151(3):1009–1020

Magalhães WF, Lima JB, Barros F et al (2009) Is *Ocypode quadrata* (Fabricius, 1787) a useful tool for exposed sandy beaches management in Bahia state (Northeast Brazil)? Braz J Oceanogr 57(2):149–152

Maia-Carneiro T, Dorigo LA, Andrade G et al (2013) Density and spatial distribution of *Ocypode quadrata* (Decapoda, Ocypodidae) in an insular environment in the state of Rio de Janeiro, southeastern Brazil. Rev Bras Zoocienc 15(1):91–96

Manoel F (2008) Biologia Populacional do caranguejo Austinixa patagoniensis na praia Suja da Ilha da Marambaia. Undergraduate Thesis. Universidade Federal do Estado do Rio de Janeiro

Matthews-Cascon H, Rocha-Barreira CA, Meirelles CAO (2011) Egg masses of some Brazilian Mollusks. Expressão Gráfica e Editora, Fortaleza

Mattos G, Cardoso RS (2012) Population dynamics of two suspension-feeding bivalves on a sheltered beach in southeastern Brazil. Helgol Mar Res 66:393–400

Mattos G, Seixas VC, Paiva PC (2019) Comparative phylogeography and genetic connectivity of two crustacean species with contrasting life histories on South Atlantic sandy beaches. Hydrobiologia 826(1):319–330

Mcallan J (2006) Dinâmica Populacional de *Callichirus major* (Decapoda – Thalassinidae). Undergraduate Thesis. Universidade Federal do Estado do Rio de Janeiro

Morsan EM, Kroeck MA (2005) Reproductive cycle of purple clam, *Amiantis purpurata* (Bivalvia: Veneridae) in northern Patagonia (Argentina). J Mar Biol Assoc UK 85:367–373

Moschetto FA, Borges RP, Duarte LFA (2020) Population structure of *Callichirus major* (Say 1818) (Crustacea: Callianassidae) and conservation considerations at Southeast coast of São Paulo, Brazil. An Acad Bras Cienc 92(1):1–13

Nakagaki JM, Pinheiro MAA (1999) Biologia populacional de *Emerita brasiliensis* Schmitt (Crustacea, Hippidae) na Praia Vermelha do Norte, Ubatuba (São Paulo, Brasil). Rev Bras Zool 16(2):83–90

Narchi W (1976) Ciclo anual da gametogênese de *Anomalocardia brasiliana* (Gmelin, 1791) (Mollusca, Bivalvia). Bol Zool 1:331–350

Neal D (2019) Introduction to population biology, 2nd edn. Cambridge University Press, Cambridge

Negreiros-Fransozo ML, Fransozo A, Bertini G (2002) Reproductive cycle and recruitment period of *Ocypode quadrata* (Decapoda, Ocypodidae) at a sandy beach in southeastern. J Crust Biol 22(1):157–161

Neves FM, Bemvenuti CE (2006) The ghost crab *Ocypode quadrata* (Fabricius, 1787) as a potential indicator of anthropic impact along the Rio Grande do Sul coast, Brazil. Biol Conserv 133(4):431–435

Oliveira FRF, Yokoyama LQ (2021) Response of *Ocypode quadrata* to storm waves on an urbanized sandy beach. Ocean Coast Res 69:e21005

Oliveira CAG, Souza GN, Soares-Gomes A (2016) Measuring burrows as a feasible non-destructive method for studying the population dynamics of ghost crabs. Mar Biodivers 46(4):809–817

Omena EP, Amaral ACZ (2000) Population dynamics and secondary production of *Laeonereis acuta* (Treadwell, 1923) (Nereididae: Polychaeta). Bull Mar Sci 67:421–431

Otegui MBP, Blankensteyn A, Pagliosa PR (2012) Population structure, growth and production of *Thoracophelia furcifera* (Polychaeta: Opheliidae) on a sandy beach in southern Brazil. Helgol Mar Res 66(4):479–488

Pauly D, Moreau J, Abad N (1995) Comparison of age-structured and length-converted catch curves of brown trout *Salmo trutta* in two French rivers. Fish Res 22(3):197–204

Peiró DF, Wehrtmann IS, Mantelatto FL (2014) Reproductive strategy of the ghost shrimp *Callichirus major* (Crustacea: Axiidea: Callianassidae) from the southwestern Atlantic: sexual maturity of females, fecundity, egg features, and reproductive output. Invertebr Reprod Dev 58(4):294–305

Pennafirme S, Soares-Gomes A (2017) Population dynamics and secondary production of a key benthic tanaidacean, *Monokalliapseudes schubarti* (Mañé-Garzón, 1949) (Tanaidacea, Kalliapseudidae), from a tropical coastal lagoon in southeastern Brazil. Crustaceana 90(11–12):1483–1499

Pennafirme S, Soares-Gomes A (2022) Population dynamics and reproductive biology of the sandhopper *Atlantorchestoidea brasiliensis* (Amphipoda: Talitridae) of a sandy beach in southwestern Atlantic Coast. J Nat Hist 56(13–16):989–1005

Petracco M, Veloso VG, Cardoso RS (2003) Population dynamics and secondary production of *Emerita brasiliensis* (Crustacea: Hippidae) at Prainha. Mar Ecol 24(3):231–245

Petracco M, Cardoso RS, Corbisier TN (2010) Population biology of *Excirolana armata* (Dana, 1853) (Isopoda, Cirolanidae) on an exposed sandy beach in southeastern Brazil. Mar Ecol 31(2):330–340

Petracco M, Cardoso RS, Turra A et al (2012) Production of *Excirolana armata* (Dana, 1853) (Isopoda, Cirolanidae) on an exposed sandy beach in southeastern Brazil. Helgol Mar Res 66(3):265–274

Petracco M, Cardoso RS, Turra A (2013) Patterns of sandy-beach macrofauna production. J Mar Biol Assoc UK 93(7):1717–1725

Petracco M, Camargo RM, Tardelli DT et al (2014) Population biology of the gastropod *Olivella minuta* (Gastropoda, Olividae) on two sheltered beaches in southeastern Brazil. Estuar Coast Shelf Sci 150:149–156

Petracco M, Cardoso RS, Martinelli Filho JE et al (2017) Effects of beach morphodynamic features on production and P/B ratio of the crab *Emerita brasiliensis* Schmitt, 1935 (Decapoda: Hippidae) in sandy beaches of South America. J Mar Biol Assoc UK 97(6):1215–1221

Petracco M, Aviz D, Martinelli Filho JE et al (2019) Effects of physical features on production of three macrofaunal species in different sandy beach zones in South America. Estuar Coast Shelf Sci 218:23–30

Pezzuto PR (1998) Population dynamics of *Sergio mirim* (RODRIGUES 1971) (Decapoda: Thalassinidea: Callianassidae) in Cassino Beach, southern Brazil. Mar Ecol 19(2):89–109

Pezzuto PR, Schio C, Almeida TCM (2010) Efficiency and selectivity of the *Anomalocardia brasiliana* (Mollusca: Veneridae) hand dredge used in southern Brazil. J Mar Biol Assoc UK 90(7):1455–1464

Pombo M, Turra A (2013) Issues to be considered in counting burrows as a measure of Atlantic ghost crab populations, an important bioindicator of sandy beaches. PLoS One 8(12):e83792

Pombo M, Turra A (2017) Variation in the body growth parameters of the ghost crab *Ocypode quadrata* from morphodynamically distinct sandy beaches. Braz J Oceanogr 65(4):656–665

Pombo M, de Oliveira AL, Xavier LY et al (2017) Natural drivers of distribution of ghost crabs *Ocypode quadrata* and the implications of estimates from burrows. Mar Ecol Prog Ser 565:131–147

Pombo M, Campagnoli M, Castilho-Martins EA et al (2018) Continuous, video-recording assessment of daily activity cycle of the ghost crab *Ocypode quadrata* Fabricius, 1787 (Brachyura: Ocypodidae) in southeastern Brazil. J Crust Biol 38(2):133–139

Ramos NIM (2014) Dinâmica populacional e produção secundária de *Talorchestia tucurauna* (Müller, 1864) (Amphipoda: Talitridae) em duas praias do litoral pernambucano. Dissertation. Universidade Federal de Pernambuco

Ricker WE (1977) Computation and interpretation of biological statistics of fish populations. Information Canada, Ottawa

Rocha-Barreira CA (2010) Reproductive cycle of *Olivancillaria vesica auricularia* (Lamarck, 1910) (Mollusca: Gastropoda: Olividae) in southern Brasil. Arq Cienc Mar 43(2):110–118

Rocha-Barreira CA, Araújo MLR (2005) Ciclo reprodutivo de *Anomalocardia brasiliana* (Gmelin, 1791) (Mollusca, Bivalvia, Veneridae) na praia do Canto da Barra, Fortim, Ceará, Brasil. Bol Inst Pesca 31(1):9–20

Rocha-Barreira CA, Batista WF, Monteiro DOM et al (2002) Aspectos da estrutura populacional de *Donax striatus* (Linnaeus, 1758) (Mollusca: Donacidae) na praia do Futuro, Fortaleza – CE. Arq Cienc Mar 35:51–55

Rodrigues A, Borges-Azevedo C, Costa R et al (2013) Population structure of the bivalve *Anomalocardia brasiliana*, (Gmelin, 1791) in the semi-arid estuarine region of northeastern Brazil. Braz J Biol 73(4):819–833

Rosa LC, Borzone CA (2008) Spatial distribution of the *Ocypode quadrata* (Crustacea: Ocypodidae) along estuarine environments in the Paranaguá Bay Complex, southern Brazil. Rev Bras Zool 25(3):383–388

Rosa LC, Freire KMF, Souza MJM (2018) Spatial distribution and population dynamics of *Callichirus major* (Crustacea, Callianassidae) in a tropical sandy beach, northeastern Brazil. Invertebr Biol 137(4):308–318

Salvador L (2001) Malacofauna da região entremarés de praias de areia com fragmentos de rochas: Distribuicão temporal e espacial; dinâmica populacional e produção secundária de *Anomalocardia brasiliana* (Gmelin, 1791) e *Corbula cubaniana* (Orbigny, 1853). PhD thesis. Universidade Estadual Paulista

Santos PJP (1994) Population dynamics and production of *Scolelepis gaucha* (Polychaeta: Spionidae) on the sandv beaches of southern Brazil. Mar Ecol Prog Ser 110:159–165

Santos PJP, Florêncio MS, Florêncio MAP (2003) Environmental control of the reproductive cycle of *Laeonereis acuta* (Annelida; Polychaeta) on a tropical intertidal sandy beach. J Coast Res 35:378–384

Schlacher TA, Schoeman DS, Dugan J et al (2008) Sandy beach ecosystems: key features, sampling issues, management challenges and climate change impacts. Mar Ecol 29:70–90

Shah Esmaeili Y, Corte GN, Checon HH et al (2021) Comprehensive assessment of shallow surf zone fish biodiversity requires a combination of sampling methods. Mar Ecol Prog Ser 667:131–144

Shimizu RM (1997) Ecologia populacional de *Scolelepis squamata* (Muller, 1806) (Polychaeta: Spionidae) e *Callichirus major* (Say, 1818) (Crustacea: Decapoda: Thalassinidae) da Praia de Barequeçaba (São Sebastião, SP). PhD Thesis. Universidade de São Paulo, São Paulo

Silva FMO (2013) Efeitos da coexistência na biologia populacional de isópodes cirolanídeos em praias arenosas. Dissertation, Universidade Federal do Estado do Rio de Janeiro

Silva CF, Corte GN, Yokoyama LQ et al (2015) Growth, mortality, and reproduction of *Tagelus plebeius* (Bivalvia: Solecurtidae) in Southeast Brazil. Helgol Mar Res 69:1–12

Silva-Cavalcanti JS, Costa MF, Alves LHB (2018) Seasonal variation in the abundance and distribution of *Anomalocardia flexuosa* (Mollusca, Bivalvia, Veneridae) in an estuarine intertidal plain. PeerJ 6:e4332

Simão DS, Soares-Gomes A (2017) Population dynamics and secondary production of the ghost shrimp *Callichirus major* (Thalassinidea): a keystone species of Western Atlantic dissipative beaches. Reg Stud Mar Sci 14:34–42

Souza JRB, Borzone CA (2000) Population dynamics and secondary production of Scolelepis squamata (Polychaeta: Spionidae) in an exposed sandy beach, Southern Brazil. Bull Mar Sci 67:221–233

Souza JRB, Borzone CA (2003) A extração de corrupto, *Callichirus major* (Say) (Crustacea, Thalassinidea), para uso como isca em praias do litoral do Paraná: as populações exploradas. Rev Bras Biol 20:625–630

Souza JRB, Borzone CA (2007) Population dynamics and secondary production of Euzonus furciferus Ehlers (Polychaeta, Ophellidae) in an exposed sandy beach of Southern Brazil. Zoologia 24:1139–1144. https://doi.org/10.1590/S0101-81752007000400034

Souza JRB, Borzone CA, Brey T (1998) Population dynamics and secondary production of *Callichirus major* (Crustacea: Thalassinidea) on a southern Brazilian sandy beach. Arch Mar Fish Res 46(2):151–164

Souza JRB, Lavoie N, Bonifácio PH et al (2008) Distribution of *Ocypode quadrata* (Fabricius, 1787) on sandy beaches of northeastern Brazil. Atlantica 30(2):139–145

Souza DGC, Petracco M, Danin APF et al (2021) Population structure and use of space by ghost crabs (Brachyura: Ocypodidae) on an equatorial, macrotidal sandy beach. Estuar Coast Shelf Sci 258:107376

Sparre P, Venema SC (1992) Introduction to tropical fish stock assessment. Part 1. Manual, FAO fisheries technical paper, 306. FAO, Rome

Tavares YAG, Borzone CA (2006) Reproductive cycle of *Mellita quinquiesperforata* (Leske) (Echinodermata, Echinoidea) in two contrasting beach environments. Rev Bras Zool 23(2):573–580

Turra A, Gonçalves MAO, Denadai MR (2005) Spatial distribution of the ghost crab *Ocypode quadrata* in low-energy tide-dominated sandy beaches. J Nat Hist 39(23):2163–2177

Turra A, Cróquer A, Carranza A et al (2013) Global environmental changes: setting priorities for Latin American coastal habitats. Glob Chang Biol 19(7):1965–1969

Turra A, Petracco M, Amaral ACZ et al (2014) Temporal variation in life-history traits of the clam Tivela mactroides (Bivalvia: Veneridae): density-dependent processes in sandy beaches. Estuar Coast Shelf Sci 150(PA):157–164

Turra A, Petracco MM, Amaral ACZ, Denadai MR (2015) Temporal variation in life-history traits of the clam Tivela mactroides (Bivalvia: Veneridae). Estuarine, Coastal and Shelf Science, [s.l.], 150:157–164, Elsevier BV. https://doi.org/10.1016/j.ecss.2013.06.004

Veloso VG, Sallorenzo IA (2010) Differences in the secondary production of *Emerita brasiliensis* (Decapoda: Hippidae) on two sandy beaches in Rio de Janeiro State, Brazil. Nauplius 18(1):57–68

Waters TF (1977) Secondary production in inland waters. Adv Ecol Res 10:91–164

Wilson EO, Bossert WH (1971) A primer of population biology. Oxford University Press, Oxford

Yin D, He F (2014) A simple method for estimating species abundance from occurrence maps. Methods Ecol Evol 5(4):336–343

Yokoyama LQ (2010) Nassarius vibex (Gastropoda, Nassariidae): crescimento e reprodução em bancos de mitilídeos de substrato areno-lamoso na costa sudeste do Brasil. 2010. Tese (Doutorado em Zoologia) - Instituto de Biociências, Universidade de São Paulo, São Paulo, 2010. https://doi.org/10.11606/T.41.2010.tde-05042010-160037. Acesso em: 2023-05-18

Yokoyama LQ, Amaral ACZ (2011) Recruitment and growth variation of *Ophionereis reticulata* (Echinodermata: Ophiuroidea). Invertebr Reprod Dev 55(2):73–81

Yokoyama LQ, Duarte LFL, Amaral ACZ (2008) Reproductive cycle of Ophionereis reticulata (Ophiuroidea, Echinodermata) on the southeast coast of Brazil. Invertebr Reprod Dev, 51:111–118

Zavarize B (2009) Biologia populacional e produção secundária de *Pagurus criniticornis* (Anomura: Decapoda) na praia das Flexeiras, Itacuruçá, RJ, Brasil. Undergraduate Thesis. Universidade Federal do Estado do Rio de Janeiro

Chapter 7
Biological Interactions

Cristina de Almeida Rocha-Barreira and José Souto Rosa-Filho

7.1 Introduction

Understanding the contribution of biological interactions to the regulation of natural communities is a key goal in ecology (Connell and Orias 1964; Hall and Kingsford 2016). In some ecosystems, pressures imposed by physical and chemical factors make it difficult to recognize the role of biological processes in structuring communities. In sandy beaches, for example, the interaction among tides, waves, and sediment size (Defeo and McLachlan 2013) shapes the morphodynamics of these ecosystems, whose effect on the fauna can overlap those of biological interactions.

Sandy beaches are dominant ecosystems on the continental shores of the world and provide habitats for well-adapted and diversified fauna (McLachlan and Defeo 2018). These environments largely lack biogenic structures and are extremely dynamic in space and time. Sandy beach fauna exhibits specific responses to the physical characteristics of the environment, thereby supporting the general validity of the autoecological hypothesis (McLachlan and Defeo 2018). This hypothesis states that biological communities are structured solely by the independent responses of individual species to environmental features.

Although sandy beaches are mostly physically controlled, biological interactions may play an important role in their biodiversity (Defeo et al. 2009; Corte et al. 2022). The intensity of biological interactions increases with decreasing environmental stress; therefore, these processes are more important toward dissipative

C. de Almeida Rocha-Barreira (✉)
Universidade Federal do Ceará, Instituto de Ciências do Mar, Laboratório de Zoobentos.
Av. da Abolição, Fortaleza, CE, Brazil

J. S. Rosa-Filho
Universidade Federal de Pernambuco, Centro de Tecnologia, Departamento de Oceanografia,
Laboratório de Bentos. Av. Prof. Moraes Rego, S/N, Cidade Universitária, Recife, PE, Brazil

The Marine Biological Association, The Laboratory, Plymouth, UK
e-mail: souto.rosa@ufpe.br

A. C. Z. Amaral et al. (eds.), *Brazilian Sandy Beaches*, Brazilian Marine Biodiversity, https://doi.org/10.1007/978-3-031-30746-1_7

beaches (Chap. 1), where conditions are relatively benign, densities of individuals are high, and resources could be limited (Defeo et al. 2003; McLachlan and Dorvlo 2005). Yet, even under stressful conditions, the species inhabiting sandy beaches can establish inter and intraspecific relationships. The importance of biological interactions in structuring macrofaunal communities, however, is extremely difficult to identify on sandy shores, since conducting field manipulative experiments in these naturally dynamic environments is very difficult.

This chapter addresses the role of biological interactions on the biodiversity of Brazilian sandy beaches, mainly in the intertidal zone, considering their effects on trophic relationships, richness, density, and distribution of the main species. We first examine biological interactions between living organisms such as competition, predation, and parasitism. Then we focus on trophic interactions commonly observed in sandy beach ecosystems.

7.2 Biological Interactions

Classically, it is assumed that species coexist in a community and establish numerous relationships, whose limits are not always easy to be established (Dajoz 2005). On sandy beaches, the intensity of biological interactions is inversely related to wave action. Thus, the influence of these interactions on benthic communities of exposed sandy beaches is generally considered negligible (Dugan et al. 2004). On the other hand, on sandy beaches where benthic communities are abundant and strongly aggregated, such as dissipative/sheltered beaches, the potential of biological interactions to control biodiversity distribution can be significant.

Although there are few studies characterizing and evaluating biological interactions on Brazilian sandy shores, some aspects are worth mentioning. Thus far, studies have pointed out that predation and interspecific and intraspecific competition are the main biological interactions modulating the structure and dynamics of benthic infaunal communities on sandy beaches (McLachlan and Jaramillo 1995). Nevertheless, other interactions, such as parasitism, may also influence the population dynamics of sandy beach fauna.

7.2.1 Competition

Competition is an interaction between individuals, brought about by a shared requirement for a resource, which may lead to a reduction in population parameters such as survival, growth, and/or reproduction of at least one of the competitors (Begon et al. 2006). Not always competing individuals interact with one another directly. Instead, individuals may be affected by changes in the amount of a resource, which can be made unavailable due to the presence and activity of other individuals (competition by interference) or be reduced due to consumption by other individuals or species (competition by exploitation) (Levine 1976).

The typical high hydrodynamic of sandy beaches makes them environments with characteristically low in situ primary production. As a consequence, sandy beach food webs are supported primarily by allochthonous supply from adjacent coastal habitats (see Michaud et al. 2019). The low productivity or the erratic and irregular supply of resources may turn food a limiting factor during some periods (e.g., during low spring tides), which may enhance competition (interspecific and/or intraspecific). This competition may be especially important in the upper zones of sandy beaches, where the accumulation of stranded macroalgae and other organic debris is even more unpredictable (Duarte et al. 2010).

These potential intra- and interspecific interactions can play an important role in structuring sandy beach populations and communities (Defeo et al. 1997). Cardoso (2002) studied the behavior and surface activity of the talitrid *Atlantorchestoidea* (= *Pseudorchestoidea*) *brasiliensis* on a beach in southeastern Brazil and recorded that while adults had nocturnal activity, most juveniles were active at sunset and sunrise. This behavior was influenced by environmental factors, such as temperature and humidity, and competition for food, given that both adults and juveniles feed on the same food source.

The cirolanid isopods *Excirolana armata* and *Excirolana braziliensis* have a wide geographical distribution along the coast of America and coexist along the Atlantic coast of South America (Dexter 1977; Castro and Brum 1969; Queiroz and Rocha-Barreira 2019). In Uruguay, Defeo et al. (1997) observed that *E. armata* only occurred on beaches with fine sands, whereas *E. braziliensis* was present on both fine and coarse sand beaches, with the highest densities in the latter. When occurring in sympatry on fine sands, both species showed maximum densities at different tidal zones, with *E. braziliensis* restricted to the upper reaches of the beach. This co-occurrence of both isopods scavengers at the intertidal zone demonstrates that partitioning the available space may be a way of reducing or avoiding potential competition of the carrion.

On dissipative sandy beaches, characterized by gentle slopes and fine sands, such as those between southeastern Brazil and northern Argentina, intertidal filter feeders species can compete for food when dense patches are established. Some studies have observed negative correlations between the density and growth rate of species due to competition (Petracco et al. 2012). The mole crab *Emerita brasiliensis* and the bivalves *Donax hanleyanus* and *Amarilladesma mactroides* occur at similar sandy beach zones and depths in the sediment, thus an intense competitive interaction for food and space can occur (Penchaszadeh and Olivier 1975; Defeo et al. 1986). After mass mortalities that sequentially occurred southward between 1993 (southern Brazil) and 2002 (Argentina), and Defeo (2003) recorded that the yellow clam *A. mactroides* was replaced by their competitors (i.e., *Emerita brasiliensis* and *Donax hanleyanus*). Recently, competition for space has also been observed between *A. mactroides* and the pea crab *Austinixa patagoniensis*, a symbiotic crab that lives in the burrows of the ghost shrimp *Sergio mirim* in southern Brazil (Carcedo et al.

2019). The high densities of *A. mactroides* reduce the area for establishment and maintenance of the ghost shrimp burrows, and this process indirectly harms the foraging behavior of the pea crab that feeds on suspended material in the proximity of these burrows.

7.2.2 Predation

Predation is the consumption of one organism by another, in which the prey is alive when attacked (Begon et al. 2006). The importance of these biological interactions has been highlighted in different coastal marine ecosystems such as rocky shores (Connell 1972; Coutinho et al. 2016), seagrass meadows (Siebert and Branch 2007; Copertino et al. 2016), seaweed beds (O'Brien et al. 2018), and coral reefs (Hall and Kingsford 2016). However, the regulation of prey populations by predators has not been well demonstrated in sandy beach ecosystems due to the physical harshness of these environments, which makes it extremely difficult to evaluate the role of predation through both observational and experimental investigations. Nevertheless, a few studies have shown that predation influence may vary with beach exposure, prey productivity, and environmental stress (McLachlan and Defeo 2018).

Three main groups of predators are present on sandy beaches: birds, arachnids, and insects from land, fishes from the sea, and resident invertebrates, notably crabs, gastropods, asteroids, and polychaetes (Pinotti et al. 2014; McLachlan and Defeo 2018). Gastropods are among the most important sandy beach predators. They usually inhabit the surf zone and shallow infralittoral of sandy beaches and make incursions into the washing zone and lower intertidal to prey upon suspension feeders during high tide. The lateral expansions of the mesopodium (the middle portion of the foot) extend and make wave movements making the animal slide over the water following the movements of the waves rising and falling (Gianuca 1983, 1985, 1991). Rocha-Barreira (2002) studied the feeding behavior of *Olivancillaria vesica auricularia* (Fig. 7.1a) on Cassino Beach (state of Rio Grande do Sul, Southern Brazil) and found that it feeds upon the bivalves *Donax hanleyanus* and *Amarilladesma mactroides* and the mole crab *Emerita brasiliensis*, the latter being the most predated, especially the largest ovigerous females (Rocha-Barreira 2002). The gastropod *Olivella minuta*, studied by several authors (e.g., Marcus and Marcus 1959; Arruda et al. 2003; Corte et al. 2019; Checon et al. 2020) on the state of São Paulo beaches, is a nonselective opportunistic species that may prey upon bivalves, foraminiferans, copepods, amphipods, scaphopods, diatoms, nematodes, and crustaceans. Both *Olivancillaria* and *Olivella* involve the prey with its foot and produce thick mucus capable of immobilizing it; in sequence, muscular movements of metapodium kill the prey by asphyxia. After the prey's death, the predator extends its proboscis and starts to ingest its tissues (Rocha-Barreira 2002).

The gastropod *Hastula cinerea* surfs expanding the foot as an underwater sail and migrates up and down the shore with the tides to find food resources. Unlike olivids, *Hastula* is a specialized predator that uses its toxoglossan radula, with a

Fig. 7.1 Examples of carnivores, scavengers, and detritivores: (**a**): *Olivancillaria vesica auricullaria*, (**b**): *Ocypode quadrata*, (**c**): *Nassarius vibex*, and (**d**): *Mellita quinquesperforata*. (Photos: Cristina Rocha Barreira)

modified harpoon-like format, to inoculate poison in the preys (generally polychaetes living in the lower intertidal zone) (Viana et al. 2005).

Naticidae gastropods feed by making holes in the prey shells, using the radula and secretions produced by the accessory boring organ (ABO) (Hughes 1986). In the intertidal zone of Brazilian sandy beaches, species of *Polinices* and *Natica* have been mentioned as predators of bivalves and other gastropods (Couto 1996; Rocha-Barreira et al. 2017; Chagas and Herrmann 2015). Martins (1996) compared laboratory experiments and field observations, citing that *Natica marochiensis* eats 17 different mollusk species on Quitéria beach (state of Ceará coast). This author observed that the selection of the prey is related to the prey population densities, with the bivalves *Tivela mactroides, Mulinia cleyriana, Strigilla pisiformes,* and *Donax striatus* being the most predated.

Veloso et al. (2006) reported that the ghost crab *Ocypode quadrata* has nocturnal habits and preys upon sandy beach species such as the bivalve *Donax hanleyanus* and the mole crab *Emerita brasiliensis* on Rio de Janeiro beaches. This species has a flexible feeding behavior, is a skilled scavenger, and can withstand starvation for long periods inside the burrow. The analyses of the stomach contents of *O. quadrata* in Santa Catarina (Branco et al. 2010) and São Paulo (Gomes et al. 2019) showed insects and other crustaceans as the most abundant items (Fig. 7.1b). Cannibalism, especially in juvenile, molting, or diseased individuals, was also observed by the authors.

The swimming crab *Arenaeus cribarius* lives superficially buried in the internal surf zone on Brazilian sandy beaches (Gianuca 1985). It has a generalist diet, feeding mainly on other crustaceans and fishes (Pinheiro and Pardal-Souza 2018). On the coast of Rio Grande do Sul, *A. cribarius* feeds on the mole crab *Emerita brasiliensis*, juveniles of the bivalves *Amarilladesma mactroides, Donax hanleyanus,* and *Donax gemmula* and also polychaetes and juvenile teleost fishes (Gianuca 1985). Rocha-Barreira (2003) also reported capsules of the gastropod *Olivancillaria vesica auricularia* consumed by *A. cribarius*.

The polychaetes *Sigalion cirrifer*, *Hemipodia californiensis*, and *Nephtys simoni* are frequent in the swash/surf-zone interface on sandy beaches, preying actively upon the meiofauna (e.g., nemerteans, ostracods, and harpacticoid copepods) or ingesting post-settled recruits of the macrofauna (e.g., bivalves, polychaetes, and crustaceans), especially during recruitment peaks (Gianuca 1997; Neves et al. 2007).

Besides invertebrates, fishes and birds are important predators on sandy beaches (Dugan et al. 2003). Predator fishes play an important role in regulating the densities of intertidal macroinvertebrates and in transferring energy to higher trophic levels (Nelson 1986; Veloso et al. 2003; Tomme et al. 2014). Most beach predators are generalists that consume both zooplankton (fish and invertebrate larvae and juveniles) and intertidal and subtidal benthic macrofauna (demersal benthic feeders).

Sandy beaches provide refuge against predators and food for juvenile fish (Félix et al. 2007), which prey mainly upon zooplankton and small benthic animals. Success in capturing prey can be related mainly to fish mobility and visual acuity. Vasconcellos et al. (2011) studied diel fluctuations in nearshore fish communities at the surf zone in Southeastern Brazil and observed that juveniles *Atherinella brasiliensis* preyed mainly upon plankton, while juveniles *Trachinotus carolinus* preyed on small invertebrates. Feeding during daylight hours in the surf zone can be a strategy employed by juvenile fishes to avoid predation since predators are mainly active at night (Vasconcellos et al. 2011).

Few studies have focused on the effect of bird predation on sandy beaches. Lopes (2003) described the structure of benthic communities used as food by wintering shorebirds on the state of Maranhão coast (Northeastern Brazil). The bill morphological adaptations for feeding and the trophic relationships of resident birds in exposed sandy shores in southernmost Brazil were reviewed by Pinotti et al. (2014). Sandpipers and plovers have short bills adapted to prey on small macrobenthic fauna buried superficially (Santos 1991; Vooren 1997). In the state of Rio Grande do Sul beaches, the American oystercatcher *Haematopus palliatus* use its specialized bill (large spoon-shaped bill) to search preferentially for the large yellow clam *Amarilladesma mactroides*, which can be buried at depths larger than 15–40 cm when adult (>50 mm) (Vooren and Chiaradia 1990). *Amarilladesma* is also preyed on by seagulls (*Larus dominicanus*) when doing tidal migrations (Vooren 1997). The ground-burrowing owl, hawks, and seagulls prey upon the ghost crab *Ocypode quadrata* on the sandy beaches of Southern Brazil (Gianuca 1997; Branco et al. 2010) and may play an important role in ghost crab population control.

7.2.3 Parasitism

Parasitism is widespread among marine animals and may regulate the population dynamics and influence the community structure of their hosts (Prinz et al. 2010). Parasites can infect sandy beach species and impose energy demands, alter behavior, affect morphology and appearance, reduce growth, and even cause mortality. Parasites may also affect the reproductive output of species. Many trematodes, for example, cause partial or total inhibition of host gamete formation in bivalves (Averbuj and Cremonte 2010).

Although parasitism may exert a strong influence on sandy beach species, this topic is still largely overlooked. Corte (2015) studied the parasitism of a digenetic trematode in the clam *Anomalocardia flexuosa* (=*brasiliana*) on two beaches with different morphodynamic states (an intermediate beach and a tidal flat) in Southeast Brazil and observed that the parasite caused castration of all hosts. Nevertheless, no difference in parasite prevalence (proportion of infected hosts) was observed between beaches (circa 7.5%). Similarly, Araújo and Rocha-Barreira (2004) observed a prevalence rate of 6% of the trematode *Bucephalus* sp. in adult bivalves of *A. flexuosa* in estuarine beaches in Ceará, Northeast Brazil.

Souza et al. (2019) described the effect of the parasitic isopod *Mothocya nana* infestation on the physiological condition and the diet of *Atherinella brasiliensis* (Brazilian silverside) in southeastern Brazil. According to these authors, general conditions of the fishes decreased according to the parasite's development and the offspring's weight. Parasitized fishes showed reduced phytoplankton intake and based their food on fewer trophic categories. Parasitism by isopods in surf-zone fish was also investigated by Shah Esmaeili et al. (2021) on 27 beaches in São Paulo, southeastern Brazil. The authors registered four species of parasitic isopods, all belonging to the Cymothoidae family, but *Livoneca desterroensis* was the most abundant (78.7%). Overall, parasite infestation was low and recorded on only seven beaches. A high infestation was found in a single site, the highly urbanized Itaguá beach (Ubatuba, state of São Paulo), where the prevalence of parasitism reached almost 70% of the Atlantic anchoveta *Cetengraulis edentulus* individuals. No effects of parasitism were detected on the general conditions (i.e., size, weight, and physiology) of affected fishes; however, according to the authors, coastal urbanization could be a factor in increasing the prevalence of isopod parasitism in surf-zone fish.

7.2.4 Positive Interactions

In sandy beaches, the availability of suitable hard substrate for oviposition and shelter is limited and can strongly influence species behavior and reproduction. Often, empty mollusk shells are the only hard substrate available on sandy beaches, as observed by Rocha-Barreira (2003), who recorded egg capsules of the gastropod *Olivancillaria vesica auricularia* fixed in fragments and empty shells of bivalves in the surf zone. The selection of an appropriate substratum for laying the egg capsules in environments with strong hydrodynamics, such as exposed beaches, can guarantee the survival of the embryos during their development, especially for species with a prolonged intracapsular period (Borzone 1995). Empty shells, however, usually remain on the sediment surface, making deposited egg capsules an easy target for predators. Furthermore, they can also be carried up the shore by wave action and currents, where the larvae might die from desiccation.

As an alternative to enhance offspring survival, some sand-dwelling marine gastropods select specific oviposition sites and attach their egg capsules to living organisms such as algae, bivalves, or conspecifics (von Dassow and Strathmann 2005). This behavior may bring several advantages when compared to laying eggs in empty shells or stones. The most obvious benefit is that predation rates may be significantly lower on egg capsules attached to "moving nurseries" (Kano and Fukumori 2010). Additionally, living organisms, especially infaunal species, provide egg capsules with considerable protection from wave action, turbulence, desiccation, and high air temperatures (Barnett et al. 1980). On Brazilian sandy shores, Corte et al. (2018) reported a massive event of egg-capsule deposition by the gastropod *Olivella minuta* on the shells of living, buried specimens of the trigonal clam *Tivela mactroides*, emphasizing the importance of biological interactions on the ecology of sandy beaches.

Low hydrodynamic environments, such as sheltered sandy beaches and tidal flats, provide conditions for the occurrence of hermit crabs, which act as prominent scavengers and deposit feeders (Negreiros-Fransozo et al. 1997; Fransozo and Mantelatto 1998). Hermit crabs generally use gastropod shells because they protect predators (Vance 1972; Hazlett 1996) and from physical stress (Reese 1969). Several species of hermit crabs have been recorded on Brazilian sandy beaches, mainly belonging to the genus *Clibanarius, Isocheles* (family Diogenidae), and *Pagurus* (family Paguridae) (Chap. 4). In Sepetiba Bay, Rio de Janeiro, Cardoso et al. (2011) observed that the abundance of *Pagurus criniticornis* may be related to the dominance of gastropods *Cerithium atratum*, which provides empty shells. In Araçá Bay, São Paulo, *P. criniticornis* occupies shells of the gastropod *Olivella minuta* (Turra and Denadai 2004).

Along the Brazilian coast, thalassinoides (Decapoda, Crustacea) build sandy or muddy burrows, which are used by several species of benthic invertebrates as a place for living, reproduction, and/or feeding (Manning and Felder 1989; Rodrigues and Shimizu 1997). One of the most common examples is the use of the burrows of ghost shrimps (*Callichirus* sp., *Sergio* sp., and *Lapidophtalmus* sp.) by pea crabs of the genus *Austinixa* (= *Pinnixa*) (Alves and Pezzuto 1998; Alves and Rodrigues 2003; Lima 2006; Lima et al. 2006; Peiró and Mantellato 2011). *Austinnixa patagoniensis* occupies the upper portion of *Callichirus major* tubes (Rodrigues and Shimizu 1997), with densities ranging from 0.9 to 3.6 crabs per tube (maximum of 26 crabs per tube) (Alves and Rodrigues 2003). Adults and juveniles commonly are observed in the same burrow, and a maximum of four mature adults occurred together. Unlike mature females, which usually are accompanied by other females, mature males rarely occur together (Alves and Rodrigues 2003).

7.3 Trophic Interactions

7.3.1 Trophic Guilds

A guild is defined as a group of species that similarly exploit the same class of environmental resources, without considering the taxonomic position, and that significantly overlap in their niche requirements (Dajoz 2005). Fauchald and Jumars (1979) considered a feeding guild as an assemblage of species with similarities in food size and type, mechanism of food capture and ingestion, and mobility during feeding activities.

Arruda et al. (2003) showed that the distribution of mollusk-feeding guilds on beaches is related to salinity, grain size, and the contents of silt clay and organic matter in the sediment. Suspension-feeding bivalves, such as *Anomalocardia flexuosa* and *Tagelus plebeius*, occurred in areas with lower salinity and moderately to well-sorted sediments with a predominance of very fine sand. Bivalve deposit feeders (e.g., *Macoma constricta* and *Tellina versicolor*) were more abundant in sediments with high silt-clay and organic matter contents. Carnivore/scavengers gastropods, such as *Olivella minuta*, were less abundant and occurred nearer the waterline, feeding on bivalves of the genus *Donax* and other food items.

Olivella minuta has been shown to have scavenger and detritus-feeder-feeding habits, with the diet varying in space and time (Corte et al. 2019). Thus, *O. minuta* can participate in different food guilds, since its generalist behavior allows individuals to shift their diet according to habitat alterations and food availability. This strategy enhances the persistence and resilience of *O. minuta* in different environments and highlights its role in the cycling of both vegetal and animal organic matter on sandy beaches (Corte et al. 2019).

Diets of several intertidal polychaete species of the state of São Paulo were studied by Checon et al. (2017). Through analysis of the stomach contents of polychaetes, the authors demonstrated that trophic guilds access different sources of food available on these beaches. Subsurface deposit feeders have a lower diet breadth and a higher dependence on organic matter detritus than surface feeders. Planktonic diatoms are of moderate importance for surface deposit feeders, which indicates that the sinking planktonic diatoms may be an important food source for these organisms, although benthic diatoms were more often consumed by all feeding guilds, which highlights the importance of local food sources for the marine food web.

Cabrini et al. (2018) described the role of feeding guilds, considering different groups of macrofauna, in the bioaccumulation of contaminants on sandy beaches with different morphodynamic states on the state of Rio de Janeiro coast. Carnivorous tissues had higher heavy metal concentrations compared to herbivorous, detritus, suspension, and filter feeders. The bioaccumulation across the feeding guild was probably related to the different magnitudes of metal contamination along the coast as to the trophic structure of each beach.

7.3.2 Scavengers

Sandy beaches are depositional environments (Barreiro et al. 2011) where wrack accumulations can often be found along and above the intertidal zone. The supralittoral zone usually has a strip of debris deposited by the action of waves and tides known as "drift line" or "stranded wrack line" (Gandara-Martins et al. 2010). However, due to the high hydrodynamics typical of sandy beaches, the occurrence and permanence of dead matter can be ephemeral.

As an essentially depositional environment, sandy beaches are favorable habitats for scavengers that occur throughout the entire intertidal zone and are dominant in the upper beach reaches (Defeo and McLachlan 2011). Scavengers are abundant on all beach types and their high diversity highlights the role of beaches in processing organic materials of marine origin (Defeo and McLachlan 2011). Scavengers are represented by invertebrates and vertebrates. Beetles, flies, talitrid amphipods, and ghost crabs dominate the scavenger fauna, while lizards, seabirds, and eventually little rodents are generally present in low densities.

Studies focusing on sandy beach scavengers are scarce. In Brazil, this theme has been addressed in a complementary way in studies on secondary production, the structure of benthic communities and feeding guilds (Arruda et al. 2003; Petracco et al. 2013; Mattos et al. 2013), and the effects of urbanization on the coast (Barbosa et al. 2017; Costa et al. 2017). On the southernmost beaches of Brazil, dense accumulations of the surf-zone diatom *Asterionellopsis glacialis* generally form dark-brown patches at the meso- or the supralittoral (Odebrecht et al. 2013) (Chap. 2), representing an important food source for terrestrial coleopterans (*Bledius hermani, B. bonariensis*, and *B. fernandezi*) and the talitrid amphipod *Atlantorchestoidea brasiliensis* (Gianuca 1983; Pinotti et al. 2014; Vianna and Borzone 2015). On the beaches of northeast Brazil, seaweed and *Halodule wrightii* accumulate in the drift line and are used as food by beetles and talitrid amphipods (Ramos 2014). The beetles *Phaleria testacea, B. caribbeanus,* and *B. hermani* are common in the drift line of sandy beaches in Northeastern Brazil (Rocha-Barreira et al. 2001; Rosa et al. 2013). The coleopteran *Phaleria* sp. is also a common inhabitant of beaches in the states of Rio de Janeiro and São Paulo, often associated with detritus in the drift line (Caldas and Almeida 1993; Veloso et al. 2003; Corte et al. 2022).

Talitrids generally are the first colonizers of stranded algae on the drift line of sandy beaches (Veloso et al. 2009). The sandhopper *Atlantorchestoidea brasiliensis* occurs from the supralittoral to the upper mid-littoral across the whole morphodynamic spectrum on exposed beaches along south and southeast sandy shores in Brazil (Cardoso and Veloso 1996; Defeo and Gómez 2005). Ramos (2014) studied populations of *Thalorchestia tucurauna* on two sandy beaches of the state of Pernambuco and found the highest densities on the beaches with more *Halodule wrightii* in the drift line.

The ghost crab *Ocypode quadrata* is a predator and also an important scavenger across the intertidal and supralittoral zones along the Brazilian coast (Alberto and

Fontoura 1999; Turra et al. 2005; Branco et al. 2010; Gomes et al. 2019; Fortaleza et al. 2019). This crab species feed on particulate organic matter (POC) found among sand grains, dead fish, and other marine animals (Wolcott 1978). After storms, empty shells of the bivalves *A. mactroides* and *D. hanleyanus* and the exoskeletons of the mole crab *E. brasiliensis* were observed around large burrows of the ghost crab in Rio Grande do Sul beaches (Pinotti et al. 2014). Gomes et al. (2019) found gut contents of ghost crabs on the state of São Paulo beaches to contain numerous arthropod groups, poriferans, plants, and even fish bones, highlighting the opportunistic nature of the species. Field observations and field experiments carried out in the city of Fortaleza (Rocha Barreira, pers. Obs.), state of Ceará, observed that *O. quadrata* feeds on several food items, such as dead animal carcasses and macroalgae, and also recorded that scavenger-feeding habit is more common than carnivory.

The intertidal zone is the area of beaches where the unpredictability of the carrion is greatest. However, in this zone, particulate debris may be the main resource of food available to scavengers. The scavenger isopods *Excirolana armata* and *E. braziliensis* inhabit the middle and the low intertidal zone and are capable of actively searching for carrion under different intensities of swash currents (Yannicelli et al. 2002). Pinotti et al. (2014) recorded that several macrobenthic species, and even large dead organisms (e.g., fishes and seabirds), are voraciously consumed by hundreds of *Excirolana* on beaches in southern Brazil.

In sheltered sand–muddy beaches, the relative greater environmental stability favors scavengers. The gastropod *Nassarius vibex* (Fig. 7.1c) is one of the most important scavenger species on sheltered beaches in the states of São Paulo and Rio de Janeiro (Denadai et al. 2005; Cardoso and Caetano 2010). Field manipulative experiments revealed that this gastropod is quickly attracted to food, with an instantaneous response to the smell of dead fishes, such that in less than 10 minutes the animals were attracted to the baits (Lucena et al. 2012).

Omnivorous-detritivorous amphipods common in Brazilian beaches are *Bathyporeiapus bisetosus* and *B. ruffoi*, *Phoxocephalopsis zimmeri* and *Puelche orensanzi,* and Platyischnopidae may use primary or secondary production products (e.g., suspended/deposited diatoms and detritus) along the swash zone as food (Escofet 1973; Lercari et al. 2010).

In the subtidal zone of beaches in south Brazil, the surf-zone whelk *Buccinanops duartei* is a scavenger frequently recorded on the beaches. This gastropod makes exploratory excursions across the swash zone/inner surf zone, employing accurate chemical receptors to detect deposited animal debris (Borzone and Gianuca 1990; Garcia and Gianuca 1997). Many fishes also use macrophyte accumulations as food. When studying two tropical Brazilian beaches with drift algae in the surf zones, Andrades et al. (2014) found that fish biomass, density, and diversity increased in the presence of drift algae. The macroalgae provide habitat/shelter and food, mainly for juvenile (young-of-the-year) fishes.

7.3.3 Detritivory

Detritivores feed upon particulate organic matter in and/or on the seafloor, distinguishing themselves from scavengers, which eat the corpses of animals and plants in different stages of decomposition lying on the seabed (Kooten and Schellekens 2015). Deposit feeders are sparse or absent on reflective beaches, where high hydrodynamics prevent the accumulation of debris. The deposit-feeding habit only prevails on fine sand beaches, as found on dissipative beaches (McLachlan and Defeo 2018).

The infaunal polychaetes *Thoracophelia furcifera* and *Spio gaucha* and the amphipods *Bathyporeiapus bisetosus, B. ruffoi, Phoxocephalopsis zimmeri, Puelche orensanzi,* and Platyischnopidae feed on the diatom *Asterionellopsis glacialis* deposited in the intertidal and supratidal zone in dissipative beaches of southern Brazil (see Pinotti et al. 2014). Although the polychaetes *Scolelepis chilensis, S. Goodbodyi,* and *S. squamata* are predominantly deposit feeders, they can shift to filter feeding on very compact, fine sand beaches on the north coast of the state of São Paulo (Pardo and Amaral 2004). The bivalves *Macoma constricta, Tellina versicolor,* and *Phacoides pectinatus* are deposit feeders in beaches with high silt-clay and organic matter contents (Fig. 7.2) (Arruda et al. 2003). These bivalves build deep tubes or galleries and can opportunistically exploit the sediment layers rich in organic matter on the subsurface (Arruda et al. 2003).

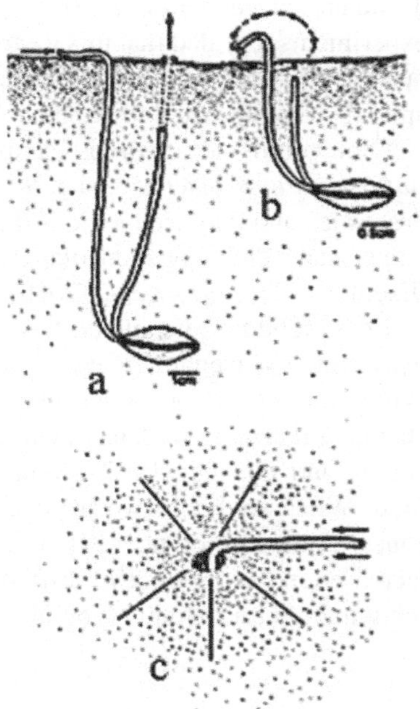

Fig. 7.2 Feeding behavior of (**a**) *Macoma constricta*, (**b**) *Tellina lineata*, and (**c**) *Tellina versicolor*. (Modified from Arruda et al. 2003). Mollusc-feeding guilds on sandy beaches in São Paulo State, Brazil. (Licensed by Springer Nature and Copyright Clearance Center)

In intertidal, sublittoral, and surf zones of sand beaches along the Brazilian coast, the sand dollar *Mellita quinsquesperforata*, an epibenthic deposit-feeder, is very common (Fig. 7.1d) (Lopes 2011). In the surf zone, this species is under high hydrodynamics and turbulence, generated by tidal currents and waves, but is still able to ingest diatoms and sand grains colonized by microorganisms (Borzone et al. 1997).

7.3.4 Trophic Webs

Food webs differ between sandy beaches due to the different food sources and beach morphodynamics (McLachlan and Defeo 2018). The main sources of organic matter in sandy beaches, such as surf-zone phytoplankton, carrion, and stranded macroalgae, vary widely in space and time (Bergamino et al. 2016; Quillien et al. 2016). Most food webs on sandy beaches include macrobenthos, zooplankton, fish, and birds. In addition, an interstitial food web, composed of bacteria, protozoans, and meiofauna, occurs commonly on sheltered beaches, and a microbial loop (bacteria, flagellates, ciliates, and other microzooplankton) is always present in the water column of the surf zone (McLachlan and Defeo 2018).

Although studies on food webs are important for understanding the relationship between energy flows and ecosystem functioning, there are still relatively few studies on sandy beaches (Thompson et al. 2012). Costa et al. (2017) applied mass-balanced models on two Brazilian sandy beaches with distinct human impact degrees on the state of Rio de Janeiro coast. In this study, the food web features did not differ between nonurbanized and urbanized sectors from Grussaí Beach (dissipative beach) and Praia Grande Beach (reflective beach), but clear differences were observed between beach morphodynamic types (Fig. 7.3). At Grussaí Beach, the majority of the Total System Throughput (TST) flowed to detritus or was exported. At Praia Grande Beach, most TST was consumed, instead of exporting or flowing to detritus. The highest trophic levels at Grussaí Beach were represented by catfish (Ariidae), while crustaceans *Emerita brasiliensis* and *Excirolana braziliensis* were the main primary consumers (Fig. 7.3a). At Praia Grande Beach, the birds (brown booby *Sula leucogaster* and the kelp gull *Larus dominicanus*) and fishes (sand drum *Umbrina coroides* and the bluefish *Pomatomus saltatrix*) were the top predators; crustaceans *Excirolana brasiliensis* and *Atlantorchestoidea brasiliensis* were the main primary consumer (Fig. 7.3b).

In a review study, Pinotti et al. (2014) outlined the main trophic relationships along subtropical sandy shores in southernmost Brazil. The authors highlighted the biological compartments of producers, macrobenthic fauna, seabirds, and surf-zone fishes (Fig. 7.4) and showed that the surf-zone diatom *Asterionellopsis glacialis* is the most important primary producer sustaining intermediate (secondary to tertiary) macrobenthic consumers like filter-feeding bivalves, hippid crabs, and scavenger isopods, forming short trophic chains. During short-term high tides, the elevated abundance and consequent high biomass of intertidal species may attract several top

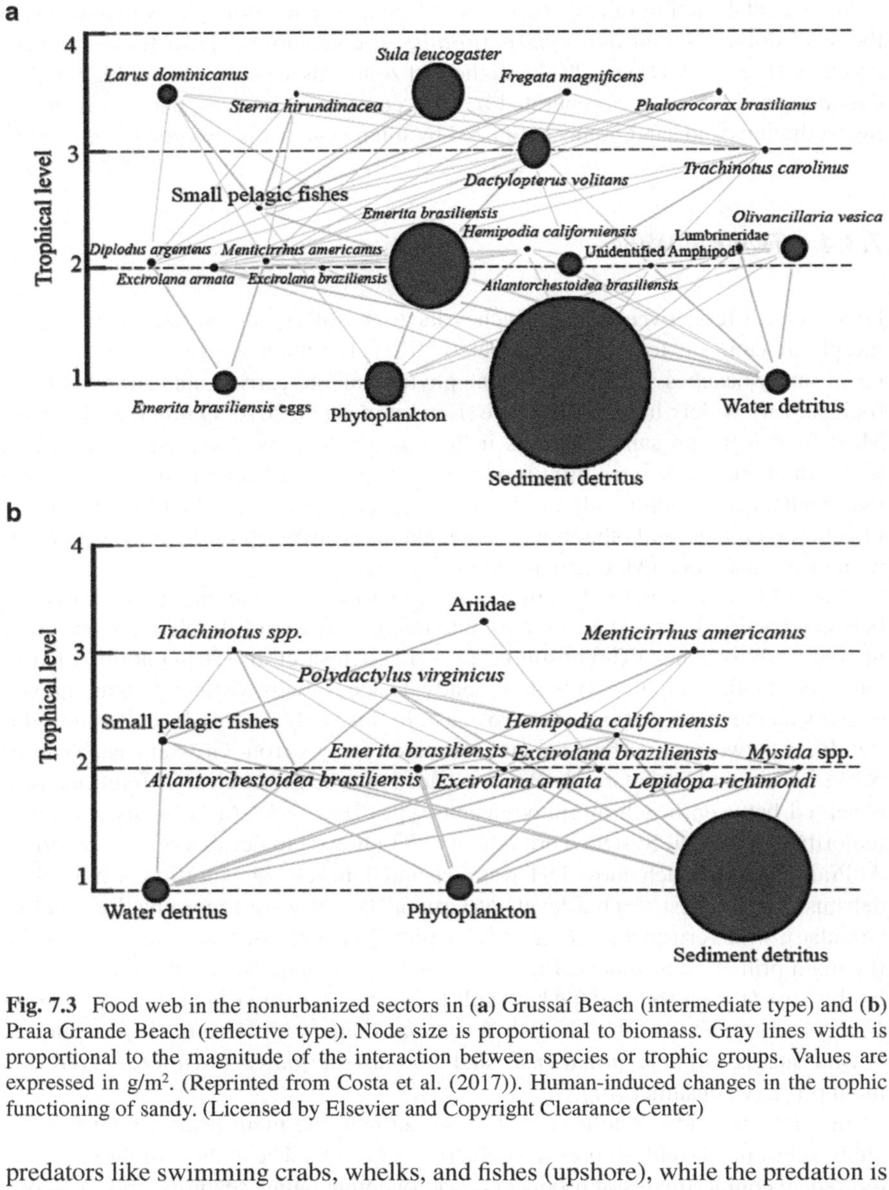

Fig. 7.3 Food web in the nonurbanized sectors in (**a**) Grussaí Beach (intermediate type) and (**b**) Praia Grande Beach (reflective type). Node size is proportional to biomass. Gray lines width is proportional to the magnitude of the interaction between species or trophic groups. Values are expressed in g/m^2. (Reprinted from Costa et al. (2017)). Human-induced changes in the trophic functioning of sandy. (Licensed by Elsevier and Copyright Clearance Center)

predators like swimming crabs, whelks, and fishes (upshore), while the predation is mainly exerted by ghost crabs and seabirds (downshore) at low tides (Gianuca 1983).

The use of methods based on the isotopic signature of C and N has become an extremely useful tool to analyze the food source, as well as the flow pathways of organic matter and the trophic level of consumers on sandy beaches (Petracco 2008). Petracco (2008) conducted the first study with this focus in Brazil and found that the trophic chain in Una Beach (São Paulo coast) seems to be based mainly on

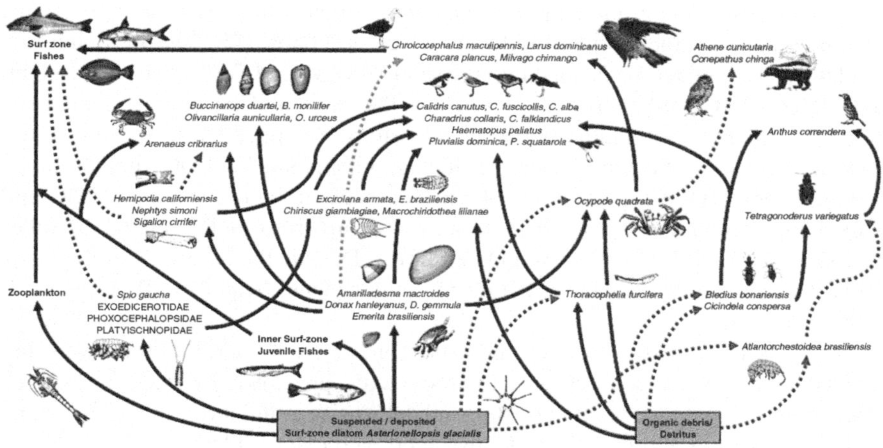

Fig. 7.4 Trophic relationships (solid lines: evidenced; dashed lines: suggested) reported for the intermediate sandy shores in southernmost Brazil. Sources of primary production and species richness within biological compartments can differ significantly through time (seasonally) and according to the morphodynamic beach state. Drawings are represented on an arbitrary scale and are copyrighted to respective owners. (Modified from Pinotti et al. 2014). A review on macrobenthic trophic relationships along subtropical sandy shores in southernmost Brazil. (Copyright Biota Neotropica – BIOTA/FAPESP Program. Licensed under a Creative Commons Attribution 4.0 International (CC BY 4.0) License)

suspended particulate matter, further suggesting that the microphytobenthos also have a relevant role and that dune plants do not contribute as primary producers in that ecosystem.

Maria et al. (2011), through laboratory experiments with labeled planktonic and benthic diatoms enriched with carbon 13C isotopes, observed that most macrofaunal animals consumed both types of diatoms, but benthic diatoms were consumed by both macrofaunae and by meiofauna. These findings reveal the importance of benthic carbon sources for macrofaunal and interstitial food webs and suggest a link between both food webs through the common use of benthic diatoms.

Almeida et al. (2019) showed differences in the stable isotopic ratio (δ13C and δ15N) of macrobenthic species between the reflective beach Praia do Farol, with a high nutrient contribution from the continent provided by the Paraíba do Sul River, and the dissipative Praia Grande Beach, which is affected by the upwelling at Cabo Frio (RJ). Their findings pointed to no significant differences between the beaches, suggesting that they have similar sources of organic matter, and the diet of organisms is based on zooplankton at both beaches. However, significant differences were found in the δ13C and δ15N isotopic signals for *Donax hanleyanus* and *Emerita brasiliensis*, indicating that these species may occupy different trophic positions at each beach. Also, the δ15N isotopic signature of the polychaete *Hemipodia californiensis* was similar to that of several benthic predators, such as carnivorous fishes and octopus (Almeida et al. 2019; Corbisier et al. 2014), revealing that it is a top predator on beach environments.

The interstitial environment is a beach compartment where bacteria, protozoans, and meiofauna interact in their food web, which is supported by dissolved and particulate organic materials flushed into the sand by wave and tide action (McLachlan and Defeo 2018). In sheltered sandy beaches, the interstitial system may include its primary producers as benthic diatoms and autotrophic flagellates. The interstitial food web consists of bacteria using particulate organic matter (POC) and dissolved organic matter (DOC), protozoans preying on bacteria (and consuming POC and DOC), meiofauna feeding on all of these, and carnivorous meiofauna taking other meiofaunal forms (McLachlan and Defeo 2018). Despite the relevance of the interstitial food web in the trophic structure of sandy beaches, studies in Brazil are still quite incipient.

Sandy beaches with well-developed surf zones have a microbial food web suspended in the water column. The hydrodynamic promoted by wave turbulence maintains surf diatoms, organic matter, and detritus accumulations in the surf zone, supporting a short and highly productive food web (McLachlan and Defeo 2018). The large amount of dissolved organic carbon produced by surf diatoms is taken up by bacteria and should fuel the surf-zone "microbial loop" (Odebrecht et al. 2013). On exposed beaches, macrobenthic filter feeders are the main consumers and form a link between diatoms and higher trophic levels (Garcia and Gianuca 1997).

7.4 Interactions Between Macrofauna and Meiofauna

The effects of macrobenthos on the meiobenthos in sandy beaches are extremely variable, species-dependent, and are often not delimited (Ólafsson 2003). Although the generally maximum density and species richness of meiofauna and macrofauna occur at different parts of the intertidal zone (Rosa-Filho et al. 2011; McLachlan and Defeo 2018), macrobenthic organisms can cause physical changes in their abiotic environment (e.g., through bioturbation and bio-irrigation) (Widdicombe et al. 2000; Volkenborn et al. 2007), which in turn may influence the distribution of meiofaunal organisms (Reise 1985). These interactions may be negative (e.g., mechanical disturbance and predation) (Sherman et al. 1983; Urban-Malinga et al. 2014) or positive (e.g., offer of shelter against predation and reduction of harmful chemical compounds in porewater) (Volkenborn et al. 2007; Aviz et al. 2019).

In Brazil, few studies have attempted to understand the interaction between macro-and meiofauna, and most studies were done with estuarine species, or on estuarine beaches. Guilherme et al. (2011) used rubble mimics to describe the role of *Diopatra* tubes in meiobenthos in Coroa do Avião (Pernambuco). The effect of the burrowing polychaete *Laeonereis culveri* on meiofauna was studied by Pinto and Bemvenuti (2003) using laboratory experiments. When reviewing the state of the art of the meiofauna of Brazilian sandy beaches, Maria et al. (2016) state that only two studies focused on biological interactions as a structuring factor of meiobenthos communities. From these studies, only Maria et al. (2013) dealt with the interaction between macrofauna and meiofauna. These authors investigated through laboratory experiments the role of *Scolelepis squamata*, a macrobenthic polychaete abundant in intertidal zones of Brazilian sandy beaches, in the colonization of defaunated sediments by sandy-beach nematodes. The main conclusion was that meiofauna recovery was not significantly affected by the presence of the polychaete. The causes of the absence of effect, differing from the results of other similar experiments conducted on temperate sandy beaches, were not clear. Authors argued that the behavior of the polychaete species (actually a species complex) used in the Brazilian experiment or differences in nematode community between tropical and temperate regions may have caused the observed differences in results.

7.5 Final Remarks

Although sandy beaches are known as physically controlled environments, intra and interspecies biological interactions may play an important role in their biodiversity even under the stressful conditions found in more reflective beaches. In Brazil, knowledge about biological interactions on sandy beaches is still limited. Efforts to understand the role of biological interactions in the structuring and functioning of benthic communities of sandy beaches have been restricted mainly to the Southern and Southeastern regions of Brazil (Amaral et al. 2016) with some important initiatives also in the Northeastern region. This situation evidences a regional gap and demonstrates the need for further studies on this topic to gain a better understanding of sandy beach ecosystems.

References

Alberto RMF, Fontoura NF (1999) Distribuição e estrutura etária de *Ocypode quadrata* (Fabricius, 1787) (Crustacea, Decapoda, Ocypodidae) em praia arenosa dolitoral sul do Brasil. Rev Bras Biol 59:95–108. https://doi.org/10.1590/S0034-71081999000100013

Almeida TC, Rocha PF, Zalmon IR et al (2019) Is there an indication of the origin of nutrient supply in different morphological structures of macrofauna at two different Brazilian southeastern sandy beaches? Comparison by C and N stable isotopes. Environ Sci Pollut Res 26(32):33023–33029. https://doi.org/10.1007/s11356-019-06376-4

Alves ES, Pezzuto PR (1998) Population dynamics of *Pinnixa patagoniensis* Rathbun, 1918 (Brachyura: Pinnotheridae) a symbiotic crab of *Sergio mirim* (Thalassinidea: Callianassidae) in Cassino Beach, Southern Brazil. Mar Ecol 19(1):37–51. https://doi.org/10.1111/j.1439-0485.1998.tb00452.x

Alves ES, Rodrigues AS (2003) Distributional patterns of *Austinixa patagoniensis* (Rathbun, 1918) (Decapoda: Pinnotheridae), a symbiotic crab of *Callichirus major* (Decapoda: Callianassidae) in Balneário Camboriú Beach, Santa Catarina, Brazil. J Coast Res 35:408–417

Amaral ACZ, Corte GN, Rosa-Filho JSR et al (2016) Brazilian sandy beaches: characteristics, ecosystem services, impacts, knowledge and priorities. Braz J Oceanogr 64(SPE2):5–16. https://doi.org/10.1590/S1679-875920160933064sp2

Andrades R, Gomes MP, Pereira-Filho GH et al (2014) The influence of allochthonous macroalgae on the fish communities of tropical sandy beaches. Estuar Coast Shelf Sci 144:75–81. https://doi.org/10.1016/j.ecss.2014.04.014

Araújo MLR, Rocha-Barreira CDA (2004) Ocurrence of *Bucephalus* sp. (Trematoda: Bucephalidae) in *Anomalocardia brasiliana* (Gmelin, 1791) (Mollusca: Veneriidae) at Canto da Barra Beach, Fortim, Ceará state, Brazil. Arquiv Cienc Mar 37(1–2):35–37

Arruda EP, Domaneschi O, Amaral ACZ (2003) Mollusc feeding guilds on sandy beaches in São Paulo State, Brazil. Mar Biol 143(4):691–701. https://doi.org/10.1007/s00227-003-1103-y

Averbuj A, Cremonte F (2010) Parasitic castration of Buccinanops cochlidium (Gastropoda: Nassariidae) caused by a lepocreadiid digenean in San José Gulf, Argentina. J Helminthol 84:381–389

Aviz D, Silva RF, Rosa-Filho JS (2019) *Sabellaria wilsoni* (Polychaeta: Sabellariidae): an ecosystem engineer and promoter of zoobenthos diversity in the Brazilian Amazon coast. J Mar Biol Assoc UK 99(5):1099–1109. https://doi.org/10.1017/S0025315418001115

Barbosa TM, Carmo RFR, Silva LP et al (2017) Diversity of *Sarcosaprophagous calyptratae* (Diptera) on sandy beaches exposed to increasing levels of urbanization in Brazil. Environ Enthomol 46(3):460–469. https://doi.org/10.1017/S0025315418001157

Barnett PRO, Hardy BLS, Watson J (1980) Substratum selection and egg-capsule deposition in *Nassarius reticulatus* (L.). J Exp Mar Biol Ecol 45(1):95–103. https://doi.org/10.1016/0022-0981(80)90072-6

Barreiro F, Gómez M, Lastra M et al (2011) Annual cycle of wrack supply to sandy beaches: effect of the physical environment. Mar Ecol Prog Ser 433:65–74. https://doi.org/0.3354/meps09130

Begon M, Townsend CR, Harper JL (2006) Ecology: from individuals to ecosystems, 4th edn. Blackwell Publishing, Oxford, p 738

Bergamino L, Martínez A, Han E et al (2016) Trophic niche shifts driven by phytoplankton in sandy beach ecosystems. Estuar Coast Shelf Sci 180:33–40. https://doi.org/10.1016/j.ecss.2016.06.023

Borzone CA (1995) Ovicapsulas de prosobrânquios (Mollusca: Gastropoda) de una playa arenosa expuesta del sur del Brasil. Iheringia Ser Zool 79:47–58

Borzone CA, Gianuca NM (1990) A zonação infralitoral em praias arenosas expostas. Pub ACIESP 71(3):280–287

Borzone CA, Tavares YAG, Soares CR (1997) Adaptação morfológica de *Mellita quinquiesperforata* (Clypeasteroida, Mellitidae) para explorar ambientes com alto hidrodinamismo. Iheringia Ser Zool 82:33–42

Branco JO, Hillesheim JC, Fracasso HA et al (2010) Bioecology of the ghost crab *Ocypode quadrata* (Fabricius, 1787) (Crustacea: Brachyura) compared with other intertidal crabs in the southwestern Atlantic. J Shellfish Res 29(2):503–512. https://doi.org/10.2983/035.029.0229

Cabrini TM, Barboza CAM, Skinner VB et al (2018) Investigating heavy metal bioaccumulation by macrofauna species from different feeding guilds from sandy beaches in Rio de Janeiro, Brazil. Ecotox Environ Safe 162:655–662. https://doi.org/10.1016/j.ecoenv.2018.06.077

Caldas A, Almeida JR (1993) Population dynamics of *Phaleria testacea* say (Coleoptera: Tenebrionidae). Coleopt Bull 47(3):221–227

Carcedo MC, Fiori SM, Scotti M et al (2019) Dominant bivalve in an exposed Sandy Beach regulates community structure through spatial competition. Estuaries Coast 42(7):1912–1923. https://doi.org/10.1007/s12237-019-00622-1

Cardoso R (2002) Behavioural strategies and surface activity of the sandhopper *Pseudorchestoidea brasiliensis* (Amphipoda: Talitridae) on a Brazilian beach. Mar Biol 141(1):167–173. https://doi.org/10.1007/s00227-002-0784-y

Cardoso RS, Caetano CHS (2010) Imposex in *Nassarius vibex*: relationship with harbor and yachting activities at five beaches in Sepetiba Bay, RJ, Brazil. Pan-Am J Aquat Sci 5(4):540–545

Cardoso RS, Meireis F, Mattos G (2011) Crustaceans composition in sandy beaches of Sepetiba Bay, Rio de Janeiro, Brazil. Check List 7(6):778–781. https://doi.org/10.15560/11024

Cardoso RS, Veloso VG (1996) Population biology and secondary production of the sandhopper *Pseudorchestoidea brasiliensis* (Amphipoda: Talitridae) at Prainha beach, Brazil. Mar Ecol Prog Ser 142:111–119. https://doi.org/10.3354/meps142111

Castro AL, Brum INS (1969) Sobre as espécies do gênero *Excirolana* Richardson do litoral atlântico das Américas (Isopoda, Cirolanidae). Bol Mus Nac Rio de Janeiro 271:164–185

Chagas RA, Herrmann MA (2015) Relação predador-presa entre *Natica marochiensis* (Gmelin, 1791)(Gastropoda: Naticidae) e *Donax striatus* Linnaeus, 1767 (Bivalvia: Donacidae) é controlada pelo tamanho da concha da presa? In: XIX Congresso Brasileiro de Engenharia de Pesca, São Luis, pp 466–470

Checon HH, Pardo EV, Amaral ACZ (2017) Breadth and composition of polychaete diets and the importance of diatoms to species and trophic guilds. Helgol Mar Res 70:19

Checon HH, Silva MO, Corte GN et al (2020) Full stomachs at empty tides: tidal cycle affects feeding activity and diet of the sandy beach gastropod *Olivella minuta*. J Mollusc Stud 86(3):219–227. https://doi.org/10.1093/mollus/eyaa007

Connell JH (1972) Community interactions on marine rocky intertidal shores. Annu Rev Ecol Evol Syst 3(1):169–192

Connell JH, Orias E (1964) The ecological regulation of species diversity. Am Nat 98(903):399–414. https://doi.org/10.1086/282335

Copertino MS, Creed JC, Lanari MO et al (2016) Seagrass and submerged aquatic vegetation (VAS) habitats off the Coast of Brazil: state of knowledge, conservation and main threats. Braz J Oceanogr 64(SI 2):53-80. https://doi.org/10.1590/S1679-875920161036064sp2

Corbisier TN, Petti MAV, Soares LS et al (2014) Trophic structure of benthic communities in the Cabo Frio upwelling system (southeastern Brazilian shelf): a temporal study using stable isotope analyses. Mar Ecol Prog Ser 512:23–38. https://doi.org/10.3354/meps10947

Corte (2015) Reproductive cycle and parasitism in the clam Anomalocardia brasiliana (Bivalvia: Veneridae). In: Invertebrate reproduction & development. https://doi.org/10.1080/07924259.2015.1007215

Corte GN, Yokoyama LQ, Denadai MR, Bessa E, Salles MCZ, Turra A (2018) Egg-capsule deposition of the marine gastropod *Olivella minuta*: the importance of an interspecific relationship with the soft-bottom bivalve Tivela mactroides. J Molluscan Stud 85(1):126–132. https://doi.org/10.1093/mollus/eyy060

Corte GN, Yokoyama LQ, Checon HH et al (2019) Spatial and temporal variation in the diet of the sandy beach gastropod *Olivella minuta*. Invertebr Biol 138(4):e12269. https://doi.org/10.1111/ivb.12269

Corte GN, Checon HH, Shah Esmaeili Y et al (2022) Evaluation of the effects of urbanization and environmental features on sandy beach macrobenthos highlights the importance of submerged zones. Mar Pollut Bull 182:113962. https://doi.org/10.1016/j.marpolbul.2022.113962

Costa LL, Tavares DC, Suciu MC et al (2017) Human-induced changes in the trophic functioning of sandy beaches. Ecol Indic 82:304–315. https://doi.org/10.1016/j.ecolind.2017.07.016

Coutinho R, Yaginuma LE, Siviero F et al (2016) Studies on benthic communities of rocky shores on the Brazilian coast and climate change monitoring: status of knowledge and challenges. Braz J Oceanogr 64(SI2):27–36. https://doi.org/10.1590/S1679-875920161015064sp2

Couto ECG (1996) Seleção por tamanho: predação de *Divaricella quadrisulcata* (Orbigny, 1824) (Bivalvia, Lucinidae) por *Polinices hepaticus* (Roding, 1798)(Gastropoda, Naticidae). Arq Biol Tecnol 39(4):815

Dajoz R (2005) Princípios de ecologia, 7th edn. Artmed, Porto Alegre, p 520

Defeo O, Brazeiro A, Alava A, Riestra G (1997) Is sandy beach macrofauna only physically controlled? Role of substrate and competition in isopods. Estuar Coast Shelf Sci 45(4):453–462. https://doi.org/10.1006/ecss.1996.0200

Defeo O, Gómez J (2005) Morphodynamics and habitat safety in sandy beaches: life history adaptations in a supralittoral amphipod. Mar Ecol Prog Ser 293:143–153. https://doi.org/10.3354/meps293143

Defeo O, Layerle C, Masello A (1986) Spatial and temporal structure of the yellow clam *Mesodesma mactroides* (Deshayes, 1854) in Uruguay. Med Amb 8(1):48–57

Defeo O, Lercari D, Gómez J (2003) The role of morphodynamics in structuring sandy beach populations and communities: what should be expected? J Coast Res SI35:352–362. https://www.jstor.org/stable/40928782

Defeo O, McLachlan A, Schoeman DS et al (2009) Threats to sandy beach ecosystems: a review. Estuar Coast Shelf Sci 81(1):1–12. https://doi.org/10.1016/j.ecss.2008.09.022

Defeo O, McLachlan A (2011) Coupling between macrofauna community structure and beach type: a deconstructive meta-analysis. Mar Ecol Prog Ser 433:29–41. https://doi.org/10.3354/meps09206

Defeo O, McLachlan A (2013) Global patterns in sandy beach macrofauna: species richness, abundance, biomass and body size. Geomorphology 199:106–114. https://doi.org/10.1016/j.geomorph.2013.04.013

Denadai MR, Amaral ACZ, Turra A (2005) Structure of molluscan assemblages in sheltered intertidal unconsolidated environments. Braz Archiv Biol Technol 48(5):825–839. https://doi.org/10.1590/S1516-89132005000600019

Dexter D (1977) Natural history of the pan-American sand beach isopod *Excirolana braziliensis* (Crustacea: Malacostraca). J Zool London 183:103–109

Duarte C, Jaramillo E, Contreras H et al (2010) Cannibalism and food availability in the talitrid amphipod *Orchestoidea tuberculata*. J Sea Res 64(3):417–421. https://doi.org/10.1016/j.seares.2010.02.009

Dugan JE, Hubbard DM, McCrary M et al (2003) The response of macroinfauna communities and shorebirds to macrophyte wrack subsidies on exposed sandy beaches of southern California. Estuar Coast Shelf Sci 58S:133–148

Dugan JE, Jaramillo E, Hubbard DM et al (2004) Competitive interactions in macroinfaunal animals of exposed sandy beaches. Oecologia 139(4):630–640. https://doi.org/10.1007/s00442-004-1547-x

Escofet A (1973) Observaciones sobre alimentación y hábitos cavadores en *Bathyporeiapus ruffoi* y *Bathyporeiapus bisetosus* (Amphipoda: Oedicerotidae). Physis 32(84):95–103

Fauchald K, Jumars PA (1979) The diet of worms: a study of polychaete feeding guilds. Oceanogr Mar Biol Annu Rev 17:193–284

Félix FC, Spach HL, Moro PS, Schwarz R Jr, Santos C, Hackradt C, Hostim-Silva M (2007) Utilization patterns of surf zone inhabiting fish from beaches in Southern Brazil. Panam J Aquat Sci 2:27–39

Fortaleza MO, Girão MML, Franklin-Junior W et al (2019) Which moon phase do we find more ghosts? Effects of the lunar cycle on the ghost crab *Ocypode quadrata* (Fabricius, 1787). Arq Ciênc Mar 52(2):85–97. https://doi.org/10.32360/acmar.v52i2.42737

Fransozo A, Mantelatto FLM (1998) Population structure and reproductive period of the tropical hermit crab *Calcinus tibicen* (Decapoda: Diogenidae) in the region of Ubatuba, São Paulo, Brazil. J Crust Biol 18(4):738–745. https://doi.org/10.1163/193724098X00610

Gandara-Martins AL, Borzone CA, Rosa LC et al (2010) Ocorrência de três espécies do gênero Bledius Leach, 1819 (Coleoptera, Staphylinidae, Oxytelinae) nas praias arenosas expostas do Paraná, Brasil. Braz J Aquat Sci Technol 14(2):23–30

Garcia VMT, Gianuca NM (1997) The beach and surf zone. In: Seeliger U, Odebrecht C, Castello JP (eds) Subtropical convergence environments: the coast and sea in the Southwestern Atlantic. Springer-Verlag, Berlin, pp 166–170

Gianuca NM (1983) A preliminary account of the ecology of sandy beaches in Southern Brazil. In: McLachlan A, Erasmus T (eds) Sandy beaches as ecosystems. W Junk, The Hague, pp 413–419

Gianuca NM (1985) The ecology of a sandy beaches in southern Brazil. Thesis. University of Southampton

Gianuca NM (1991) Comportamiento predatorio y alimenticio de los gasteropodos de la zona de barrido de la Playa de Cassino, Rio Grande, Brasil. In: I Congreso Latinoamericano de Malacologia, Caracas, 61

Gianuca NM (1997) A fauna das dunas costeiras do Rio Grande do Sul. Oecologia Aust 3(1):121–133

Gomes TT, Gheler-Costa C, Rinaldi CA et al (2019) Natural diet of *Ocypode quadrata* (Fabricius, 1787)(Crustacea, Decapoda, Brachyura) from the Northern Coast of São Paulo. Brazil Pap Avulsos Zool 59:e20195957. https://doi.org/10.11606/1807-0205/2019.59.57

Guilherme BC, Silva GB, El-Deir ACA et al (2011) Meiofauna associada ao tubo de *Diopatra cuprea* Bosc, 1902 (Polychaeta: Onuphidae). Rev Bras Zool 5:37–52

Hall A, Kingsford M (2016) Predators exacerbate competitive interactions and dominance hierarchies between two coral reef fishes. PlosOne 11(3):e0151778. https://doi.org/10.1371/journal.pone.0151778

Hazlett BA (1996) Recent experience and the shell-size preference of hermit crabs. Marine Freshw Behav Physiol 28(3):177–182. https://doi.org/10.1080/10236249609378988

Hughes RN (1986) A functional biology of marine gastropods. Croom Helm Ltd, London, p 245

Kano Y, Fukumori H (2010) Predation on hardest molluscan eggs by confamilial snails (Neritidae) and its potential significance in egg-laying site selection. J Molluscan Stud 76(4):360–366. https://doi.org/10.1093/mollus/eyq018

Kooten T, Schellekens T (2015) A food web model of invertebrate subtidal soft-bottom communities Part A: model derivation and effects of productivity. IMARES, Wageningen

Lercari D, Bergamino L, Defeo O (2010) Trophic models in sandy beaches with contrasting morphodynamics: comparing ecosystem structure and biomass flow. Ecol Model 221(23):2751–2759. https://doi.org/10.1016/j.ecolmodel.2010.08.027

Levine SH (1976) Competitive interactions in ecosystems. Am Nat 110(976):903–910

Lima JF (2006) Larval development of *Austinixa bragantina* (Crustacea: Brachyura: Pinnotheridae) reared in the laboratory. Fortschr Zool 26(1):143–154. https://doi.org/10.1590/S1984-46702009000100020

Lima JF, Abrunhosa F, Coelho PA (2006) The larval development of *Pinnixa gracilipes* Coelho (Decapoda, Pinnotheridae) reared in the laboratory. Rev Bras Zool 23(2):480–489. https://doi.org/10.1590/S0101-81752006000200023

Lopes ATL (2003) Estrutura das comunidades de organismos bentônicas em regiões entre-marés utilizadas por aves limícolas migratórias na costa norte do Brasil. Thesis, Universidade Federal do Pará

Lopes RP (2011) Fossil sand dollars (Echinoidea: Clypeasteroida) from the southern Brazilian coast. Rev Bras Paleontol 14(3):201–214. https://doi.org/10.4072/rbp.2011.3.01

Lucena JM, Meirelles CAO, Matthews-Cascon H (2012) Feeding behavior of *Nassarius vibex* (Gastropoda: Nassariidae). Arq Ciênc Mar 45(2):60–66

Manning RB, Felder DL (1989) The *Pinnixa cristata* complex in the western Atlantic, with descriptions of two new species (Crustacea: Decapoda: Pinnotheridae). Smithsonian Contrib Zool 473:1–36

Marcus E, Marcus E (1959) Studies on Olividae. Bolm Fac Filos Ciênc Univ São Paulo 22:100–188

Maria TF, Esteves AM, Vanaverbeke J et al (2013) Is nematode colonisation in the presence *of Scolelepis* in tropical sandy-beach sediment similar to the colonisation process in temperate sandy beaches? Braz J Biol 73(1):19–28. https://doi.org/10.1590/S1519-69842013000100004

Maria TF, Troch M, Vanaverbeke J et al (2011) Use of benthic vs planktonic organic matter by sandy-beach organisms: a food tracing experiment with 13C labelled diatoms. J Exp Mar Biol Ecol 407(2):309–314. https://doi.org/10.1016/j.jembe.2011.06.028

Maria TF, Wandeness AP, Esteves AM (2016) State of the art of the meiofauna of Brazilian Sandy Beaches. Braz J Oceanogr 64(SPE2):17–26. https://doi.org/10.1590/S1679-875920160946064sp2

Martins IX (1996) Predação de Natica marochiensis (Gmelin, 1791) (Mollusca: Gastropoda: Naticidae) da praia de Quitéria, município de Icapuí-Ceará. Dissertation, Universidade Federal da Paraíba

Mattos G, Cardoso RS, Santos AS (2013) Environmental effects on the structure of polychaete feeding guilds on the beaches of Sepetiba Bay, South-Eastern Brazil. J Mar Biol Assoc UK 93(4):973–980. https://doi.org/10.1017/S0025315412000707

McLachlan A, Defeo O (2018) The ecology of sandy shores, 3rd edn. Academic Press, London, p 560

McLachlan A, Dorvlo A (2005) Global patterns in sandy beach macrobenthic communities. J Coast Res 214:674–687. https://doi.org/10.2112/03-0114.1

McLachlan A, Jaramillo E (1995) Zonation on sandy beaches. Oceanogr Mar Biol Ann Rev 33:305–335

Michaud KM, Emery KA, Dugan JE et al (2019) Wrack resource use by intertidal consumers on sandy beaches. Estuar Coast Shelf Sci 221:66–71. https://doi.org/10.1016/j.ecss.2019.03.014

Negreiros-Fransozo ML, Fransozo A, Mantelatto FLM et al (1997) Anomuran species (Crustacea, Decapoda) in their ecological distribution at Fortaleza Bay sublittoral, Ubatuba, São Paulo, Brazil. Iheringia Ser Zool 83:187–194

Nelson WO (1986) Predation and prey population variation in a high energy sand beach macrofaunal community. Ophelia 26:305–316. https://doi.org/10.1080/00785326.1986.10421996

Neves LP, Silva PSR, Bemvenuti CE (2007) Zonation of benthic macrofauna on Cassino Beach, southernmost Brazil. Braz J Oceanogr 55(4):293–307. https://doi.org/10.1590/S1679-87592007000400006

O'Brien BB, Mello K, Litterer A et al (2018) Seaweed structure shapes trophic interactions: a case study using a mid-trophic level fish species. J Exp Mar Biol Ecol 506:1–8. https://doi.org/10.1016/j.jembe.2018.05.003

Odebrecht C, Du Preez DR, Abreu PC et al (2013) Surf zone diatoms: a review of the drivers, patterns and role in sandy beaches food chains. Estuar Coast Shelf Sci 150:24–35. https://doi.org/10.1016/j.ecss.2013.07.011

Ólafsson E (2003) Do macrofauna structure meiofauna assemblages in marine soft-bottoms? A review of experimental studies. Vie Milieu 53(4):249–265

Pardo EV, Amaral ACZ (2004) Feeding behavior of *Scolelepis* sp (Polychaeta: Spionidae). Braz J Oceanogr 52(1):74–79. https://doi.org/10.1590/S1679-87592004000100007

Peiro DF, Mantelatto FL (2011) Population dynamics of the pea crab *Austinixa aidae* (Brachyura, Pinnotheridae): a symbiotic of the ghost shrimp *Callichirus major* (Thalassinidea, Callianassidae) from the southwestern Atlantic. Iheringia Ser Zool 101(1–2):5–14. https://doi.org/10.1590/S0073-47212011000100001

Penchaszadeh PE, Olivier SR (1975) Ecologia de una población de "berberecho" (*Donax hanleyanus*) en Villa Gesell, Argentina. Malacologia 15(1):133–146

Petracco M (2008) Produção secundária da macrofauna bentônica da zona entremarés no segmento norte da praia do Una, litoral sul do estado de São Paulo. Thesis, Universidade de São Paulo

Petracco M, Cardoso RS, Corbisier TN et al (2012) Brazilian sandy beach macrofauna production: a review. Braz J Oceanogr 60(4):473–484. https://doi.org/10.1590/S1679-87592012000400006

Petracco M, Cardoso RS, Turra A (2013) Pattern of sandy-beach macrofauna production. J Mar Biol Assoc UK 93(7):1717–1725. https://doi.org/10.1017/S0025315413000246

Pinheiro MAA, Pardal-Souza AL (2018) Historical background of the population biology of the swimming crab *Arenaeus cribrarius* (Crustacea: Portunidae) in the Ubatuba coast, Southeast Brazil. Bol Inst Pesca 42(2):419–429. https://doi.org/10.20950/1678-2305.2016v42n2p419

Pinotti RM, Minasi DM, Colling LA et al (2014) A review on macrobenthic trophic relationships along subtropical sandy shores in southernmost Brazil. Biota Neotrop 14(3). https://doi.org/10.1590/1676-06032014006914

Pinto TKO, Bemvenuti CE (2003) Effects of burrowing macrofauna on meiofauna community structure. Acta Limnol Bras 15(3):41–51

Prinz K, Kelly TC, O'Riordan RM, Culloty SC (2010) Occurrence of macroparasites in four common intertidal molluscs on the south coast of Ireland. Marine Biodiversity Records 3:1–7

Queiroz LR, Rocha-Barreira CA (2019) Morphodynamics and macrofauna communities in 12 sandy beaches of Brazil northeast: a semiarid tropical study. In: Mochel FR (ed) Gerenciamento Costeiro e Gerenciamento Portuário, vol 2. Atena Editora, Ponta Grossa, pp 85–106

Quillien N, Nordström MC, Schaal G et al (2016) Opportunistic basal resource simplifies food web structure and functioning of a highly dynamic marine environment. J Exp Mar Biol Ecol 477:92–102. https://doi.org/10.1016/j.jembe.2016.01.010

Ramos NLM (2014) Dinâmica populacional e produção secundária de *Talorchestia tucurauna* (Müller, 1864) (Amphipoda: Talitridae) em duas praias do litoral pernambucano. Thesis, Universidade Federal de Pernambuco.

Reese ES (1969) Behavioral adaptations of intertidal hermit crabs. Am Zool 9:343–355

Reise K (1985) Macrofauna promotes meiofauna. Tidal flat ecology. An experimental approach to species interactions. In: Billings WD, Golley F, Lange OL et al (eds) Ecological studies: analysis and synthesis, vol 54. Springer-Verlag, Berlin, pp 119–145

Rocha-Barreira CA (2002) Feeding behavior of *Olivancillaria vesica auricularia* (Lamarck, 1810):(Mollusca, Olividae). Thalassas 18(2):83–89

Rocha-Barreira CA (2003) Desenvolvimento embrionário e larval de *Olivancillaria vesica auricularia* (Lamarck, 1810) (Mollusca, Gastropoda, Olividae) em condições de laboratório. Atlântica (Rio Grande) 25(2):107–116

Rocha-Barreira CA, Barros KVS, Matthews-Cascon H et al (2017) Ecology of Mollusc communities in marine environments: central region of the Semiarid Coast of Brazil. In: Jenkins OP (ed) Advances in animal science and zoology, 1st edn. Nova Science Publishers, New York, pp 165–208

Rocha-Barreira CA, Monteiro DO, Franklin-Júnior W (2001) Macrofauna bentônica da faixa entremarés da Praia do Futuro, Fortaleza, Ceará, Brasil. Arq Ciênc Mar 34(1–2):23–38

Rodrigues AS, Shimizu RM (1997) Autoecologia de *Callichirus major* (say, 1818). Oecologia Bras 3(1):155–170

Rosa LC, Caron E, Sousa-Souto L (2013) First record of *Bledius caribbeanus* Blackwelder, 1943 (Coleoptera: Staphylinidae: Oxytelinae) from Brazil and distributional extension of *B. hermani* Caron and Ribeiro-Costa, 2007. Check List 9(2):450–451

Rosa-Filho JS, Gomes TP, Almeida MF, Silva RF (2011) Benthic fauna of macrotidal sandy beaches along a small-scale morphodynamic gradient on the Amazon coast (Algodoal Island, Brazil). J Coast Res SI64:435–439. www.jstor.org/stable/26482209

Santos PJP (1991) Morphodynamical influence of a temporary freshwater stream on the population dynamics of *Scolelepis gaucha* (Polychaeta:Spionidae) on a sandy beach in Southern Brazil. Bull Mar Sci 48(3):657–664

Shah Esmaeili Y, Checon HH, Corte GN et al (2021) Parasitism by isopods in sandy beach fish assemblages: role of urbanization and environmental characteristics. Hydrobiologia 848:4885–4901. https://doi.org/10.1007/s10750-021-04680-0

Sherman KM, Reidenauer JA, Thistle D et al (1983) Role of a natural disturbance in an assemblage of marine free-living nematodes. Mar Ecol Prog Ser 11:23–30. www.jstor.org/stable/24815063

Siebert T, Branch GM (2007) Influences of biological interactions on community structure within seagrass beds and sandprawn-dominated sandflats. J Exp Mar Biol Ecol 340(1):11–24. https://doi.org/10.1016/j.jembe.2006.08.007

Souza JS, Canellas BGC, Sakabe R et al (2019) The parasitic isopod *Mothocya nana* drives dietary shifts and poorer condition of Brazilian silversides *Atherinella brasiliensis*. Dis Aquat Org 132(3):229–239. https://doi.org/10.3354/dao03307

Thompson RM, Brose U, Dunne JA et al (2012) Food webs: reconciling the structure and function of biodiversity. Trends Ecol Evol 27(12):689–697. https://doi.org/10.1016/j.tree.2012.08.005

Tomme J, Degraer S, Vincx M (2014) Role of predation on sandy beaches: predation pressure and prey selectivity estimated by laboratory experiments. J Exp Mar Biol Ecol 451:115–121. https://doi.org/10.1016/j.jembe.2013.11.009

Turra A, Denadai MR (2004) Interference and exploitation components in interespecific competition between sympatric intertidal hermit crabs. J Exp Mar Biol Ecol 310(2):183–193. https://doi.org/10.1016/j.jembe.2004.04.008

Turra A, Gonçalves MAO, Denadai MR (2005) Spatial distribution of the ghost crab Ocypode quadrata in low-energy tide-dominated sandy beaches. J Natl Hist 39(23):2163–2177. https://doi.org/10.1080/00222930500060165

Urban-Malinga B, Drgas A, Gromisz S et al (2014) Species-specific effect of macrobenthic assemblages on meiobenthos and nematode community structure in shallow sandy sediments. Mar Biol 161:195–212. https://doi.org/10.1007/s00227-013-2329-y

Vance RR (1972) Role of shell adequacy in behavioral interactions involving hermit crabs. Ecology 53:1075–1083

Vasconcellos RM, Araújo FG, Santos JNS et al (2011) Diel seasonality in fish biodiversity in a sandy beach in South-Eastern Brazil. J Mar Biol Assoc UK 91(6):1337–1344. https://doi.org/10.1017/S0025315410000652

Veloso VG, Caetano CHS, Cardoso RS (2003) Composition, structure and zonation of intertidal macrofauna in relation to physical factors in microtidal sandy beaches at Río de Janeiro State, Brazil. Sci Mar 67(4):393–402. https://doi.org/10.3989/scimar.2003.67n4393

Veloso VG, Sallorenzo IA, Ferreira BCA et al (2009) *Atlantorchestoidea brasiliensis* (Crustacea: Amphipoda) as an indicator of disturbance caused by urbanization of a beach ecosystem. Braz J Oceanogr 58(1):13–21. https://doi.org/10.1590/S1679-87592010000100002

Veloso VG, Silva ES, Caetano CH et al (2006) Comparison between the macroinfauna of urbanized and protected beaches in Rio de Janeiro State, Brazil. Biol Conserv 127(4):510–515. https://doi.org/10.1016/j.biocon.2005.09.027

Viana MG, Rocha-Barreira CA, Grossi-Hijo CA (2005) Macrofauna bentônica da faixa entremarés e zona de arrebentação da praia de Paracuru (Ceará-Brasil). Braz J Aquat Sci Technol 9(1):75–82

Vianna AXMR, Borzone CA (2015) Distribuição espacial e temporal de espécies de *Bledius* (Coleoptera, Staphylinidae) em praias arenosas no sul do Brasil. Iheringia Ser Zool 105(1):122–132. https://doi.org/10.1590/1678-476620151051122132

Volkenborn N, Polerecky L, Hedtkamp SIC et al (2007) Bioturbation and bioirrigation extend the open exchange regions in permeable sediments. Limnol Oceanogr 52(5):1898–1909. https://doi.org/10.4319/lo.2007.52.5.1898

von Dassow YJ, Strathmann RR (2005) Full of eggs and no place to lay them: hidden cost of benthic development. Mar Ecol Prog Ser 294:23–34

Vooren CM (1997) Sea and shore birds. In: Seeliger U, Odebrecht C, Castello JP (eds) Subtropical convergence environments: the coast and sea in the Southwestern Atlantic. Springer-Verlag, Berlin, pp 154–159

Vooren CM, Chiaradia A (1990) Seasonal abundance and behavior of coastal birds on Cassino Beach, Brazil. Ornit Neotrop 1:9–24

Widdicombe S, Austen MC, Kendall MA et al (2000) Bioturbation as a mechanism for setting and maintaining levels of diversity in subtidal macrobenthic communities. Hydrobiol 440:369–377. https://doi.org/10.1007/978-94-017-1982-7_34

Wolcott TG (1978) Ecological role of ghost crabs *Ocypode quadrata* (Fabricius) on an ocean beach: scavengers or predators? J Exp Mar Biol Ecol 31(1):67–82. https://doi.org/10.1016/0022-0981(78)90137-5

Yannicelli B, Palacios R, Gimenez L (2002) Swimming ability and burrowing time of two cirolanid isopods from different levels of exposed sandy beaches. J Exp Mar Biol Ecol 273:73–88. https://doi.org/10.1016/S0022-0981(02)00146-6

Chapter 8
Ecosystem Services

Helio Herminio Checon, Luciana Yokoyama Xavier, and Leandra Regina Gonçalves

8.1 Ecosystem Services

Ecosystems are broadly defined as systems that encompass the interaction between living organisms with the abiotic features of a particular environment. For much time, scientists have framed ecosystems as natural entities detached from human influence. However, as the need to understand the intrinsic interaction between natural systems and human societies as components of a single system emerged, science has moved to a broader approach. The social-ecological systems frameworks describe an integrated system of people and nature in which feedback occurs between human and biophysical system elements. Among them, the ecosystem services framework translates the idea that ecosystems provide unreplaceable benefits to human populations (Daily and Matson 2008).

The ecosystem services classifications and approaches have been developed throughout the end of the last century, engaging scholars from different areas (Braat and de Groot 2012; Costanza et al. 2017). Although the term can be traced back to 1981 (Costanza et al. 2017), the Millennium Ecosystem Assessment (MEA 2005)

H. H. Checon (✉)
Departamento de Oceanografia Biológica, Instituto Oceanográfico da Universidade de São Paulo, São Paulo, SP, Brazil

Departamento de Biologia Animal, Instituto de Biologia, Universidade Estadual de Campinas, Campinas, SP, Brazil

L. Y. Xavier
Departamento de Oceanografia Biológica, Instituto Oceanográfico da Universidade de São Paulo, São Paulo, SP, Brazil

L. R. Gonçalves
Departamento de Oceanografia Biológica, Instituto Oceanográfico da Universidade de São Paulo, São Paulo, SP, Brazil

Departamento de Ciências do Mar, Instituto do Mar da Universidade Federal de São Paulo, Santos, SP, Brazil

A. C. Z. Amaral et al. (eds.), *Brazilian Sandy Beaches*, Brazilian Marine Biodiversity, https://doi.org/10.1007/978-3-031-30746-1_8

promoted global recognition of ecosystem services and their importance. This assessment categorized and defined ecosystem services to assess the impacts on human economics and well-being caused by ecosystem degradation (Carpenter et al. 2009). The findings that anthropic activities impacted 60% of ecosystem services worldwide were a warning to politicians and scientists that boosted studies on ecosystem services and fostered their application as a conservation tool (Vihervaara et al. 2010; Costanza et al. 2017).

The MEA also proposed the currently most applied framework to classify ecosystem services, distinguished among provisioning, regulating, supporting, and cultural services (MEA 2005). Provisioning services are the goods generated by ecosystem functioning (e.g., food, raw materials, energy, and others). Regulating services are those responsible for maintaining quality or preventing damages to goods (e.g., erosion prevention, water purification, carbon sequestration, and others). Supporting services are related to processes derived from the ecosystem functioning that maintain the ecosystem goods (e.g., nutrient cycling, photosynthesis, pollination, and others). Cultural services are related to the immaterial benefits resulting from the connection of humans with the ecosystem (e.g., tourism, religious practices, topophilia, and others) (MEA 2005).

Building on the ecosystem services approach, other frameworks emerged (see Costanza et al. 2017 for a detailed review of the most prominent ones). The Economics of Ecosystems and Biodiversity project (TEEB) stresses the economic benefits of nature (TEEB 2008). The Common International Classification of Ecosystem Services (CICES) distinguishes ecosystems' structure, processes, services, benefits, and values, organized in a linear cascade model that highlights the influence of the social component on the natural capital (feedback loop; e.g., value and governance systems) (Haines-Young and Potschin 2018). The Intergovernmental Platform on Biodiversity and Ecosystem Services (IPBES) understands ecosystems services as part of broader "nature's contributions to people" and highlights the intrinsic value of nature, the direct connection between nature and good quality of life, and the need to include traditional views on human–nature relations (e.g., traditional and indigenous views on nature) in decision-making (Pascual et al. 2017).

As the discussion on the concept evolves, it is unlikely that a single definition of ecosystem services is possible, even though it would facilitate global comparisons (Fisher et al. 2008). Despite this, management and conservation policies are increasingly applying the concept of ecosystem services to guide the decision-making process and translate the importance of the conservation of ecosystem functioning (Costanza et al. 1997; Fisher et al. 2008). In the next few years, the concept of ecosystem services, its frameworks, and approaches shall undoubtedly be further developed and applied by different areas of knowledge and management sectors, which will improve it and connect approaches through interdisciplinarity and integrated management.

Concerns regarding the application of the ecosystem services concept exist. One of the main criticisms lies in what is perceived as an anthropocentric view of nature, which results in the understanding of ecosystems as a commodity that needs to be attributed a monetary value, whereas the valuing process in itself may be subject to ethical questioning (Schröter et al. 2014; Costanza et al. 2017). Attributing monetary value to cultural services has proven to be especially tricky and leads to their undervaluation (Daniel et al. 2012). Other critics point to the misconception that ecosystem processes only provide benefits (McCauley 2006). Despite criticisms, the concept has been valuable for communicating nature's values and implementing management policies. The positive view of ecosystems represents a utility rather than a hindrance to their protection (Schröter et al. 2014).

It is well known that the ocean and coastal areas support the livelihoods of hundreds of millions of people and contribute to more than 60% of the total economic value of the biosphere (Martínez et al. 2007; Costanza et al. 2014). However, ecosystem services studies for these environments are still incipient, and the complexity and the dynamic among coastal ecosystems and well-being have not been comprehensively assessed (Blythe et al. 2020). Given the crescent centrality of the ocean in international sustainability discussions, the research topic is growing rapidly while academic and nonacademic publications are quantitatively increasing.

Among coastal ecosystems, sandy beaches have distinctive cultural, economic, and environmental importance. They provide habitat and food for many marine species and attract people who enjoy their beauty. The activities associated with their direct use support the economy of coastal cities worldwide. Sandy beach ecosystem services, however, still have much to be explored and studied. In Brazil, a few articles have addressed the topic (see Amaral et al. 2016), and many are descriptive and focus on single ecosystem services. In this chapter, we present an overview of ecosystem services provided by sandy beaches worldwide and review examples of studies on the Brazilian coast to promote a more comprehensive understanding of beach ecosystems. For this chapter, we adopt the broader definition of ecosystem services as "the ecological characteristics, functions, or processes that directly or indirectly contribute to human well-being" (Costanza et al. 2017).

8.2 Sandy Beaches Ecosystem Services

Sandy beaches are widespread ecosystems, occupying around a third of the unfrosted coastlines worldwide (Luijendijk et al. 2018). Like other coastal depositional ecosystems (such as mangroves, estuaries, and seagrasses), sandy beaches are dynamic environments influenced by continental (i.e., riverine input, winds, organic matter exportation) and oceanic processes (i.e., tides, waves). Thus, the functioning of this ecosystem depends on the stability and conservation of terrestrial and marine environments. Similarly, the threats also come from terrestrial and marine activities, including human occupation on coastlines worldwide.

Sandy beaches are the coastal ecosystem most used by human societies (Defeo et al. 2009), as they are the most accessible coastal habitat, providing a gateway to the sea for economic and recreational activities. They carry significant cultural and economic importance for human societies. However, sandy beaches were for a long time seen as desertic systems in terms of biodiversity. The works of Anton McLachlan during the 1980s were among the first to consider them as singular ecosystems and fostered the development of sandy beach science. Nowadays, sandy beaches still are the least studied coastal ecosystem – number of publications from 1950 to 2013: sandy beach = 2936 studies; rocky shores = 3157 studies; mangroves = 11,149 studies; coral reefs = 20,065 studies; and estuaries = 36,358 studies (Nel et al. 2014) – but an exponential increase in the number of studies has been registered since the mid-1980s (Nel et al. 2014), a trend that continues up to date. Similarly, sandy beach services have been more recognized and valued only in the last decades. Due to this recent recognition, efforts to better understand them, their role, distribution, impacts, and values are still needed. Several services have been attributed to sandy beaches, encompassing all MEA-established categories (Fig. 9.1).

In Brazil, sandy beaches are an integral part of society. Similar to the worldwide trend, the country's highest population densities are in coastal states and cities. Thus, the urban development of cities fosters occupation and subsequent changes in coastal environments. Brazilian sandy beaches provide fundamental ecosystem goods and services, playing an essential role in human population health and in promoting biodiversity conservation. Among the practices and activities developed on Brazilian beaches, shellfish harvesting and fishing modalities are the main activities carried out by coastal populations. Sandy beaches also host cultural manifestations and support fishing (e.g., the launching and beaching of small boats), tourism, and recreation activities. Although less common, Brazilian beaches provide other goods and services, including aquaculture and the exploitation of sea salt (especially in the northeastern region of the country) (Amaral et al. 2016).

Despite their ecological and social importance, human activities broadly impact sandy beaches, both by direct (e.g., dredging, overexploitation of species, coastal engineering) or indirect impacts (e.g., organic enrichment, plastic pollution, heavy metals contamination) (Defeo et al. 2009; Ivar do Sul and Costa 2014). Owing to these impacts and threats, the management of coastal areas has garnered much attention, especially with the more recent concerns regarding the climate change scenarios to which the coastal ecosystems are highly vulnerable (Amaral et al. 2016; Xavier et al. 2022). Thus, a comprehensive understanding of the sandy beaches' ecosystem services is necessary to provide insights into where management could be improved to maintain the natural condition of these ecosystems. In the next sections, we will review some of the most recognized and well-studied services on beaches, summarized in Fig. 8.1, using examples of studies carried out on shorelines around the world and the Brazilian coast. Some of these services are summarized within other chapters of this book (e.g., Chaps. 2, 4, and 5).

Fig. 8.1 Ecosystem services provided by beaches around the world

8.3 Provisioning Services

Fisheries are one of the most common coastal activities around the world and support the economy of coastal cities and the economy and subsistence of local communities (Vasconcellos et al. 2011; Hanazaki et al. 2013). Small-scale and artisanal fisheries comprise the majority of fishing activities and support the livelihood and

Fig. 8.2 The provision of food is the main provisioning service generated by sandy beaches. (**a**) Small-scale fishery by hand collection of mollusks in the intertidal area (state of São Paulo, photo: Álvaro Migotto/Cifonauta); (**b**) fisheries vessels equipped with shrimp fishing apparatus (state of Rio de Janeiro state, photo: Guillermo Bresciano/Unsplash); (**c**) fish catching on the surf zone is usually made using seine nets, with or without the use of vessels (state of Santa Catarina, photo: Cassiano Psomas/Unsplash); examples of commercially relevant sandy beach invertebrates: (**d**) the blue crab of the genus *Callinectes* (photo: Álvaro Migotto/Cifonauta); (**e**) the ghost shrimp *Callichirus major* (state of São Paulo; photo: Francisco Buchmann) and (**f**) the heart clam *Anomalocardia flexuosa* (state of São Paulo, photo: Márcia Denadai)

culture of many families in traditional fishing communities (McGoodwin 2001; Begossi 2010). In Brazil, it responds to 53–70% of total fish landings (Begossi 2010; Vasconcellos et al. 2011). Beaches are an integral part of this activity. On the one hand, they provide the resource to be fished, support food webs of fishing resources, and provide sites for foraging and nursery. On the other hand, they provide easy access to the sea for small-scale fisheries and sites where fish landing and initial commercialization occur (Vasconcellos et al. 2011; Carrilho and Sinisgalli 2018) (Fig. 8.2).

Sandy beaches house many vertebrate and invertebrate species that are explored for food, especially by small-scale fisheries (for profit, subsistence, or recreational purposes). Fisheries of marine invertebrates are a common practice on sandy beaches around the world. These are almost always carried out by small-scale, artisanal methods, and capture is used for human consumption, bait for other fisheries, or handicraft purposes (Defeo 2003; Turra et al. 2016). A study conducted on nine sandy beaches of the state of Rio Grande do Norte (Northeast Brazil) found nine molluscan species explored for subsistence and sale, with a selling price ranging from US$ 1.2 to US$ 3 kg^{-1} (Silva and Martins 2017).

The most commonly exploited sandy beach species in Brazil is the heart clam *Anomalocardia flexuosa* (known previously as *Anomalocardia brasiliana*) (Chap. 4), a mollusk found mainly on estuarine beaches (Fig. 8.2e). This species, known popularly as berbigão, vongole, or búzio (depending on the regional context), is extensively gathered and represents an integral part of the economy of traditional gathering communities. In northeast Brazil, where *A. flexuosa* is the most relevant exploited species, mollusk gathering represents up to 5.8% of the total coastal fisheries income (Silva-Cavalcanti and Costa 2011). The clam *Tivella mactroides* (also popularly known as *berbigão* or *sapinhauá*) is often captured and sold as an alternative to *A. flexuosa* (Turra et al. 2016; Silva and Martins 2017). On the state of São Paulo (southeast Brazil), recreational exploration of *T. mactroides* by residents and tourists is estimated to range from 8.8 to 24.6 tons/year (Denadai et al. 2015). On the southernmost temperate coast of Brazil, the yellow clam *Amarilladesma mactroides* represents an economic and food resource (Ortega et al. 2012).

Although associated mainly with rocky shores and mangroves, mussel beds may also grow on sandy beaches with low hydrodynamics and very fine sands. Among them, the *Mytella* species are harvested for human consumption and sold for US$ 1.21/unit in the state of Rio de Janeiro (Silva and Martins 2017). Besides food provisioning, beached shells of mollusks are raw materials used for commercial purposes, sold as a whole, or used to craft souvenirs (Turra et al. 2016).

Sandy beach crustaceans are also captured for commercial purposes. The ghost shrimp *Callichirus major* (popularly known as *corrupto*) is a commonly captured crustacean (Fig. 8.2f), taken out for its burrows on the lower intertidal using suction tools. This crustacean is commonly used as bait for small-scale and recreational fishing activities (Souza and Borzone 2003). A study carried out at beaches in south Brazil found that *C. major* harvesting was responsible for depleting ~10% of the total stock, which may threaten these populations (Chap. 6). Impacts on these organisms may affect the whole ecosystem and its services since they are important bioturbators whose activity enhances the nutrient cycling services provided by beaches (Machado et al. 2004).

Blue crabs are also economically relevant crustaceans that inhabit sandy beaches. In Brazil, the *Callinectes danae*, *C. sapidus,* and *C. ornatus* (Severino-Rodrigues et al. 2001) are fished by small traditional communities. Anacleto et al. (2016) reported a mean commercial value of US$ 11.20 kg^{-1} in the state of Paraná state

(South Brazil). Fisheries of these crustacean species, however, have seen a decline during the last decades, which can be attributed to environmental degradation and changes, threatening the food security of many traditional communities (Bail and Branco 2007; Carrilho and Sinisgalli 2018).

Shrimp fishing represents another valuable economic resource for traditional coastal communities (Bail and Branco 2007). The most common species associated with shrimp fishing are the Atlantic bob (*Xiphopenaeus kroyeri*, known locally as camarão-sete-barbas), the white shrimp (*Litopenaues schmitii*), and the pink shrimp (*Farfantepenaeus brasiliensis* and *F. paulensis*) (Souza et al. 2009). Although shrimp fishery occurs in areas farther from the coast or inside estuaries, juvenile stages can be found occupying the surf zone of beaches and depend on the conservation of beach ecosystems to grow.

Fish captures present the same scenario. Large-scale fish captures occur outside the beach area but are, in many cases, connected to the beach ecosystem, which has a fundamental role in fish life cycles. Many commercially important fish species captured in other marine ecosystems (e.g., estuaries and open ocean) occur on beach surf zones during their juvenile stages (nursery function, see Supporting Services).

Table 8.1 Commercially important fish species whose juvenile stages occupy the surf zone of sandy beaches in Brazil

Popular name (English)	Popular name (Portuguese)	Scientific denomination	Source
Anchovies	Anchova	*Anchoa januaria*	Pereira et al. (2015)
Anchovies	Anchova	*Anchoa lyolepsis*	Félix et al. (2007)
Anchovies	Anchova	*Anchoa tricolor*	Pereira et al. (2015)
			Félix et al. (2007)
Catfish	Bagre	*Genides genides*	Shah Esmaeili et al. (2021)
Catfish	Bagre	*Genides barbus*	Shah Esmaeili et al. (2021)
Croaker	Corvina	*Micropogonias furnieri*	Pereira et al. (2015)
False herring	Sardinha-cascuda	*Harengula clupeola*	Pereira et al. (2015)
Herrings	Sardinha	*Sardinella brasiliensis*	Félix et al. (2007)
Leatherjacket	Guavira	*Oligoplites saliens*	Shah Esmaeili et al. (2021)
Lebranche mullet	Parati	*Mugil liza*	Pereira et al. (2015)
Pompano	Pampo	*Trachinotus carolinus*	Andrades et al. (2012)
Pompano	Pampo	*Trachinotus falcatus*	Andrades et al. (2012)
Silverside	Peixe-rei	*Atherinella brasiliensis*	Del Favero and Dias (2015)
Swordspine snook	Robalo-flecha	*Centropomus ensiferus*	Shah Esmaeili et al. (2021)
Swordspine snook	Carapeba-listrada	*Eugerres brasilianus*	Shah Esmaeili et al. (2021)

Studies on the Brazilian coast reveal economically important species whose juvenile or adult stages occupy the surf zones of sandy beaches (Table 8.1).

Fish captures may also occur within the beach area. However, due to the characteristics of sandy beaches (e.g., shallow habitats, wave action), fish captures in the beach area are usually small-scale, carried out using small vessels or trawling using beach seine nets. These small-scale fisheries are necessary for the economy and subsistence of local communities in many areas of the Brazilian coast (Nascimento and Sassi 2007). Vertebrate fisheries have long been practiced on Brazilian shores as a recreational activity, and fishing contests increase in number yearly (Freire et al. 2014) (Fig. 8.2).

Although characterized by small-scale practices, sandy beach fisheries can be a threat to the beach environment. Many fished species are vulnerable to overexploitation, and studies show a decrease in stocks of beach invertebrate species worldwide (Castilla and Defeo 2001). Intense shell removal may also negatively impact ecosystem processes. Shells are important for epibiontic algae and organisms, offer protection for hermit crabs, and their degradation is important to the maintenance of the calcium carbonate cycle within the marine environment (Kowalewski et al. 2014). Other activities on sandy beaches can also affect fisheries. For instance, urbanization and tourism may impact the richness and abundance of fish in surf zones, leading to a stock decrease both for coastal and oceanic fisheries (Vargas-Fonseca et al. 2016; Costa et al. 2017). The occupation of beach areas by tourists or tourism infrastructure (such as restaurants, kiosks, and bungalows) reduces species richness and biomass of invertebrates (Corte et al. 2022) and impacts fisheries activities (Hoefle 2014). Sandy beaches should be understood as habitats whose conservation is essential for the whole artisanal fishing sector and managed as such to promote a more extensive set of ecosystem services.

As for the other provisioning services mentioned in Fig. 8.1 (i.e., clean water, genetic and ornamental resources, biodiversity), there is a less known recognition of these services, at least in the ecosystem services framework. For instance, biodiversity assessments and ecological studies recognizing beach diversity, which has strong endemism to this ecosystem (Harris et al. 2014), are common on Brazil (Amaral et al. 2016); however, the recognition of the importance of biodiversity as a service, or its contribution to other services, is poorly explored (Checon et al. 2022).

8.4 Supporting Services

Sandy beaches are substantial foraging and nursery grounds for many fish species. Because of the protection provided by waves and shallow environments, juvenile stages find shelter from predators in these areas, and they may feed and grow with higher success before leaving for estuaries and open ocean to complete their life cycles, hence the nursery function (Beck et al. 2001; Able et al. 2013). Worldwide,

Fig. 8.3 Examples of beaches as nurseries, nesting, and feeding grounds for nonresident species. (**a**) A juvenile pompano (*Trachinotus carolinus*) in the beach surf zone; (**b**) the green turtle *Chelonia mydas* on a beach in Northeast Brazil (Praia do Forte, state of Bahia); (**c**) the loggerhead sea turtle (*Caretta caretta*) laying eggs in a nesting beach site in Southeast Brazil (Comboios, state of Espírito Santo); (**d**) release of baby turtles promoted by the Projeto TAMAR (photos b, c, and d; Banco de imagens Projeto Tamar|Fundação Pró-Tamar); (**e**) mixed flocks of Sternidae (*Onychoprion fuscatus* and *Anous stolidus*) and Sulidae (*Sula dactylatra*) occupying the dune area of a beach in Northeast Brazil (Atol das Rocas, state of Rio Grande do Norte) (photo: Maíra Pombo); (**f**) mixed flocks of Scolopacidae and Sternidae in the lower intertidal area of a beach in Brazil (Massambaba-Arraial do Cabo, Região dos Lagos, state of Rio de Janeiro) (photo: Eduardo Gomes Pimenta)

many commercially important fish species use beaches' surf zones as nursery habitats (Able et al. 2013; Olds et al. 2018).

The nursery function has been identified in studies developed in different areas along the Brazilian coast (Andrades et al. 2012; Pereira et al. 2015) (Fig. 8.3a). However, nurseries are not provided equally by all morphodynamics types of beaches. Two studies on the northeast and southeast Brazil reported a higher richness of species and abundance of juvenile individuals in dissipative beaches (i.e., those with wide surf zones that dissipate wave energy) than reflective ones (i.e., those with reduced surf zones and harsh swash climate) (Vasconcellos et al. 2007; Oliveira and Pessanha 2014). A higher abundance of planktonic food sources and lower stress from hydrodynamic conditions explained richness and abundance variations (Vasconcellos et al. 2007). It does not mean that reflective beaches are unimportant as nursery grounds. Morphodynamic conditions affect fish assemblages' composition in surf zones (Olds et al. 2018), and fish species with a higher capacity to withstand harsh environmental conditions may rely on reflective conditions to feed and protect from predators (Shah Esmaeili et al. 2021). Understanding the links between morphodynamics and nursery habitat can indicate which sandy beaches may be more suitable for protection to secure fish stocks.

Aside from the diverse benthic invertebrate fauna, sandy beaches also support the life cycles of resident and nonresident species of reptiles, birds, and mammals (Chap. 5). These organisms use the supra and intertidal areas of beaches for foraging, resting, and nesting, and act in exporting and importing material and connecting sandy beach food webs to other ecosystems. Sea turtles are one of the most emblematic examples of the use of sandy beaches for nesting (Fig. 8.3b, c). All seven existing species migrate from the ocean toward beach areas during reproduction. The nesting of sea turtles influences sandy beach processes by inputting organic matter and increasing organism turnover (Madden et al. 2008), which affects nutrient cycling and food web support services. Sea turtles transport a varied number of epibiont organisms, contribute to the transfer of energy from nutrient-rich to nutrient-poor beaches, and affect the marine food web, especially as consumers of small invertebrates (Bjorndal and Jackson 2002).

As sea turtles depend on sandy beaches, preserving supporting service is essential for their conservation. Given the ecological role of sea turtle species, their protection also affects the stability of coastal processes and represents one of the services related to biodiversity. The International Union for Conservation of Nature Red List categorizes six of the seven sea turtle species as threatened: leatherback (*Dermochelys coriacea*), loggerhead (*Caretta caretta*), and olive ridley (*Lepidochelys olivacea*) sea turtles are vulnerable; green sea turtle (*Chelonia mydas*) is endangered; hawksbill (*Eretmochelys imbricata*) and Kemp's ridley (*Lepidochelys kempii*) sea turtles are critically endangered (IUCN 2020). The vulnerability of such species tends to rise with the reduction of beach areas by coastal urbanization and sea-level rise.

In Brazil, turtles are likely the most known flagship species for coastal preservation. Five sea turtle species use Brazilian sandy beaches as nesting sites – *Caretta caretta*, *Chelonia mydas*, *Eretmochelys imbricata*, *Lepidochelys olivacea*, and *Dermochelys coriacea* – all listed as threatened. Most of these species nest on beaches of the northeast (especially the states of Bahia, Sergipe, Alagoas, and Rio Grande do Norte) and southeast (states of Espírito Santo and Rio de Janeiro) (Reis and Goldberg 2017). An exception is *Chelonia mydas*, the species with more abundance of juveniles along the coast, which nests preferably in oceanic islands (Reis and Goldberg 2017) (Chap. 5). The dependence on beach sites makes turtles vulnerable to egg poaching, net trampling, and capture by fishing activities. A total of 2515 turtles from 4 different species were accidentally captured from 1991 to 1998 on an important nesting site in Ubatuba, state of São Paulo (Southeast Brazil) (Santos et al. 2016). Many efforts for turtle conservation are in place in the country. The nongovernmental organization Projeto TAMAR has a prominent role in protecting nesting sites and promoting environmental education, having more than 20 units along the Brazilian coast, from the state of Santa Catarina (South Brazil) to the state of Ceará (Northeast Brazil) states (Fig. 8.3d).

Beaches of the state of Espírito Santo (Southeast Brazil) exemplify the effectiveness of conservation measures. There, the annual number of leatherback turtle nests increased from 6 (1993–1994) to 92 (2002–2003) due to nests control and egg poaching prevention (Thomé et al. 2007). The same trend was observed for the olive ridley sea turtle on beaches of the states of Sergipe and Bahia (Northeast Brazil) where annual nests increased from 252 in 1991–92 to 2606 in 2002–03 (Silva et al. 2007).

Resident and migratory seabird species also use sandy beaches for nesting and foraging (Barbieri and Paes 2008) (Fig. 8.3e, f). Nesting usually occurs where sparse vegetation cover is present at the dune level, which acts as a camouflage against predators (Swaisgood et al. 2018). Feeding may occur throughout the entire beach area. One of the first studies to address beach use by birds was carried out by McLachlan et al. (1980) in South Africa. They estimated that the consumption of invertebrates by 15 species of birds during 1 year totaled 8% of invertebrate biomass. Schlacher et al. (2013) found that birds are the most efficient scavengers to feed upon stranded marine carrion. This finding highlights the importance of beaches as a feeding ground for bird species and also the importance of birds as top predators and scavengers of these ecosystems.

Most studies regarding bird use of beaches in Brazil are concentrated on its south and southeast coast, where the presence of migratory birds from the colder southernmost areas of the continent is more pronounced. Vooren and Brusque (1999) estimated that 111 bird species occur in coastal habitats along the Brazilian coast (with 59% migratory species); 86 of those use sandy beaches, especially waders (suborder Charadrii). In the state of Rio Grande do Sul (South Brazil), Leal and coauthors found that the abundance of migratory shorebirds is higher during austral summer and on nonurbanized beaches (Leal et al. 2013). A 2-year survey on a single beach in Ilha Comprida, state of São Paulo (Southeast Brazil) registered 205,339 individuals of 52 species (with 14 migratory species), which increased beach use by birds during summer months (Barbieri and Paes 2008).

The importance of beaches as a nesting and feeding site for birds is threatened by an array of human activities, such as vehicle trafficking, the presence of domestic pets, beach cleaning, and occupation of the supralittoral area (Meager et al. 2012) (Chap. 9). In addition, the beach used by migratory birds during the summer coincides with the season when tourism is at its highest, making these organisms more susceptible to anthropic impacts. Plastic pollution is also a growing concern for birds. Plastic is an abundant item of marine litter found on Brazilian beaches and is easily available for foraging birds as small pieces or pellets on the strandline (Moreira et al. 2016). Of 115 dead birds (mostly albatrosses and petrels) found on beaches of the state of Rio Grande do Sul (South Brazil), 38.3% had plastic inside their digestive tract, with two species having more plastic than prey items (Colabuono et al. 2009). Thus, urbanization processes can harm the population of birds that

Fig. 8.4 Protection against extreme events and coastal erosion is the main regulating service provided by sandy beaches. When coastal engineering structures are built close to or upon the dune and supratidal area, many damaging processes can occur. Examples of (**a**) coastal squeezing in Balneário Camboriú beach (state of Santa Catarina, photo: Gaspar Rocha/Pixabay) and (**b**) Juquehy beach (state of São Paulo, photo: Fernanda Terra Stori); (**c**) damages to the beachfront at Macumba beach (state of Rio de Janeiro, Brazil, Tomaz Silva/Agência Brasil Fotografias/CCBY (https://creativecommons.org/licenses/by/2.0)); (**d**) Santos municipality (state of São Paulo, photo: Fernanda Terra Stori); and (**e**) flooding of the streets caused by storm waves at Santos municipality (state of São Paulo, photo: Luiz Henrique Alves do Pateo)

depend on sandy beaches for feeding and nesting, highlighting the importance of having beaches with low anthropic pressure to guarantee their use by migratory birds.

8.5 Regulating Services

Coastal protection against storms, waves, and wind action is among the essential services provided by coastal ecosystems, as they act as a buffer to alleviate damages (Barbier et al. 2011; Barbier 2015) (Fig. 8.4). The sloping profile of sandy beach face and sublittoral zones serves as a buffer that reduces wave energy that would otherwise fall upon the adjacent back beach areas. Buffering considerably reduces the damages that would be caused, especially in urbanized areas. For instance, Lins-de-Barros et al. (2016) estimated losses totaling US$ 59,171.60 km^{-1} due to property damages caused by storm waves on an urbanized beach in the state of Rio de Janeiro, Southeast Brazil. However, anthropic interventions can compromise this protection promoted by beaches, which often comes at the cost of environmental damage (Fig. 8.4).

Storms can severely impact sandy shores, mainly due to the erosion process that removes sand from the beach zones, shortening their width (Roelvink et al. 2009) (Chap. 9). Provided that hydrodynamics and sediment budgets remain constant, beaches can naturally recover from such impacts over time. Beaches with preserved vegetated dunes at the back beach and a wide beach face are more resilient to erosion from storm impacts because of the constant sand budget and thus provide enhanced protection against damages (Taylor et al. 2015). However, several natural and anthropic causes contribute to the global coastal erosion concern. The sea-level rise, increase in storm events, armoring of coast shoreline, trapping of sediment by artificial structures, sand mining, and beach cleaning are among the processes and activities that contribute to the negative erosion impacts (Dugan et al. 2008; Defeo et al. 2009).

Worldwide, sandy beaches are shrinking worldwide, with nearly a quarter experiencing an erosion of >0.5 m/year (Luijendijk et al. 2018). For Brazil, this number is higher, with 40% of the entire coastline facing erosion processes (Brasil 2018). This value is highly variable along the coast, ranging from as low as 4% of eroding beaches in the state of São Paulo to 65% of eroding beaches in the state of Amapá (Brasil 2018). Despite the low number of eroding beaches, nearly 52% of beaches in the state of São Paulo are at high or very high risk of erosion (Souza 2012). Coastal erosion also makes beaches less attractive to tourists, as observed in the state of Ceará (Northeast Brazil) (Silva et al. 2018), which further damages the provision of cultural services. Thus, there is a need to map and monitor erosion impacts on shorelines to maintain sandy beach services as well as reduce the socio-economic impacts of extreme events.

The rising number of studies addressing coastal erosion has to do with the growing concern about climate change. Predictions of sea-level rise have caused alarm surrounding the potential negative impacts on coastal settlements and ecosystems. Sandy beaches can have a positive effect on alleviating these impacts. As dynamic ecosystems, sandy beaches can exert a "backward" migration toward dune areas with the sea-level rise (Defeo et al. 2009), maintaining their coastline protection service. However, for such a process to occur, beaches must be able to migrate, which means that areas with the presence of coastal armoring and occupied dunes are limited in maintaining sandy beach protection (Dugan et al. 2008). An assessment of the northeast Brazilian coast estimated that coastal vulnerability to sea-level rise is much higher in areas with high population densities, a high number of buildings, and the ongoing process of coastal squeezing due to the construction of hard structures (Martins et al. 2017).

Besides coastal protection, other relevant processes of nutrient storage, cycling, and export affect food webs in most ecosystems. Although this process is not as comprehensively investigated on sandy beaches as for terrestrial or estuarine ecosystems, beaches are important sites for cycling organic matter input. The organic input on sandy beaches mainly occurs from macroalgae and macrophytes brought by tides and waves from nearby rocky shores and mangroves. Dugan et al. (2011) quantified the inorganic and organic nitrogen of pore and surf zone water. They found that beaches store dissolved nutrients and export the re-mineralized nutrients from macrophyte inputs. Andrades et al. (2014) also found that the macroalgae strands on the surf zone increase the function of nursery habitats due to increased nutrient availability. The presence of organisms, especially those related to bioturbation, is important for cycling these nutrients, as evidenced by the increased nutrient concentration in the water from *Callichirus major* burrows in south Brazil (Machado et al. 2004). Due to the importance of nutrient cycling services and the role of macroalgae inputs, beach cleaning should not be recommended for areas where such environmental services are to be preserved.

Beaches and dunes are also beneficial for regulating services of water provision and purification, due to the groundwater storage, and for carbon storage on beaches where foredune vegetation is preserved (Barbier et al. 2011). While these regulating services exist on many beaches in Brazil (Ferreira et al. 2015; Oliveira 2015a; Silva and Paula 2017), very few studies attempt to quantify and evaluate their relative importance. In Brazil, Carrilho and Sinisgalli (2018) valued the stored carbon on organisms of a tidal flat area in southeast Brazil; it was, however, less than 1% of the value of the total services identified for the area. Whether these figures can reach higher values on other beach types remains a knowledge gap that deserves to be investigated.

8.6 Cultural Services

While regulating services related to coastline protection are essential to the maintenance of the beach area, the most acknowledged services are cultural ones. In Brazil, Romeiro et al. (2014) estimated that cultural ecosystem services respond to 77% of the services identified for beaches by the local population. It is not hard to understand the reason. Among coastal environments, sandy beaches are the most welcoming in terms of access to sea and sunbathing and for the development of recreational activities. Tourists who travel to the beach are generally motivated by a sense of well-being and quality of life: escaping from daily routines and seeking interactions, inner self-experience, relaxation, and personal attachment (Van der Merwe et al. 2010; Bell et al. 2015) (Fig. 8.5). Recreational activities include sunbathing and sea bathing, which is performed by most visitors; the gathering of species, such as the case of the clam *Tivella mactroides* on beaches of the state of São Paulo, southeast Brazil (Turra et al. 2016); and sports practice.

Beach tourism is a significant income source for many coastal cities and a booster of international tourism. It supports a range of commercial activities, such as sellers working within the beach area (including beach hawkers, kiosks, and restaurants), real estate markets, hotel businesses, and supermarkets. Yepes and Medina (2005) showed that Spanish beaches were responsible for about 11.4% of the country's total gross domestic product and 11.2% of direct employment. In the Caribbean, tourism contribution to the gross domestic product may reach values as high as 74% (Antigua & Barbuda) and 64% (Anguilla), and even the lowest contributions reach 18% (Bahamas) and 7% (Trinidad & Tobago) (Scott et al. 2012). In the US coastal states, beaches respond to 85% of the total tourism income (Houston 2008). In the touristic area of the Great Barrier Reef, Australia, beach-related tourism generates almost US$ 600 million per year, with an average expenditure of US$ 35.09 per person on a single visit (Rolfe and Gregg 2012).

Expectedly, the relevance of sandy beaches for tourism and their impact on a country's economy will depend on quality, climate, and tourism promotion. Brazilian beaches are among the most famous in the world and are a primary destination for tourists. Most national and international travelers go to coastal cities (MTur 2019). The highest demand for Brazilian tourism agencies is for "sun and beach" destinations (45.9%) (MTur 2019). Internationally, beaches are the second most used reference to describe the country (Chagas and Dantas 2009) and the image most associated with touristic activities in Brazil (Rezende-Parker et al. 2003). Leisure is the motivation for 58% of the total travels to Brazil, of which 71% are motivated by an interest in visiting Brazilian beaches (FIPE 2018). Beach tourism has great importance to the gross domestic product in Brazil, especially for the northeast coastal states and states of Rio de Janeiro and São Paulo (FIPE 2018). In 2018, tourism generated US$ 152.5 billion, responsible for 8.1% of the country's gross domestic product (WTTC 2019).

The potential for beaches to serve as a ground for tourists and recreational activities depends on beaches characteristics and users profile. Users tend to value safe and clean beaches (i.e., lack of litter and wrack) with easy access and relevant facilities (Phillips and House 2009; Botero et al. 2013). Domestic tourists seem to favor characteristics such as the total length of the beach face, whereas international tourists also value cultural activities (Onofri and Nunes 2013). Surfers choice are based on their hydrodynamics characteristics, preferring exposed beaches with higher waves, while families prefer beaches with wide intertidal areas (Phillips and House

Fig. 8.5 Examples of cultural services provided by sandy beaches. (**a**) Tourism is the main activity related to sandy beaches and the source of economic income for many coastal cities (state of Rio de Janeiro, photo: Martins Cardoso/Unsplash) (**b**) an array of tourism-related economic activities are carried out in sandy beaches, such as boat ride trips (Arraial do Cabo municipality, state of Rio de Janeiro, photo: Jonathan Borba/Unsplash); (**c**) among the sports practiced in beaches, surfing is likely the most common in Brazil (Ipanema beach, state of Rio de Janeiro, photo: Rafael Cerqueira/Unsplash); (**d**) religious activities are also commonly practiced, especially those linked to the Candomblé spiritual entity named Yemanjá (Praia das Cabeçudas, Itajaí municipality, state of Santa Catarina, photo: Marcus Polette); (**e**) social projects such as those involving environmental education are developed in beaches (beach cleaning made by school kids in Bahia, photo: Banco de imagens Projeto Tamar|Fundação Pró-Tamar); (**f**) the scenic beauty of beaches are one of their main assets that trigger tourism and serve as a source for artistic inspiration and improvement of well-being (photo: Álvaro Migotto/Cifonauta)

2009). In Brazil, a deeper understanding of the link between beach choice and visitor's profile is still required. However, it is reasonable to assume a similar pattern and that the choice depends on the user's motivation. Preliminary studies in Ubatuba (Pombo et al. 2009) and Guarujá (Silva and Souza 2018) municipalities (state of São Paulo, southeast Brazil) showed that dissipative beaches attract families with children and people who value infrastructure as kiosks and restaurants, while reflective beaches attract younger people who engage in sport more frequently.

One of the biggest challenges of beach management is dealing with the trade-off between tourism development and beach conservation. The urbanization required to enhance tourist activities (e.g., construction of buildings and hard infrastructure on the sand) alters the landscape and may damage ecosystem services (Dugan et al. 2008; Defeo et al. 2009). Tourism and urbanization have been related to an increase in the input of contaminants and marine debris on the sea (Tourinho and Fillmann 2011; Suciu et al. 2017). Tourism impacts also affect beach fauna (Defeo et al. 2009), assemblage richness, and population density (Veloso et al. 2006; Machado et al. 2017), compromising other ecosystem services related to biodiversity. As proposed by McLachlan et al. (2013), the ideal scenario for managers is to identify, based on the services provided and the potential for recreation or conservation, what is the "vocation" of particular sandy beaches. This way, beaches can be individually selected to be classified and managed as recreational, primarily for conservation, or as a mixed-use beach (McLachlan et al. 2013).

Beach management must also consider the impacts of other activities or events on tourism and related economic activities. For example, in late 2019, Brazilian northeast and southeast beaches were washed by a massive oil spill, considered the most significant environmental disaster in Brazil (Araújo et al. 2020). Although the economic effects are uncertain, impacts on coastal areas were ten times higher than in inland regions, and the gross domestic product was estimated to decrease by up to 0.2% (Ribeiro et al. 2020). Additionally, given that the fish trade and tourism sectors were paralyzed, the unemployment of professionals in these sectors increased (Araújo et al. 2020). The COVID-19 pandemic aggravated the impact. Less than a year after the oil spill, the local business and the fisheries communities have to deal with the pandemic, which led to a massive decrease in the number of people traveling to the coastal regions (Botero et al. 2020). Although these examples are catastrophic and unpredictable, the effects on the provision of cultural services by beaches caused profound impacts on the economy and people's livelihoods, and management must develop strategies to circumvent or reduce these problems.

Aside from tourism, beaches also host sports activities and events, including beach soccer, beach volley, foot volley, squash, surf, and jogging, which populate wide intertidal zones. Sports such as surfing attract many tourists and represent annual values as high as US$ 4,229,551 for single beaches (Bosqueti and Souza 2019). Beach sports can also promote social inclusion. Surf schools are established in many areas and represent a tool for social projects. On the Portuguese coast, for instance, surf teaching schools are offered to kids and teenagers from foster homes, with positive effects on the development of social skills and inner knowledge (Matos et al. 2016). In Brazil, public surf schools promote social inclusion also by using surfing boards adapted for disabled people (Siqueira et al. 2020). Similar to surfing, beach soccer, and volleyball also promote social inclusion and combat harmful practices (e.g., littering and drug use).

Beaches may also support environmental education activities. On the southeast Brazilian coast, students and teachers from public schools were presented with pedagogical methods to include local beaches as part of environmental activities, aiming to raise awareness and environmental citizenship and engage citizens in the management of local coastal areas (Santos et al. 2018). Exposing people to activities in natural ecosystems increases their connection to nature and sense of well-being (leisure satisfaction) (Rosa et al. 2019). Furthermore, greater exposure can lead to increased involvement with beach conservation and additional value to beach ecosystem services. In this sense, beaches can be important places for the development of ocean literacy principles (i.e., understanding the relationships between oceans and humans), an experience that has been carried out in Brazilian schools (Costa et al. 2021).

Beaches are also grounds for the development of practices of religious and spiritual significance. Shells of sandy beach mollusks are culturally relevant for religious purposes in Brazil. Religious rituals and ceremonies of the Afro-Brazilian religion, Candomblé, use sandy beach species (nonexclusive) or empty shells, such as *Monetaria moneta* and *M. annulus*, *Cassia tuberosa*, *Anomalocardia flexuosa*, and *Stramonita rustica* (Léo-Neto et al. 2018). Nationwide people send miniature boats with candles and white roses into the sea in an offering to Yemanjá (the Candomblé spiritual entity of the sea) as part of the ritual of New Year's Eve.

8.7 Identification and Valuation of Beach Services

Information regarding the economic value of ecosystem services can be useful for decisions that involve understanding how much is being lost or gained by different services (Costanza et al. 2017). Even with uncertainties surrounding the monetary estimates, valuation offers a more transparent framework for the decision-making process (Costanza et al. 2017). Costanza et al. (1997) were not the first, but certainly one of the most essential earlier studies aiming to evaluate the monetary benefits of ecosystem services. Although beaches were not directly valued, coastal systems were estimated to provide US$ 4052/ha/year, mostly from nutrient cycling services, but also from recreation, culture, disturbance regulation, food provision, and biological control. From there, the number of studies increased, and an array of techniques has been developed to assess the monetary and nonmonetary value of ecosystem services (see review by Christie et al. 2012).

Worldwide, scientists and management actors have tried to identify and stipulate monetary values for market and nonmarket services generated by sandy beaches, implement policies based on scientific information, and raise awareness of the importance of preserving this ecosystem. The influential global scale valuation works by Costanza et al. (1997) (with updates in Costanza et al. (2014)) and de

Groot et al. (2012) highlight the importance of services provided by coastal systems and wetlands; however, neither work evaluated sandy beaches individually. To this end, we need to look into studies focusing specifically on this ecosystem.

Examples of beach service evaluation exist worldwide. On the Central Coast of Portugal, Alves et al. (2009) estimated that dunes and sandy beaches are responsible for 47% of the 193 million euros provided by the country's ecosystem services per year. They also projected that coastal erosion predicted for 2058 would result in the loss of 25% of the total ecosystem value, with the most impact being on the services provided by sandy beaches and dunes (Alves et al. 2009). On the Mexican Gulf coast, sandy beaches and dunes services of coastline protection and recreation were estimated to provide US$ 80,459/ha/year, the highest estimated value among other coastal ecosystems in the area (Mendoza-González et al. 2012). The same scenario was found in the state of New Jersey, in the United States, where beaches were the most valuable ecosystem among marine and terrestrial ones, providing coastline protection and stabilization, esthetical, recreational, cultural, and spiritual services worth U$ 42,147/ha/year (Liu et al. 2010).

Brazil is the third country in the number of publications on sandy beaches worldwide (8% of the total), only behind the USA and South Africa (Nel et al. 2014). However, few studies provide a complete characterization and valuation of beach ecosystem services. The more common approach in Brazil is the investigation of a single ecosystem service, such as coastline protection (e.g., Lins-de-Barros et al. 2016; Martins et al. 2017) and tourism potential (e.g., Silva et al. 2018). The majority of these studies were published in local journals or are part of the "gray literature" (i.e., thesis, reports, and conference annals), which, although essential for beach management, do not reach the general public and scientific community. There is also a focus on touristic beaches and services related to economic activities, which limits the comprehension of the provision of ecosystem services on beaches with low urbanization and the interaction among services of different categories (Checon et al. 2022). Nonetheless, there is an increasing trend in the number of studies (Checon et al. 2022) and some significant contributions assessed services on Brazilian beaches.

Souza-Filho et al. (2014) investigated the services provided by several sandy beaches in the state of Bahia, northeast Brazil, and found that natural conditions affect their services. Beaches associated with adjacent ecosystems such as estuaries and mangroves were more likely to provide services such as water filtration and nutrient cycling, and urbanization negatively affected the provision of services such as coastline protection and food provision (Souza-Filho et al. 2014). The combined valuation of two beaches in the state of Rio Grande do Norte (Northeast Brazil) found a maximum value of US$ 62,247/ha/year (Ferreira et al. 2015). Carrilho and Sinisgalli (2018) identified and estimated the ecosystem services provided by a coastal bay tidal flat in São Paulo (Southeast Brazil), including the four sandy beaches that surround the tidal flat. The authors identified 13 ecosystem services provided by this single area. The economic value was estimated at US$ 340, 610.29/year, mainly due to services of waste depuration, harboring vessels, access to the sea, and hosting scientific and educational activities (Carrilho and Sinisgalli 2018).

Economic loss due to degradation and damages have also been estimated. Lins-de-Barros and Parente-Ribeiro (2018) estimated that sea-level rise and coastal erosion in two highly touristic beaches in the state of Rio de Janeiro (Southeast Brazil) would result in a monthly loss of US$ 316,622.69 during the nonvacation period, reaching US$ 1,055,409.00 during the high season. In April 2016, the loss of sand on Arpoador beach, state of Rio de Janeiro, during a storm resulted in a monthly loss of US$ 42,216.00 (Lins-de-Barros and Parente-Ribeiro 2018). Based on the willingness to pay for beach quality, Oliveira (2015b) estimated an annual loss of US$ 796,660.16 for residents of a neighborhood in the state of Bahia, northeast Brazil, due to the current degraded status of local beaches. Krelling et al. (2017) estimated annual losses from US$ 880.000 to US$ 8.5 million in three different marine litter concentration scenarios in Pontal do Sul municipality (South Brazil). These values show the importance of understanding the benefits and monetary worth brought by ecosystem services and what can be lost if the services are compromised. Table 8.2 summarizes additional studies that identified and valued general or specific beach services (e.g., coastal erosion, tourism) in Brazil.

The willingness to pay (WTP) is one of the metrics commonly used to value ecosystem services (Christie et al. 2012). The method estimates whether and how much an individual is willing to pay to protect a given ecosystem and maintain its services. Understanding this willingness can guide managers to identify and stipulate acceptable prices for nonmarket ecosystem services (Nielsen-Pincus et al. 2017). In a study on the state of Ceará (Northeast Brazil), Leite et al. (2019) found that 54.05% of the local population would be unwilling to pay for ecosystem services provided by local beaches, despite recognizing its importance, most citing that the government should be held accountable for such issues. Those numbers are lower than studies carried out on beaches of Colombia (70%, Enriquez-Acevedo et al. 2018), South Florida (USA) (67%, Lindsay et al. 1992), and Turkey (92%, Birdir et al. 2013); but higher than those obtained for beaches of Liguria (Italy) (36%, Marin et al. 2009) and Cadiz (Spain) (12.9%, Alves et al. 2015). The willingness to pay is a metric that depends on the political and social context and may also vary with time (Matthews et al. 2017). Studies on Brazilian beaches with this method need to be developed considering local particularities, not only to implement potential management policies but also to understand whether citizens perceive the importance of beaches to human populations.

8.8 Final Remarks

We presented in this chapter an extensive but not exhaustive overview of ecosystem services provided by sandy beaches, using examples from around the world as well as in Brazil. The development of this topic is motivated by the need for management practices to prevent the impacts of the increased pressure on coastal ecosystems. Ecosystem services have a vital role in the development of such practices and in raising public awareness about the importance of sandy beaches and other marine

Table 8.2 Summary of studies that estimated and/or valued sandy beach services on the Brazilian coast

Reference	State	Ecosystem services identified	Estimated value	Method
Abreu et al. (2008)[1]	Alagoas	Nonspecified	>US$ 14,767,000/ month	Travel cost/ WTP
Araújo et al. (2018)[2]	Rio Grande do Norte	Food supply, groundwater retention, high biodiversity, gene fluxes, maintenance of soil and water chemistry, pollination, carbon sequestration, dune stabilization, pollination, tourist attractions, well-being, hosting scientific activities, scenic beauty, leisure, and recreation	–	–
Bosqueti and Souza (2019)[3,*]	Santa Catarina	Surf tourism	Revenue from surf tourism (2018): US$ 2,978,018 (Summer); US$ 239.119 (Fall+Winter); US$ 1,012,416 (Spring)	Direct expenditure method
Carrilho and Sinisgalli (2018)[2,*]	São Paulo	Food supply, raw materials, harboring vessels, sea access, protection against sea-level rise, scenic beauty, hosting scientific and educational activities, cultural heritage and topophilia, leisure and recreation, Satisfaction from conservation, waste depuration, carbon stocking	US$ 340,610.29/year	Market price/ cost of replacement/ project valuation
Carvalho and Mondo (2010)[2]	Santa Catarina	Tourism and surfing	Individual expenses for surfing and tourism: Fuel: US$ 17.75/day (tourist), US% 5.92/day (locals) Food: US$ 17.75/day Rent: US$ 266.27–473.37 monthly (locals); R$ 29.58/ day (tourists)	Questionnaire for assessing daily expenditure

(continued)

Table 8.2 (continued)

Reference	State	Ecosystem services identified	Estimated value	Method
Ferreira (2015)[2]	Santa Catarina	Organic matter supply, habitat diversity, sailing area, nursery grounds, biomass stock, hydric resources, nutrient cycling, sediment trapping, shoreline protection, climate control, cultural significance, scenery, historical value	–	Questionnaire for assessing daily expenditure
Ferreira et al. (2015)[2]	Rio Grande do Norte	Food supply, organic matter input, genetic resources, raw materials, carbon sequestration, water filtration, groundwater retention, climate regulation, sediment trapping, marine nursery, leisure and recreation, tourist attractions, religious practices, cultural heritage, pollination, nutrient cycling, primary production	US$ 62,247.00 ha/year	Not informed
Finco and Abdallah (2002)[1]	Rio Grande do Sul	Nonspecified	US$ 190,692.24/month	WTP
Freire et al. (2008)[1]	Bahia	Tourism activities	US$ 107.65/tourist (high season) US$ 191.38/tourist (low season)	Travel cost
Krelling et al. (2017)[2,*]	Paraná	Tourism activities (daily expenditure)	US$ 96.83/trip by second houses owners; US$ 83.75/trip by tourists	Questionnaire for assessing daily expenditure
Lima and Paula (2017)[2]	Ceará	Food supply, sediment trapping, wave dissipation, genetic resources, raw material, ecotourism, tourist attractions, leisure and recreation, cultural activities/value	–	–

(continued)

Table 8.2 (continued)

Reference	State	Ecosystem services identified	Estimated value	Method
Lins-de-Barros and Parente-Ribeiro (2018)[2,*]	Rio de Janeiro	Tourism activities	US$ 316,622.69/month (low season) US$ 1,055,409.00/month (high season)	Market value
Martins and Peres (2005)[2]	Ceará	Tourism and cultural services	US$ 21.84/person; US$ 764,343.98/year	WTP/WTA
Medeiros and Araújo (2010)[1]	Ceará	Coastal and building protection, recreational and sports activities, tourism, surf competitions, scenic beauty	US$ 918,253.88/year (erosion control and beach maintenance)	WTP
Oliveira and Berkes (2014)[2,*]	Rio de Janeiro	Inspiration for traditional songs and dances; leisure and recreation; traditional knowledge (related to canoe making by the locals – abandoned practice); religious ceremonies; scenery for TV and cinema; scenic beauty	–	Multimethod approach – Questionaries for services perception
Oliveira (2015a)[2]	Bahia	Nonspecified. Losses estimated due to poor beach quality	US$ 796,685.92/year	Contingent valuation
Oliveira (2015b)[4]	Santa Catarina	Sediment trapping, food supply, shoreline protection, nursery grounds, wave dissipation, nutrient cycling and pollutant assimilation, groundwater recharge, genetic resources, ecotourism, leisure, and recreation, scenic beauty	–	–
Oliveira et al. (2017)[2]	Rio Grande do Sul	Recreational and tourism activities (valuation related to the maintenance of the beach for recreational use)	WTP: US$ 10.70/person.year. Total: US$ 2,149,063.63/year	WTP

(continued)

Table 8.2 (continued)

Reference	State	Ecosystem services identified	Estimated value	Method
Reis (2018)[5]	Pernambuco	Cultural services: scientific research, scenic beauty, religion, recreation, environmental education, cultural heritage, ecotourism, contemplative activities, sports.	US$ 2.64/person fee for the quality and conservation of environmental and cultural services	WTP
Romeiro et al. (2014)[2]	São Paulo	18 nonspecified services, related to the food supply, fish stocks and leisure and well-being – 77.7% belonging to the cultural category	–	–
Santos (2017)[5]	Rio Grande do Norte	Tourism and recreation activities	US$ 798,774,390.24/year	Travel cost
Souza-Filho et al. (2014)[2,*]	Bahia	Food supply, raw material, tourist attractions, aquifer recharge, sediment trapping, groundwater storage, pollution recycling and assimilation, marine refuge and nursery, erosion control, nutrient cycling, wave dissipation, genetic resources, site for food production	–	–
Santos et al. (2010)[1]	Paraná	Nonspecified	US$ 166,480.47/day (high season)	WTP
Santos-Frossard et al. (2015)[2]	Bahia	Nonspecified	US$ 30,430.89/month	WTP/WTA
Silva and Nascimento (2016)[2]	Rio Grande do Norte	Regulation of terrestrial and marine processes, coastline protection, flood control, support for coastal buildings, habitat for marine invertebrates, landscape aesthetics, sports practice, inspiration for arts, site for social development, research, education and extension	–	–

(continued)

Table 8.2 (continued)

Reference	State	Ecosystem services identified	Estimated value	Method
Silva and Paula (2017)[1]	Ceará	Food supply, sediment trapping, groundwater recharge, wave dissipation, coastline protection, site for food production, ecotourism, leisure and recreation, cultural activities/value, scenic beauty, raw material, water resources	–	–
Toste et al. (2011)[1]	Rio de Janeiro	Tourism activities	US$ 173,043,010.75/year (potential revenue for the municipality); Beach: US$ 17,502,904.30/year (high season: 12.3 mi; low season: 5.2 mi).	–
Uchikawa (2008)[5]	Pernambuco	Nonspecified	US$ 4.00/person. month; US$ 6,325,878.06/month for the municipality	WTP

Note: [1]Conference Proceedings; [2]Journal; [3]Report; [4]Thesis; [5]Monography; (*) denotes English-language publications. Estimated values are converted to dollars using the currency of the last day of the year of the publication

ecosystems. As we have seen throughout this chapter, the ecosystem services approach has been applied to beaches. However, efforts are still needed, especially considering the need to account for local and temporal variations, which may play a significant role in the perception of ecosystem value (Matthews et al. 2017).

In regard to the Brazilian coasts, studies assessing the ecosystem services provided by beaches need to move from the gray literature into the scientific community. This would enhance the accessibility and application of this knowledge to foreigners and even researchers from other regions of Brazil. Another concern is the notable regional imbalance in beach studies in Brazil. Although ecological assessments of biodiversity are more geared toward the south and southeastern coast, studies assessing and valuing beach services are carried out mainly on the northeast coast of Brazil. Such effort is extremely important given the tourism potential and associated risks of beaches in this region. However, other areas, mainly from the southeast coast, are under crescent pressure from the development of industrial activities in coastal areas, but there is a clear gap in the evaluation of ecosystem services in this region.

We hope that studies along the Brazilian coast follow the increasing worldwide trend to guide management strategies for protecting these valued, yet threatened ecosystems. The United Nations Decade of Ocean Science for Sustainable

Development (2021–2030)[1] is an opportunity for this. To boost ocean science, promote worldwide cooperation, and engage society in ocean protection, the decade can provide the momentum for innovative and applied scientific and technological development. It is up to those engaged (as scientists and society) to guarantee that beaches catch this wave!

Acknowledgments The authors would like to acknowledge the Fundação de Amparo à Pesquisa do Estado de São Paulo (FAPESP) for financial support (FAPESP HHC: 2018/22036-0; LYX: 2017/21797-5 and 2019/13851-5; LRG: 2018/00462-8 and 2019/04481-0). The work is part of the activities of the ongoing thematic project, "Environmental Governance in the São Paulo Macrometropolis, due to climate variability" (FAPESP: 2015/03804-9), linked to the FAPESP Global Climate Change Research Program) and the project "Fostering the ecosystem-based approach in beach spatial planning and conservation," supported by FAPESP (2018/19776-2, linked to the FAPESP Biodiversity Research Program).

References

Able KW, Wuenschel MJ, Grothues TM et al (2013) Do surf zones in New Jersey provide "nursery" habitat for southern fishes. Environ Biol Fish 96:661–675

Abreu EAP, Silva AG, Silva Junior GG (2008) Valoração econômica: aplicação do método de custo de viagem para a Praia da Avenida, Maceio. In: Annals of the XXXVI National Economy Meeting

Alves F, Roebeling P, Pinto P et al (2009) Valuing ecosystem services losses from coastal erosion using a benefits transfer approach: a case study for the central Portuguese Coast. J Coast Res 56:1169–1173

Alves B, Torrent RR, Ballester R et al (2015) Coastal erosion perception and willingness to pay for beach management (Cadiz, Spain). J Coast Conserv 19:269–280

Amaral ACZ, Corte GN, Rosa-Filho JS et al (2016) Brazilian sandy beaches: characteristics, ecosystem services, impacts, knowledge, and priorities. Braz J Oceanogr 64:5–16

Anacleto A, Broetto B, Xavier FP et al (2016) Retail trade of caged crabmeat (*Callinectes* sp.) on the coast of Paraná-Brazil. Bus Manag Dyn 5:15

Andrades R, Bolzan MS, Contaifer LS et al (2012) Evidence of sandy beaches as growth grounds for commercial fish in the south-western Atlantic. Pan Am J Aquat Sci 7:107–110

Andrades R, Gomes MP, Pereira-Filho GH et al (2014) The influence of allochthonous macroalgae on the fish communities of tropical sandy beaches. Estuar Coast Shelf Sci 144:75–81

Araújo IS, Maciel ABC, Ferreira JCV et al (2018) Identificação de serviços ecossistêmicos na Praia de Ponta Negra, Natal-RN. Rev Geocienc Nordeste 4:298–312

Araújo ME, Ramalho CWN, Melo PW (2020) Artisanal fishers, consumers and the environment: immediate consequences of the oil spill in Pernambuco, Northeast Brazil. Cad Saude Publica 36:e00230319

Bail GC, Branco JO (2007) Pesca artisanal do camarão sete-barbas: uma caracterização socioeconômica na Penha, SC. Braz J Aquat Sci Technol 11:25–32

Barbier EB (2015) Valuing the storm protection service of estuarine and coastal ecosystems. Ecosyst Serv 11:32–38

Barbier EB, Hacker SD, Kennedy C et al (2011) The value of estuarine and coastal ecosystem services. Ecol Monogr 81:169–193

Barbieri E, Paes ET (2008) The birds at Ilha Comprida beach (São Paulo State, Brazil): a multivariate approach. Biota Neotrop 8:41–50

[1] https://www.oceandecade.org/ (Accessed 15 July 2020).

Beck MW, Heck KL, Able KW et al (2001) The identification, conservation and management of estuarine and marine nurseries for fish and invertebrates. Bioscience 51:633–641

Begossi A (2010) Small-scale fisheries in Latin America: management models and challenges. MAST 9:7–31

Bell SL, Phoenix C, Lovell R et al (2015) Seeking everyday well-being: the coast as a therapeutic landscape. Soc Sci Med 142:56–67

Birdir S, Unal O, Birdir K et al (2013) Willingness to pay as an economic instrument for coastal tourism management: cases from Mersin, Turkey. Tour Manag 36:279–283

Bjorndal KA, Jackson JBC (2002) Roles of sea turtles in marine ecosystems: reconstructing the past. In: Lutz PL, Musick JA, Wyneken J (eds) The biology of sea turtles. CRC Press, Boca Raton

Blythe J, Armitage D, Alonso G et al (2020) Frontiers in coastal well-being and ecosystem services research: a systematic review. Ocean Coast Manag 185:105028

Bosqueti MA, Souza MA (2019) Surfonomics Guarda do Embaú, Brazil: the economic impact of surf tourism on the local economy. UFSC, Florianópolis. https://www.savethewaves.org/surfonomics/guarda-do-embau/UFSC. Accessed 20 Jun 2020

Botero CM, Anfuso G, Williams AT et al (2013) Reasons for beach choice: European and Caribbean perspectives. J Coast Res 65:880–885

Botero CM, Mercadé S, Cabrera JA et al (eds) (2020) O turismo de sol e praia no contexto da Covid-19: cenários e recomendações. Rede Ibero-americana de Gestão e Certificação de Praias – PROPLAYAS, Santa Marta

Braat LC, de Groot R (2012) The ecosystem services agenda: bridging the worlds of natural science and economics, conservation and development, and public and private policy. Ecosyst Serv 1:4–15

BRASIL (2018) Panorama da erosão costeira no Brasil. Available at: https://www.mma.gov.br/images/arquivo/80342/Panorama_erosao_costeira_Brasil.pdf. Accessed 15 Mar 2020

Carpenter SR, Mooney HA, Agard J et al (2009) Science for managing ecosystem services: beyond the millennium ecosystem assessment. Proc Natl Acad Sci 106:1305–1312

Carrilho CD, Sinisgalli PAA (2018) Contribution to Araçá Bay management: the identification and valuation of ecosystem services. Ocean Coast Manag 164:128–135

Carvalho ACB, Mondo TS (2010) O valor das ondas: um estudo de caso sobre a Praia do Campeche-Florianópolis na perspectiva de surfistas, moradores e visitantes. Patrim Laz Tur 7:75–98

Castilla JC, Defeo O (2001) Latin American benthic shellfisheries: emphasis on co-management and experimental practices. Rev Fish Biol Fish 11:1–30

Chagas MM, Dantas AVS (2009) A imagem do Brasil como destino turístico nos websites das operadoras de turismo europeias. Observ Inov Tur 2:1–19

Checon HH, Xavier LY, Gonçalves LR et al (2022) Beach market: what have we been computing in Brazil. Ocean Coast Res 69:e21038

Christie M, Fazey I, Cooper R et al (2012) An evaluation of the monetary and non-monetary techniques for assessing the importance of biodiversity and ecosystem services to people in countries with developing economies. Ecol Econ 83:67–78

Colabuono FI, Barquete V, Dominguez BS et al (2009) Plastic ingestion by Procellariformes in Southern Brazil. Mar Pollut Bull 58:93–96

Corte GN, Checon HH, Shah Esmaeili Y, Defeo O, Turra A (2022) Evaluation of the effects of urbanization and environmental features on sandy beach macrobenthos highlights the importance of submerged zones. Mar Poll Bull 182:113962

Costa LL, Landmann JG, Gaelzer LR et al (2017) Does human pressure affect the community structure of surf zone fish in sandy beaches. Cont Shelf Res 132:1–10

Costa DA, Lucena RFP, Christoffersen ML et al (2021) Improving environmental awareness and ocean literacy through hands-on activities in the tropics. Appl Environ Educ Commun 21:1–20

Costanza R, d'Arge R, de Groot R et al (1997) The value of the world's ecosystem services and natural capital. Nature 387:253–260

Costanza R, de Groot R, Sutton P et al (2014) Changes in the global value of ecosystem services. Glob Environ Change 26:152–158

Costanza R, de Groot R, Braat L et al (2017) Twenty years of ecosystem services: how far have we come and how far do we still need to go. Ecosyst Serv 28:1–16
Daily GC, Matson PA (2008) Ecosystem services: from theory to implementation. Proc Natl Acad Sci 105:9455–9456
Daniel TC, Muhar A, Arnberger A et al (2012) Contributions of cultural services to the ecosystem services agenda. Proc Natl Acad Sci 109:8812–8819
De Groot R, Brander L, van der Ploeg S et al (2012) Global estimates of the value of ecosystems and their service in monetary units. Ecosyst Serv 1:50–61
Defeo O (2003) Marine invertebrate fisheries in sandy beaches: an overview. J Coast Res SI35:56–65
Defeo O, McLachlan A, Schoeman DS et al (2009) Threats to sandy beach ecosystems: a review. Estuar Coast Shelf Sci 81:1–12
Del Favero JM, Dias JF (2015) Juvenile use of the shallow zone of beaches of the Cananéia-Iguape coastal system, southeastern Brazil. Braz J Oceanogr 63:103–114
Denadai MR, Pombo M, Bernardochi LC (2015) Harvesting the beach clam *Tivella mactroides:* short and long-term dynamics. Mar Coast Fish 7:103–115
Dugan JE, Hubbard DM, Rodil IF et al (2008) Ecological effects of coastal armoring on sandy beaches. Mar Ecol 29:160–170
Dugan JE, Hubbard DM, Page HM et al (2011) Marine macrophyte wrack inputs and dissolved nutrients in beach sands. Estuaries Coast 34:839–850
Enriquez-Acevedo T, Botero CM, Cantero-Rodelo R et al (2018) Willingness to pay for beach ecosystem services: the case study of three Colombian beaches. Ocean Coast Manag 161:96–104
Félix FC, Spach HJ, Moro PS et al (2007) Utilization patterns of surf-zone inhabiting fishes from beaches in Southern Brazil. Pan Am J Aquat Sci 2:27–39
Ferreira W (2015) Os serviços ecossistêmicos e a gestão do uso público nas unidades de conservação na Ilha de Santa Catarina, SC. Uso Público Unidad Conserv 3:94–102
Ferreira JCV, Silva EES, Amaro VE et al (2015) Serviços ambientais fornecidos pelas praias da Ponta Negra e Via costeira: perspectiva de valoração. Rev CERES 1:23–29
Finco MVA, Abdallah PR (2002) Valoração econômica do meio-ambiente: o método do custo de viagem aplicado ao litoral do Rio Grande do Sul. Rev Teor Evid Econ 10:49–63
Fisher B, Turner K, Zylstra M et al (2008) Ecosystem services and economic theory: integration for policy relevant research. Ecol Appl 18:2050–2067
Freire CRF, Cerqueira CA, Guimarães Junior GS (2008) Valoração econômica de serviços ambientais como estratégia de conservação: o caso do Litoral de lheus-BA. In: XLVI Congress of the Brazilian Society of Economy, Administration and Rural Sociology. Rio Branco, Brasil
Freire KMF, Bispo MCS, Luz RMC (2014) Competitive marine fishery in the state of Sergipe. Acta Fish Aquat Resour 2:59–72
Fundação Instituto de Pesquisas Econômicas (FIPE) (2018) Caracterização e dimensionamento do turismo internacional no Brasil – 2013–2017 – Relatório Descritivo. Ministério do Turismo, Brasil
Haines-Young R, Potschin MB (2018) Common International Classification of Ecosystem Services (CICES) V5.1 and guidance on the application of the revised structure. Nottingham, UK
Hanazaki N, Berkes F, Seixas CS et al (2013) Livelihood diversity, food security and resilience among the Caiçara of coastal Brazil. Hum Ecol 41:153–164
Harris L, Campbell EE, Nel R, Schoeman D (2014) Rich diversity, strong endemism, but poor protection: addressing the neglect of sandy beach ecosystems in coastal conservation planning. Diver Distrib 20:1120–1135
Hoefle SW (2014) Fishing livelihoods, seashore tourism, and industrial development in coastal Rio de Janeiro: conflict, multi-functionality, and juxtaposition. Geogr Res 52:198–211
Houston JR (2008) The economic value of beaches – a 2008 update. Shore Beach 76:22–26
IUCN (2020) The IUCN red list of threatened species. Version 2020-2. Available at: www.iucncredlist.org. Accessed 15 Mar 2020
Ivar do Sul JA, Costa MF (2014) The present and future of microplastic pollution in the marine environment. Environ Pollut 185:352–364

Kowalewski M, Domenèch R, Martinell J (2014) Vanishing clams on an Iberian Beach: local consequences and global implications of accelerating loss of shells to tourism. PLoS One 9:e83615
Krelling AP, Williams AT, Turra A (2017) Differences in perception and reaction of tourist groups to stranded marine debris influence the loss of tourism revenue in coastal areas. Mar Policy 85:87
Leal GR, Efe MA, Ozorio CP (2013) Use of sandy beaches by shorebirds in Southern Brazil. Ornithologia 6:14–21
Leite JS, Moraes CRF, Medeiros DHM et al (2019) Avalação da disposiçao a pagar dos frequentadores de uma praia pelos benefícios ambientais. Geotemas 9:70–86
Léo-Neto NA, Voeks RA, Dias TLP et al (2018) Mollusks of Candomblé: symbolic and ritualistic importance. J Ethnobiol Ethnomed 8:10
Lima JC, Paula DP (2017) Serviços ecossistêmicos em litorais urbanos: o caso de Fortaleza, Ceará, Brasil. In: XVII Brazilian Symposium of Applied Physical Geography. Campinas, Brasil
Lindsay BE, Halstead JM, Tupper HR et al (1992) Factors influencing the willingness to pay for coastal beach protection. Coast Manag 20:291–302
Lins-de-Barros FM, Parente-Ribeiro L (2018) How much is a beach worth: economic use and vulnerability to coastal erosion: the case of Ipanema and Arpoador beaches, Rio de Janeiro (Brazil). In: Leal-Filho W, Freitas LE (eds) Climate change adaptation in Latin America. Springer, pp 207–222
Lins-de-Barros FM, Zeidan F, Lima RF (2016) Adaptações e percepção da população a eventos de ressaca do mar no litoral de Maricá, Rio de Janeiro. Rev Gestão Cost Integr 16:147–161
Liu S, Constanza R, Troy A et al (2010) Valuing New Jersey's ecosystem services and natural capital: a spatially explicit benefit transfer approach. Environ Manag 45:1271–1285
Luijendijk A, Hagenaars G, Ranasinghe R et al (2018) The state of the world's beaches. Sci Rep 8:6641
Machado E, Souza JRB, Borzone CA et al (2004) Effects of bioturbation of the sandprawn *Callichirus major* on nutrient fluxes from sediments of a subtropical sandy beach, Brazil. In: Lacerda LD, Santelli RE, Duursma EK et al (eds) Environmental geochemistry in tropical and subtropical environments. Springer, Berlin, Heidelberg
Machado PM, Suciu MC, Costa LL et al (2017) Tourism impacts on benthic communities of sandy beaches. Mar Ecol 38:e12440
Madden D, Ballestero J, Calvo C et al (2008) Sea turtle nesting as a process influencing a sandy beach ecosystem. Biotropica 40:758–765
Marin V, Palmisani F, Ivaldi R et al (2009) Users perception analysis for sustainable management in Italy. Ocean Coast Manag 52:268–277
Martínez ML, Intralawan A, Vázquez G et al (2007) The coasts of our world: ecological, economic and social importance. Ecol Econ 63:254–272
Martins EC, Peres FC (2005) O turismo como alternativa de desenvolvimento sustentável: o caso de Jericoacoara, Ceará. Rev Econ Nordeste 36:228–241
Martins KA, Pereira PS, Silva-Casárin R et al (2017) The influence of climate change on coastal erosion vulnerability in Northeast Brazil. Coast Eng J 59:1–25
Matos M, Santos A, Fauvelet C et al (2016) O surfe na promoção da saúde e bem-estar de jovens em contexto de acolhimento – avaliação de uma intervenção. Rev Psicol Criança Adolesc 7:433–447
Matthews YS, Scarpa R, Marsh D (2017) Stability of willingness-to-pay for coastal management: a choice experiment across three time periods. Ecol Econ 138:64–73
McCauley DJ (2006) Selling out on nature. Nature 443:27–28
McGoodwin JR (2001) Understanding the cultures of fishing communities: a key to fisheries management and food security. FAO Fisheries and Aquaculture Technical Paper 401
McLachlan A, Wooldridge T, Schramm M et al (1980) Seasonal abundance, biomass, and feeding of shore birds on sandy beaches in the Eastern Cape, South Africa. Ostrich 51:44–52
McLachlan A, Defeo O, Jaramillo E et al (2013) Sandy beach conservation and recreation: guidelines for optimizing management strategies for multi-purpose use. Ocean Coast Manag 71:256–268
Meager JJ, Schlacher TA, Nielsen T (2012) Humans alter habitat selection of birds on an ocean-exposed sandy beach. Divers Distrib 18:294–306

Medeiros ECS, Araujo RCP (2010) Valoração ambiental dos serviços ambientais de controle da erosão da Praia do Icaraí, Caucaia-CE. In: 48° Congresso SOBER (Sociedade Brasileira de Economia, Administração e Sociologia Rural). Campo Grande, Mato Grosso do Sul, Brazil, pp. 25–28

Mendoza-González G, Martínez ML, Lithgow D et al (2012) Land use change and its effects on the value of ecosystem services along the coast of the Gulf of Mexico. Ecol Econ 82:23–32

Millennium Ecosystem Assessment (MEA) (2005) A report of the millennium ecosystem assessment. Ecosystems and human well-being. Island Press, Washington, DC

Moreira FT, Prantoni AL, Martini B et al (2016) Small-scale temporal and spatial variability in the abundance of plastic pellets on sandy beaches: methodological considerations for estimating the input of microplastics. Mar Pollut Bull 102:114–121

MTur – Ministério do Turismo (2019) Pesquisa de Sondagem Empresarial: Agências e organização de viagens 2ndo semestre. Brasília

Nascimento MSV, Sassi R (2007) Análise da atividade pesqueira e das condições socioeconômicas dos pescadores artesanais de Cajueiro da Praia, Estado do Piaui, Brasil. Gaia Sci 1:141–154

Nel R, Campbell EE, Harris L et al (2014) The status of sandy beach science: past trends, progress and future trends. Estuar Coast Shelf Sci 150:1–10

Nielsen-Pincus M, Sussman P, Bennett DE et al (2017) The influence of place on the willingness to pay for ecosystem services. Soc Nat Resour 30:1423–1441

Olds AD, Vargas-Fonseca E, Connolly RM et al (2018) The ecology of fish in the surf zones of sandy beaches: a global review. Fish Fish 19:78–89

Oliveira KTLL (2015a) Análise de valoração contingente das praias do bairro Rio Vermelho, Salvador-BA: uma aplicação dos instrumentos econométricos logit e probit. Nexos Econ 9:135–162

Oliveira TCR (2015b) 5. Uso e qualidade de praias arenosas da Ilha de Santa Catarina, SC, Brasil: bases para seu planejamento ambiental. Dissertation, Universidade Federal de Santa Catarina

Oliveira LEC, Berkes F (2014) What value São Pedro's procession? Ecosystem services from local people perception. Ecol Econ 107:114–121

Oliveira REMCC, Pessanha ALM (2014) Fish assemblages along a morphodynamic continuum on three tropical beaches. Neotrop Ichthyol 12:165–175

Oliveira CR, Machado GC, Gonçalves RR et al (2017) Castelos de lama: uma proposta de valoração ambiental da Praia do Cassino, Rio Grande do Sul, Brasil. Rev Espac 38:5

Onofri L, Nunes PALD (2013) Beach "lovers" and "greens": a worldwide empirical analysis of coastal tourism. Ecol Econ 88:49–56

Ortega L, Castillo JC, Espino M et al (2012) Effects of fishing, market price, and climate on two South American clam species. Mar Ecol Prog Ser 469:71–85

Pascual U, Balvanera P, Díaz S et al (2017) Valuing nature's contributions to people: the IPBES approach. Curr Opin Environ Sustain 26–27:7–16

Pereira HH, Neves LM, Costa MR et al (2015) Fish assemblage structure on sandy beaches with different anthropogenic influences and proximity of spawning grounds. Mar Ecol 36:16–27

Phillips MR, House C (2009) An evaluation of priorities for beach tourism: case studies from South Wales, UK. Tour Manag 30:176–183

Pombo M, Fisner M, Xavier LY et al (2009) Estudo da variação do Perfil de usuários e usos de praias em Ubatuba, São Paulo, Brasil, conforme a morfodinâmica e a intensidade turística. In: 13° Congresso Latino Americano de Ciências do Mar. Havana, Cuba

Reis JV (2018) Valoração dos serviços ambientais culturais das praias do município de Tamandaré, Litoral Sul do Estado de Pernambuco, Brasil. Dissertation, Federal University of Pernambuco

Reis EC, Goldberg DW (2017) Pesquisa e conservação de tartarugas marinhas no Brasil e as recentes contribuições da telemetria e genética. In: Reis EC, Curvello-Fernandez MP (eds) Mamíferos, quelônios e aves: caracterização ambiental regional da Bacia de Campos, Atlântico Sudoeste. Elsevier, Rio de Janeiro, pp 91–120

Rezende-Parker AM, Morisson AM, Ismail JA (2003) Dazed and confused? An exploratory study of the image of Brazil as a travel destination. J Vacat Market 9:243–259

Ribeiro LCS, Souza KB, Domingues EP et al (2020) Blue water turns black: economic impact of oil spill on tourism and fishing in Brazilian Northeast. Curr Issues Tour 24:1042–1047

Roelvink D, Reniers A, van Dongeren A et al (2009) Modelling storm impacts on beaches, dunes and barrier islands. Coast Eng 56:1133–1152

Rolfe J, Gregg D (2012) Valuing beach recreation across a regional area: the Great Barrier Reef in Australia. Ocean Coast Manag 69:282–290

Romeiro A, Sinisgalli P, Jericó-Daminello C et al (2014) A comunidade do Marujá e o Parque Estadual da Ilha do Cardoso: co-manejo, serviços ecossistêmicos e processo socio-tecnico. Rev Virtual Redesma 7:75–85

Rosa CD, Collado S, Profice CC et al (2019) Nature-based recreation associated with connectedness to nature and leisure satisfaction among students in Brazil. Leis Stud 38:682–691

Santos DK (2017) Valor econômico do turismo em uma praiaurbana. Monography, Universidade Federal do Rio Grande do Norte

Santos SR, Almeida CR, Silva LF (2010) A disposição a pagar (DAP) e a valoração dos ativos ambientais: uma aplicação do método na praia do Atalaia em Salinópolis – PA. In: XIV Encontro Latino Americano de Iniciação Científica

Santos AJB, Neto JXL, Vieira DHG et al (2016) Individual nest site selection in hawksbill turtle within and between nest seasons. Chelonian Conserv Biol 15:109–114

Santos CR, Grilli NM, Ghilardi-Lopes NP et al (2018) A collaborative work process for the development of coastal environmental education activities in a public school in São Sebastião (São Paulo State, Brazil). Ocean Coast Manag 164:147–155

Santos-Frossard TC, Passos HDB, Silva MS et al (2015) Valoração ambiental da Praia da Coroinha, Itacaré/BA. Inf Gepec 19:90–108

Schlacher TA, Strydom S, Connolly RM et al (2013) Donor-control of scavenging food webs at the land-ocean interface. PLoS One 8:e68221

Schröter M, van der Zanden EH, van Oudenhoven APE et al (2014) Ecosystem services as a contested concept: a synthesis of critique and counter-arguments. Conserv Lett 7:514–523

Scott D, Simpson MC, Sim R (2012) The vulnerability of Caribbean coastal tourism to scenarios of climate change related sea-rise. J Sustain Tour 20:883–898

Severino-Rodrigues E, Pita JB, Graça-Lopes R (2001) Pesca artesanal de siris (Crustacea, Decapoda, Portunidae) na região estuarina de Santos-São Vicente, Brasil. Bol Inst Pesca 27:7–19

Shah Esmaeili YML, Corte GN, Checon HH, Gomes TRC, Lefcheck JS, Amaral ACZ, Turra A (2021) Comprehensive assessment of shallow surf zone fish biodiversity requires a combination of sampling methods. Mar Ecol Prog Ser 667:131–144

Silva EJ, Martins IX (2017) A pesca de moluscos em ambientes intermareais no Oeste do Estado do Rio de Janeiro, Brasil. Arq Cienc Mar 50:110–118

Silva AES, Paula DP (2017) Avaliação dos serviços ecossistêmicos oferecidos pelas praias do município de Caucaia, Ceará, Brasil. In: XVII Simpósio Brasileiro de Geografia Física Aplicada

Silva JJ, Souza CRG (2018) Perfil dos usuários em praias do município do Guarujá (SP) e sua relação com a morfodinâmica praial. In: Anais do XI Encontro Nacional de Gerenciamento Costeiro & II Simpósio Brasileiro de Praias Arenosas. Santa Catarina, Brasil

Silva ACCD, Castilhos JC, Lopez GG et al (2007) Nesting biology and conservation of the olive ridley sea turtle (*Lepidochelys olivacea*) in Brazil, 1991/1992 to 2002/2003. J Mar Biol Assoc UK 87:1047–1056

Silva MLN, Nascimento MAL (2016) Os valores dageodiversidade de acordo com os serviços ecossistêmicossensu murray gray aplicados a estudos in situ na cidade doNatal (RN). Cad Geogr 26:338–354

Silva JS, Leal MMV, Araújo MCB et al (2018) Spatial and temporal patterns of the use of Boa Viagem beach, Northeastern Brazil. J Coast Res 24:79–86

Silva-Cavalcanti JS, Costa MF (2011) Fisheries of *Anomalocardia brasiliana* in tropical estuaries. Pan Am J Aquat Sci 6:86–99

Siqueira DS, Peres LFB, Bosquetti MA (2020) Praias Acessíveis e Surf Adaptado no Brasil: inovação social baseado no Design Universal. Cuad Cent Estud Diseñ Comun 83:145–162

Souza CRG (2012) Praias oceânicas do Estado de São Paulo: síntese do conhecimento sobre morfodinâmica, sedimentologia, transporte costeiro e erosão costeira. Rev Depart Geogr 30:308–317

Souza JRB, Borzone CA (2003) A extração do corrupto, *Callichirus major*, para uso como iscas em praias do litoral do Paraná: as populações exploradas. Rev Bras Zool 20:625–630

Souza KM, Casarini LM, Henriques MB et al (2009) Viabilidade econômica da pesca do camarão-sete-barbas com embarcação de pequeno porte na praia do Perequê, Guarujá, Estado de São Paulo. Inf Econ 39:30–37

Souza-Filho JR, Santos RC, Silva IR (2014) Evaluation of recreational quality, carrying capacity and ecosystem services supplied by sandy beaches of the municipality of Camaçari, northern coast of Bahia, Brazil. J Coast Res 70:527–532

Suciu MC, Tavares DC, Costa LL et al (2017) Evaluation of environmental quality of sandy beaches in southeastern Brazil. Mar Pollut Bull 119:133–142

Swaisgood RR, Nordstrom LA, Schuetz JG et al (2018) A management experiment evaluating nest site selection by beach-nesting birds. J Wildl Manag 82:192–201

Taylor EB, Gibeaut JC, Yoskowitz DW et al (2015) Assessment and monetary valuation of the storm protection function of beaches and foredunes on the Texas Coast. J Coast Res 31:1205–1216

TEEB (2008) The economics of ecosystem & biodiversity: an interim report. A Banson Production, Cambridge

Thomé JCA, Baptistotte C, Moreira LMP et al (2007) Nesting biology and conservation of the leatherback sea turtle (*Dermochelys coriacea*) in the State of Espírito Santo, Brazil, 1988–1989 to 2003–2004. Chelonian Conserv Biol 6:15–27

Toste R, Caiuby G, Freitas M et al (2011) Vulnerabilidade costeira as mudanças climáticas Armação de Búzios (RJ, Brasil). In: XIV IWRA (World Water Congress). Porto de Galinhas, Pernambuco, Brasil, pp 25–29

Tourinho PS, Fillmann G (2011) Temporal trend of litter contamination at Cassino Beach, Southern Brazil. Rev Gestão Cost Integr 11:97–102

Turra A, Xavier LY, Pombo M et al (2016) Assessment of recreational harvesting of the trigonal clam *Tivella mactroides*: socioeconomic aspects and environmental perception. Fish Res 174:58–67

Uchikawa D (2008) Verificando a importância da praia de Boa Viagem utilizando métodos de valoração ambiental. Dissertation, Federal University of Pernambuco

Van der Merwe P, Slabbert E, Saayman M (2010) Travel motivations of tourists to selected marine destinations. Int J Tour Res 13:457–467

Vargas-Fonseca E, Olds AD, Gilby BL et al (2016) Combined effects of urbanization and connectivity on iconic coastal fishes. Divers Distrib 22:1328–1341

Vasconcellos RM, Santos JNS, Silva MA et al (2007) Efeito do grau de exposição às ondas sobre a comunidade de peixes juvenis em praias arenosas do município do Rio de Janeiro, Brazil. Biota Neotrop 7:93–100

Vasconcellos M, Diegues AC, Kalikoski DC (2011) Coastal fisheries of Brazil. In: Salas R, Chuenpagdee AC, Seijo JC (eds) Coastal fisheries of Latin America and the Caribbean, FAO Fisheries and Aquaculture Technical Paper, No 544. FAO, Rome, pp 73–116

Veloso VG, Silva ES, Caetano CHS et al (2006) Comparison between the macroinfauna of urbanized and protected beaches in Rio de Janeiro. Biol Conserv 127:510–515

Vihervaara P, Kumpula T, Tanskanen A et al (2010) Ecosystem services – a tool for sustainable management of human-environment systems. Case study Finnish Forest Lapland. Ecol Complex 7:410–420

Vooren CM, Brusque LF (1999) As aves do ambiente costeiro do Brasil: biodiversidade e conservação. Programa Nacional da Diversidade Biológica/Laboratório de Elasmobrânquios e Aves Marinhas, Fundação Universidade do Rio Grande, Rio Grande (RS), Brasil, 139 pp

WTTC (2019) Travel and tourism economic impact 2019. Available at: https://wttc.org/Research/Economic-Impact. Accessed 23 Apr 2020

Xavier LY, Guilhon M, Gonçalves LR et al (2022) Waves of change: towards ecosystem-based management to climate change adaptation. Sustainability 14:1317

Yepes V, Medina JR (2005) Land use tourism models in Spanish coastal areas: a case study of the Valencia region. J Coast Res SI49:83–88

Chapter 9
Threats and Impacts

Abílio Soares-Gomes, Ilana R. Zalmon, Phillipe Mota Machado, and Leonardo Lopes Costa

9.1 Introduction

Sandy beaches are found anywhere along the Brazilian coast, with over 4000 beaches found in the country, comprising 2% of all coastal ecosystems (Muehe 2003; Klein and Short 2016). About 46 million people live on the Brazilian coast, which extends for more than 9000 km, from 34° S to 4° N latitudes (Dominguez 2009). This huge occupation of the coast and the establishment of industries nearby the littoral and its watershed threaten Brazilian beaches.

In Brazil, beaches have been particularly jeopardized by oil spills (Marques et al. 2017; Soares et al. 2020), pollution (Teixeira et al. 2012), sewage discharges (Oliveira et al. 2010; Roth et al. 2016), marine litter (Ivar do Sul and Costa 2007), overfishing (Bender et al. 2014), heavy metal contamination (Cabrini et al. 2017; Baeyens et al. 2019), and loss of coastal vegetation (Rêgo et al. 2018) and dune habitats (Soares et al. 2022). Coastal development inherent to economic progress has resulted in extensive changes in coastal ecosystems, especially on sandy beaches, due to their tourist and recreational importance. These anthropogenic disturbances threaten the maintenance of functions, goods, and environmental services provided by these coastal ecosystems (Defeo et al. 2009, see Chap. 8).

A. Soares-Gomes
Laboratório de Ecologia de Sedimentos, Departamento de Biologia Marinha, Universidade Federal Fluminense, Niterói, RJ, Brazil

I. R. Zalmon (✉) · L. L. Costa
Centro de Biociências e Biotecnologia, Universidade Estadual do Norte Fluminense, Campos, RJ, Brazil
e-mail: ilana@uenf.br

P. M. Machado
Departamento de Biologia, Centro de Ciências Exatas, Naturais e da Saúde, Universidade Federal do Espírito Santo, Alegre, ES, Brazil

A. C. Z. Amaral et al. (eds.), *Brazilian Sandy Beaches*, Brazilian Marine Biodiversity, https://doi.org/10.1007/978-3-031-30746-1_9

Besides direct anthropic impacts, climate change consequences, such as higher frequency and intensity of storm wave events, cold fronts, and sea-level rise (IPCC 2013), threaten sandy beaches and their biodiversity (Machado et al. 2016, 2019). This emphasizes that local, regional, and global cumulative stressors impact biodiversity synergistically, challenging the prediction of changes in the biodiversity during the Anthropocene (Defeo et al. 2021).

Threats to beaches can cause chronic or acute impacts depending on the duration of their effects. Acute impacts usually result from disaster events, such as oil spills, but could also happen in programmed interventions such as beach nourishment and cleaning services. Chronic impacts can be spatially diffuse or restricted. Still, they are long-standing, usually associated with industrial activities and lack of sanitation, such as the discharge of raw domestic sewage and untreated industrial effluents, widespread occurrence of marine litter, and effects of human trampling. Although impacts are usually defined as acute or chronic, they are generally hard to fit in one or another class or have both attributes. For example, marine litter may be characterized as an acute impact on a specific beach after a social event such as New Year's Eve; however, it is also a chronic impact observed on most sandy beaches worldwide, even those without human occupation. Similarly, vehicle traffic can be acute or chronic, depending on intensity and frequency.

In this chapter, we review the main threats and impacts affecting sandy beach ecosystems. First, we focus on the direct effects of human disturbance on sandy beach biological integrity. Second, we discuss how impacts related to climate change may affect sandy beaches. Coastal erosion and its countermeasures, such as coastal armoring and beach nourishment, are examined next. Then, we move on to evidence of how indicator species have emerged as an important tool for impact assessments and monitoring, providing examples of their use along the Brazilian coast. Finally, we discuss initiatives to mitigate the adverse effects of human activities and climate change on the biodiversity of these ecosystems.

9.2 Direct Human Impacts

9.2.1 Human Traffic

Continuous human population growth in coastal zones influences the occupation and recreational use of this environment, promoting considerable changes and anthropogenic impacts at various temporal and spatial scales (McLachlan and Defeo 2018). Tourism- and recreation-related pressures on sandy beaches include mainly trampling, vehicle traffic, marine debris through direct littering, and noise pollution (Schlacher and Thompson 2008; Machado et al. 2017; Leduc et al. 2021). An increasing number of beachgoers result in cumulative and synergic stressors for both resident and transient fauna (Corte et al. 2022) (Fig. 9.1a–c).

Fig. 9.1 Human disturbances on sandy beaches. (**a**) Tire tracks from vehicle traffic (Grussaí Beach, state of Rio de Janeiro); (**b**) beach cleaning (Grussaí Beach, state of Rio de Janeiro); (**c**) harvesting of mole crabs (Praia Seca Beach, state of Rio de Janeiro) (photos: Leonardo Costa); (**d**) angling tournament (Peruíbe Beach, state of São Paulo) (photo: Rosangela Costa)

Tourism associated with recreational activities is one of the main sources of chronic disturbances on Brazilian sandy beaches, particularly on the Southeast coast (Amaral et al. 2016). Due to the intrinsic relationship between benthic fauna and sediment and the sedentary habit of most species, intense trampling triggers spatial and temporal changes in these communities (Haynes and Quinn 1995). Although distinguishing the effects of trampling from generalized impacts is challenging, several authors have attributed changes in species richness, the abundance of individuals, and biomass to trampling on Brazilian beaches (Veloso et al. 2006, 2008; Vieira et al. 2012; Machado et al. 2017; Corte et al. 2022), especially for species that have no external hard structures like shells and resistant carapaces. For robust and high-mobile species, such as ghost crabs (Crustacea: Ocypodidae), trampling seems to not result in lethal effects (Costa et al. 2018a, b), although the increase in the number of beachgoers may be associated with a reduction in crab abundance because of possible changes in sediment features (Schlacher et al. 2016). Trampling effects on macroinvertebrates also have negative consequences for energy flow efficiency and trophic functioning on Brazilian sandy beaches, since they are preyed on by several fish and bird species and have an important role in detritus cycling (Costa et al. 2017).

Among uninterrupted recreational impacts on sandy beaches, vehicle traffic has negative effects on physical features (e.g., sediment compactness, moisture, erosion, and vegetation loss) (Schlacher and Thompson 2008), beach biodiversity (Schlacher et al. 2007), and species behavior (e.g., home range and excavation)

(Merwe et al. 2012). Consequently, macroinvertebrates are usually less abundant in areas with intense vehicle traffic compared to beaches with low traffic (Moss and McPhee 2006). In Brazil, some studies have shown a reduction in burrow abundance of the ghost crab *Ocypode quadrata* (Fabricius 1787) and changes in community patterns attributed to vehicle traffic (Neves and Bemvenuti 2006; Magalhães et al. 2009; Bom and Colling 2020). Vehicles can cause direct mortality of both vertebrates (e.g., turtle hatchlings and shorebirds), and invertebrates (e.g., ghost crabs) found on the sediment surface (Costa et al. 2020; Santos et al. 2022). However, studies about spatial and temporal patterns of animal-vehicle collisions on sandy beaches are scarce in Brazilian sandy beaches.

Human traffic rarely ceases on sandy beaches, but in some cases, touristic activities are acute, and biodiversity benefits from the partial removal of anthropogenic stressors (Costa et al. 2017; Costa and Zalmon 2019a). The restriction of beach use during COVID-19 lockdowns provided a unique opportunity to design observational experiments aiming to assess the recovery potential of beach biodiversity from human activities and pollution (Neves et al. 2022). Because of this “Anthropause” (sensu Rutz et al. 2020), it has been reported an exceptionally low frequency of beach users, which in turn reduced litter and noise pollution, and increased invertebrate densities (Soto et al. 2021; Costa et al. 2022a, b).

9.2.2 Harvesting

Harvesting is a common practice for recreational, subsistence, or commercial purposes on sandy beaches worldwide (Defeo 2003). Macroinvertebrates are usually picked up in intertidal and infralittoral fringe by hand gathering, skin diving, and with traps and other devices (Fig. 9.1c). The main targets of beach harvesting are polychaetes, mole crabs (*Emerita* spp.), clams (Donacidae, Veneridae, and Mesodesmatidae), and ghost shrimps (Callianassidae) used mainly as bait for angling activity or food. There is an increasingly global market based on the use of macroinvertebrates as bait (Pombo et al. 2018); polychaetes worms from intertidal zones, for example, can achieve a trading value estimated at US$7 billion year^{-1} (Watson et al. 2018). This activity is predominantly artisanal and for noncommercial features in Brazil and, at least for now, harvesting has not been considered a major local disturbance for beach communities (Rech et al. 2021). However, there are scarce studies about harvesting impacts on Brazilian beaches, although this activity is commonly seen in YouTube videos showing local anglers teaching how to collect crustaceans, polychaetes, and mollusks. Scientific studies about the impacts of bait capture on Brazilian beaches only exist for the ghost shrimp *Callichirus major* on the southern and southeast coast, where the species is also traded as bait (Rosa and Guilheme 2014). In the municipality of Santos (state of São Paulo), the capture has been forbidden since 1992 by the enforcement of a municipality law after a decrease in the local population probably caused by overfishing. Indeed, studies on Brazilian beaches showed a decreasing density of ghost shrimps on beaches with extractive activity, which achieves up from 10% to 71% of the total standing stock (Souza and Borzone 2003; Hernáez et al. 2019).

In some cases, artisanal extraction for subsistence or commercial purposes is considered obsolete by local anglers due to changes in fishery technologies and market targets, as shown by Owczarzak (2015) with the reduction in the harvesting of the polychaete *Hemipodia californiensis* in Santa Catarina. Denadai et al. (2015a, b) studied the short- and long-term dynamics of harvesting of the bivalve mollusk *Tivella mactroides* in Caraguatatuba Bay, southeastern Brazil, and showed that this clam is intensely harvested by both residents and tourists, amounting to about 25-ton year^{-1}. According to the authors, due to the high local abundance of *T. mactroides*, this harvested volume does not negatively affect the clam population (Chap. 6). However, the authors warned that since the size limit for harvesting is unknown, an increase in capture pressure might compromise the local beach ecology in the future.

9.2.3 *Fishing*

The surf zones of sandy beaches are important sites for recreational angling and commercial net fisheries (Fig. 9.1d). However, there are scarce data about capture rates during this activity and the impacts on the fish communities from surf zones, even when the catch-release approach is applied. Barrella et al. (2016) estimated a catch of ~4 fish (488 g) per angler per day in the Baixada Santista, state of São Paulo, corresponding to only 2% of commercial landings. Besides angling, removing large predators from surf zones in shark control programs to improve swimmer safety has been performed worldwide. In Brazil, increasing shark accidents after the construction of a port complex in Recife, state of Pernambuco, led the government to create specific monitoring and research programs in 2004. Although most of the programs focused on capturing only potentially aggressive sharks and removing them from the hazardous area, there is no study about the effects of these strategies on surf zone fauna. Other harmful threats to sharks and large vertebrates are bycatch and "ghost fishing," when nontarget species are incidentally captured by fishing gears, including lost or abandoned ones (Matsuoka et al. 2005).

9.2.4 *Urbanization*

Coastal urbanization is usually associated with intense recreational pressures, but it also imposes other chronic impacts on sandy beaches. This includes habitat loss due to the removal of dunes and vegetation for the construction of paved roads, buildings, and artificial lighting, which is required for nocturnal touristic activities. A meta-analysis at continental scales, but including mainly Brazilian beaches in the dataset, showed that the conservation status of the coastal vegetation is closely related to higher densities of the ghost crab *O. quadrata* (Barboza et al. 2021). Accordingly, despite having some infrastructure facilities, beaches with well-preserved dunes and backshore vegetation generally support a higher number of individuals and species (Cardoso et al. 2016). Another problem of coastal urbanization is the crossing roads, where vehicles crush various species such as crabs and turtles attracted by artificial light (Costa et al. 2020).

The introduction of artificial lighting on sandy beach habitats has been considered an important chronic threat to biodiversity (Duarte et al. 2019; Lynn et al. 2021). Light from natural sources (i.e., sun and moon) plays a central role in ecological processes, including migration, predator–prey interactions, and circadian rhythms of invertebrates and vertebrates (Gaston et al. 2012). For example, artificial light imposes negative effects by inhibiting and disrupting sea turtles' nesting and hatchlings orientation (Peters and Verhoeven 1994), facilitating the detection of prey by predators (Rydell 1992; Silva et al. 2017), and changing foraging and behaviors (Bird et al. 2004). Beach resident fauna includes invertebrates displaying circadian activities, and recent studies have shown disrupted consumption and growth rates or lower density of amphipods induced by artificial light (e.g., Luarte et al. 2016; Lynn et al. 2021). Artificial light also reduced the foraging activity of the gastropod *Olivella minuta* in manipulative experiments (Checon et al. 2021).

Similar to the global tendency, most studies about the effects of artificial lighting on Brazilian beaches have focused on vertebrates, mainly turtles (Kamrowski et al. 2012). As an initiative of the Brazilian National Sea Turtle Conservation Program (TAMAR), created in 1980, geospatial tools have been developed for the identification of critical sea turtle habitats jeopardized by several emerging threats such as coastal development, including artificial lighting (Lopez et al. 2015). Serafini et al. (2010) described a simple and efficient method to identify emerged nests with turtle hatchlings disrupted by artificial light on urbanized sites in Bahia, northeast Brazil. The Brazilian National Environmental Agency (IBAMA) has established some laws to protect recognized nesting beaches from coastal urbanization, but in some regions, environmental laws often fail to protect beaches from this stressor (Mascarenhas et al. 2004).

Human occupation in dune and berm areas disrupts the interchange of sediment between the sea and the land through wind- and wave-driven transport, reducing sediment supply and increasing coastal erosion. On the coast of Espírito Santo, Southeast Brazil, Albino et al. (2001) reported an intensification of coastal erosion after the construction of houses and touristic infrastructure on the local dunes. According to the authors, erosive processes are further intensified in the region due to its high degree of wave exposure and low presence of rocky outcrops. Consequently, erosion processes have intensified and threatened the maintenance of urban infrastructure, stimulating the implementation of management actions that seem ineffective and ecologically adverse.

9.2.5 Marine Litter

Sandy beaches are common sites of litter stranding, even in nonurbanized areas, because of the passive transport by wind, currents, and tides (Corcoran et al. 2009). Thus, although local urbanization and direct littering by beachgoers are typically decisive factors for the loss of environmental quality, the sea and landscape characteristics are generally preponderant to local pollution. Distance to land run-offs and ports, fishery activities, navigation routes, and ocean circulation patterns are among the regional factors driving the amount of marine litter on Brazilian sandy beaches (Andrades et al. 2020).

The accumulation of litter in marine ecosystems is devastating, causing various sublethal dysfunctions or leading many organisms to death, including invertebrates and megafauna (Derraik 2002). Nowadays, assessing the litter impact on biodiversity, tourism, and even human health is a major global challenge (Rochman et al. 2015). Ingestion, entrapment, and entanglement have been reported as the main deleterious impact on marine fauna (Derraik 2002). The drift litter is also a pathway to the invasion of alien species (Barnes 2002).

Brazil produces more than 11 million metric tons of plastic annually and 2.4 million are discharged irregularly (WWF 2019). From this mismanaged litter, almost 1 million metric tons reach the ocean transported by rivers or are directly discarded in coastal areas (Jambeck et al. 2015). As a result, Brazilian sandy beaches are severely impacted by marine litter. This situation is perhaps most dire on the central coast of the state of Rio de Janeiro, which has sandy beaches with a higher density (4 items/m^2) of litter, mainly plastics (>80%), than many beaches worldwide (Suciu et al. 2017) (Fig. 9.2). This high concentration of marine litter on sandy beaches can discourage beach users, resulting in high economic losses. For example, extreme scenarios of pollution (15 items/m^2) result in US$8.5 million in local losses per year on beaches of Paraná, South Brazil (Krelling et al. 2017). Besides economic losses, marine litter also affects sandy beach resident fauna through the ingestion of microplastic (<1 mm) (Costa et al. 2019a). The interaction between beach biota and marine debris in Brazil was first reported by Costa et al. (2018a, b). The authors pointed out, with an innovative experimental design, the possibility of marine debris usage by the ghost crab *O. quadrata*, with selectivity by some types, such as soft plastic, straw, rope, and foam. This interaction occurs because ghost crabs probably misidentify (stimulated by food smell) marine debris with food sources (Costa et al. 2019b).

9.2.6 Microplastics

A large proportion of plastics on sandy beaches have been found as microplastics (MPs) (<5 mm synthetic polymer particles). MPs are classified as primary when manufactured initially to a small size (e.g., beads from cosmetic and hygienic

Fig. 9.2 Different types of plastic are found on Brazilian beaches, threatening the fauna (**a**) and (**b**) cigarette butts and bottle caps found in a single-day survey in Grussaí Beach, state of Rio de Janeiro (photo: Marjorie Suciu); (**c**) marine turtle found entangled with a rope in Grussaí Beach (photo: Leonardo Costa); (**d**) ghost crab holding a plastic spoon in Praia Seca Beach, state of Rio de Janeiro (photo: Danilo Rangel)

products and pellets for industrial applications) and secondary if fragmented from larger plastic items (GESAMP et al. 2016). Plastic fragmentation into small particles occurs effectively due to constant wave abrasion, photodegradation, and turbulence on beaches (Cole et al. 2011), resulting in an enormous variety in size, shape, color, and polymer type among the secondary marine MPs. Similar to global reports, MPs are stranded and buried in the sand of Brazilian beaches as fibers, fragments, styrofoam, and pellets (Costa et al. 2010; Carvalho and Neto 2016). There are evidence that pellets from shallow sediment have a higher concentration and toxicity potential of polycyclic aromatic hydrocarbons (PAHs) and persistent organic pollutants (POPs), imposing serious environmental risks (Fisner et al. 2013). Although further integration of beach ecology paradigms into beached plastics studies is still a global demand (Fanini et al. 2021), some results from Brazilian beaches suggest that MPs can be mainly accumulated in urban, sheltered, and dissipative beaches (Martinelli-Filho and Monteiro 2019; Pinheiro et al. 2019; Tsukada et al. 2021).

Metals are also found adsorbed in plastic pellets from Brazilian beaches closer to ports (Vedolin et al. 2018). Unsurprisingly, beach invertebrates, both macrofauna (Costa et al. 2019a), meiofauna (Gusmão et al. 2016), and surf zone fishes (Dantas et al. 2020) on Brazilian beaches, ingest these MPs, mainly fibers. Synthetic fiber occurrence in the water and sediment has been attributed mainly to the release from fabrics during washing and the fragmentation of fishing materials (e.g., trawl nets and ropes) (Cole 2016). MPs can also act as a vector for the assimilation of adsorbed contaminants (PAHs, POPs, and metals) into beach organisms, potentially responsible for bioaccumulation and biomagnification in the beach food webs (Chua et al. 2014).

9.2.7 Beach Cleaning and Grooming

The presence of anthropogenic litter and wrack (natural materials, such as macroalgae, seagrass, and invertebrates, that wash onto the beach) on Brazilian sandy beaches is a reason for beachgoers to visit or not visit a certain beach (Krelling et al. 2017; Suzuki et al. 2021). Consequently, beach cleaning and grooming are common management actions to maintain the aesthetic values and recreational potential of urban sandy beaches (Zielinski et al. 2019). For example, Krelling et al. (2017) reported that the municipality of Pontal do Paraná, South Brazil, spent US$200,000 to cleaning a 25 km long stretch of a beach for 53 days.

While beach cleaning and grooming may increase the adequacy of sandy beaches for tourism, it may also compromise sandy beach biodiversity and functioning.

Beach cleaning promotes acute adverse effects on sandy beaches by removing wrack and carrion, which are habitat and food sources for several detritivorous organisms (Dugan et al. 2003). In addition, heavy machines used in cleaning may directly kill both vertebrates and invertebrates. Thus, in terms of environmental conservation, cleaning beaches is not entirely desirable (Defeo et al. 2009). Two studies reported the potential effects of detritus removal on beach macroinvertebrates in Brazil. On the coast of Paraná, Vieira et al. (2016) found a density reduction in detritivorous macroinvertebrates, such as the coleopteran *Phaleria Testacea*, after removing mangrove propagules from a beach. Similarly, in the state of Rio de Janeiro, Souza et al. (2017) found that sandy beaches frequently cleaned had a lower number of burrows of the ghost crab *O. quadrata* and other detritus-feeder species. Detritus-feeders such as talitrid amphipods and ghost crabs are important components of the cycle of organic matter and nutrients on sandy beaches, and their absence may seriously compromise sandy beach functioning. Furthermore, besides removing wrack and carrion, beach cleaning and grooming often remove a large volume of beach sediments, thereby contributing to coastal erosion.

9.2.8 Organic Pollution

Organic pollutants on beaches generally derive from human settlements and interventions, such as construction, agricultural activities, and urbanization along the watershed (Islam and Tanaka 2004). Sewage from industrial and domestic waste, as well as runoff from agriculture and landscaping fertilizers, is considered one of the main pollutants in coastal waters, affecting mainly estuarine beaches at regional scales (Barile 2004; Roth et al. 2016). The organic enrichment of coastal waters favors the proliferation of bacteria and the reduction of dissolved oxygen near the shore (eutrophication), affecting the structure and function of aquatic communities (Valiela et al. 1992). It also fuels harmful algal blooms (HABs), which may release toxic substances or deplete water oxygen depletion (Chap. 2), thereby compromising public health, ecosystem function, fisheries, and tourism industries.

More punctual sources of organic pollution, such as domestic and synanthropic animals (e.g., dogs and pigeons), can contaminate the environment with feces hosting some pathogens (e.g., *Escherichia coli* and *Enterococcus* spp.) (Aragonés et al. 2016). Although sandy beaches are dynamic ecosystems with constant water and sediment exchange, there are several Brazilian beaches considered inappropriate for recreational activities, according to threshold values for coliform concentrations in the sand (e.g., *E. coli*) (e.g., Pinto et al. 2012; Davino et al. 2015). Part of these bacteria is resistant to antibiotics and can lead to serious public health problems (Cardonha et al. 2004; Oliveira et al. 2010). The consumption of shellfish from contaminated beaches is a huge concern to food safety in Brazil, as demonstrated by Denadai et al. (2015a, b) for Caraguatatuba sandy beach.

9.2.9 Trace Elements and Persistent Organic Pollutants

Sandy beaches that receive significant organic input usually have other contaminants, such as heavy metals and persistent organic pollutants. Concentrations of arsenic above the Brazilian standard threshold established by the Brazilian National Health Agency (ANVISA) for marine sediments (70 mg kg^{-1}) have been found on some beaches in Brazil, related to anthropogenic sources from metallurgical plants, phosphate fertilizer plants, and gold and iron mining (Baeyens et al. 2019). On the other hand, concentrations of other metals (Cr, Zn, Pb, Ni, Cu, Cd, Co, and V) on Brazilian beaches have been reported under the limits established by ANVISA. Trace element concentrations in macroinvertebrates were reported as spatially variable along the Rio de Janeiro coast (Cabrini et al. 2017) because of the diffuse source of contamination, but the biomagnification has been evidenced in carnivorous species (Cabrini et al. 2018).

Persistent organic compounds from biocides used in agriculture activities and antifouling paints can impose adverse effects on beach biodiversity and public health due to their toxicity. The imposex occurrence on the beach snail *Hastula cinerea* in São Paulo beaches with nautical activities is a further example of how beach pollution by organic compounds from seascape sources (e.g., Tributyltin-TBT) threats the resident fauna at the organism level (Ragagnin and Turra 2022).

9.2.10 Oil Spillage

Oil spillage is a typical acute threat that can prompt severe, long-term consequences for marine organisms on a large spatial scale if cleanup efforts are not properly adopted. Beaches are vulnerable to oil spillage, but the magnitude of the consequences for biota and cleanup possibilities depends on the oil type (light or heavy), the state of the oil reaching the beaches (fresh or weathered), and the beach morphodynamic features. Depending on the characteristics of the oil, it can stay in the sediment surface (<15 cm) or penetrate deeper into the sediment column (>25 cm), hampering the cleanup efforts, persisting in the environment, and prolonging changes in the abundance and composition of beach assemblages for weeks to decades (Benjarano and Michel 2016).

Oil spillages are catastrophic events that shock public opinion for the biological, social, and economic consequences, but scientific literature on the impacts of oil spills on beach fauna is scarce (Benjarano and Michel 2016). In Brazil, three studies were published about the ecological consequences of oil spillage on beach invertebrates. Borzone and Rosa (2009) studied the consequences of a spillage that took place in January 2004 in Paranaguá Bay, Southern Brazil, after the Chilean vessel Vicuña exploded and sank, leaking 291 tons of bunker fuel oil. Since the authors had data from the previous survey, it was possible to make a before-after (30 and 120 days) approach to study the impacts on talitrid amphipods. Manual removal of stranded macrodetritus (oiled or clean) and oil patches from the sand surface was the cleanup method adopted. Their results showed that the amphipods' assemblage structure was not affected by the spillage. Curiously, the increased abundance and geographical distribution of the talitrid *Platorchestia monodi* after an oil incident were attributed to the removal of wrack debris. Before the oil spill, the species was restricted to three beaches, and after 3 months of the spillage, it expanded to six beaches. Shimizu and Rosso (2000) studied the influence of oil spill on the abundance of the ghost shrimp *C. major* in the Baraqueçaba beach, state of São Paulo. Compared to previous data sampled on the same beach, the authors reported a significant reduction in densities after the accident that lasted 19 months and was attributed to the impact of the oil.

In August 2019, thousands of tons of crude oil residue from mysterious sources began to wash up along almost a 3000-km stretch of shoreline and >980 beaches in Brazil (Soares et al. 2020). The range of this large-scale acute impact overlapped the distribution of many vertebrate species, including elasmobranchs, bony fishes, turtles, birds, and marine mammals (see Chap. 5). However, the consequences of this impact on marine vertebrate biodiversity remain to be discovered. Monitoring and research strategies to investigate the effects of this disaster have been proposed by the Brazilian scientific workgroup of sandy beaches (GT5), supported by the Scientific Coordination of the Assessment and Accompaniment of the Oil Spill in the Brazilian Littoral (GAA). Among the main results, da Rosa (2022) argued that changes in the macrofauna community after the oil spillage reflected natural fluctuation rather than a response to the disturbance.

9.3 Climate Change Impacts

Coastal ecosystems are among the most vulnerable to climate change effects as they are subject to terrestrial, atmospheric, and marine environmental impacts. Climate change may modify the natural features of sandy beach ecosystems, as well as change the phenology, physiology, distribution, assemblage composition, and species interactions of beach biodiversity (Defeo et al. 2009). Nevertheless, the number of studies investigating the effects of this long-lasting and large-scale stressor on beach ecosystems is still limited, and most predictions are derived from other systems. Here, we highlight three main climate-related drivers that are expected to strongly affect sandy beach biodiversity: temperature, changes in water salinity and pH, and extreme weather events.

9.3.1 Rising Temperatures

Changes in temperatures may have direct and indirect consequences on coastal ecosystems. Rising temperatures may directly affect the physiology, distribution, and interaction of sandy beach organisms (Defeo et al. 2009). The direct influence of temperature changes may be particularly important on intertidal ecosystems as they are one of the most thermally variable and stressful habitats on the planet (Seuront and Cribb 2017). For example, long-term monitoring has revealed that rocky shore intertidal biota distribution limits have advanced toward the poles at a rate of over 50 km per decade (Southward et al. 1995), and similar results may be expected for sandy beach species. Nevertheless, the effects of rising temperature may be attenuated in sandy beach ecosystems since most species inhabiting this environment may seek thermal refugia by burrowing in the sediment (Chap. 4).

Shifts in temperature may particularly affect sandy beach species lacking dispersive larval stages, such as peracarid crustaceans, which could have their rates of range extension outpaced by changes in temperature and become locally extinct (Defeo et al. 2009). Despite the compelling need to better comprehend the direct ecological effects of temperature changes on sandy beach biodiversity, a few studies have investigated this topic, which still represents a significant knowledge gap.

Sandy beach organisms may also be indirectly affected by temperature changes through changes in planktonic assemblages and sea-level rise. For example, temperature increases have caused major changes in the distribution and structure of planktonic assemblages (Southward et al. 1995; Richardson and Schoeman 2004). Given that planktonic organisms form the base of the coastal trophic web, changes in planktonic communities are likely to significantly affect sandy beach ecosystems. Furthermore, rising temperatures are directly associated with sea-level rise, which may change beach sediments, intensify erosive processes, and alter habitat structure (Scapini et al. 2019). These changes, in turn, influence the distribution and foraging behavior of benthic invertebrates and increase the risk of local extirpations (Hubbard et al. 2014). Sea-level rise also pushed the high-water mark landward, causing beaches to migrate (Defeo et al. 2009). In areas where this migration is not possible, such as sites where dunes and berms have been suppressed by human modifications on the coastal zone, sandy beach habitats will have their area reduced and may even disappear. A recent global analysis of sandy shoreline dynamics showed that, if current greenhouse gas emissions are not reduced, almost half of the world's sandy beaches can disappear by the end of the century due to coastal recession driven by sea-level rise (Vousdoukas et al. 2020).

9.3.2 Changes in Water Salinity and pH

Rainfall increase, predicted in the global climate change scenario (IPCC 2021), promotes the input of freshwater to coastal regions, which may intensify freshwater floods, reduce water salinity, and change beach morphodynamic and sediment organic contents (Laurino et al. 2020). These changes, in turn, can affect larval settlement, recruitment, reproduction, habitat selection, vertical displacement, and behavior of beach species (Naylor 2010; Laurino et al. 2020). In addition, further invasion by species tolerant to salinity variations, which compete with resident ones, can be responsible for decreasing endemic invertebrates' populations (Fanini et al. 2017; Scapini et al. 2019). On the coast of São Paulo, Laurino et al. (2020) simulated saltwater and freshwater floods in manipulative experiments to examine their effects on the vertical distribution of the semi-terrestrial crustacean *Excirolana armata* within the sediment. The authors found that both saltwater and freshwater floods induced a movement of *E. armata* toward deeper layers of the sediment; however, fewer individuals reached deeper sediment layers during freshwater floods. Still, on the São Paulo coast, Laurino and Turra assessed small-scale spatial changes in macrobenthic assemblages related to freshwater influence on three subtropical beaches. Changes in salinity were the only environmental factor explaining the macrofaunal variations, with lower macrobenthic abundance and richness in sites close to freshwater sources.

Climate change also impacts water acidity. Human-driven increased levels of carbon dioxide in the atmosphere lead to more CO_2 dissolving into the ocean and lower pH. Carbonate saturation and pH decline can decrease calcification rates and affect the metabolism and anatomy of sandy beach invertebrates, especially mollusks, and crustaceans (Schoeman et al. 2014; Scapini et al. 2019). Also, pH fluctuations affect the reproductive behavior, sensory capacity, larval settlement, and recruitment and foraging activities in marine invertebrates (Lucey et al. 2015; Duarte et al. 2016; Espinel-Velasco et al. 2018; Scapini et al. 2019). These effects combined with anthropogenic stressors can exert further impacts, but no studies have investigated their possible consequences on the beach fauna and ecosystem functioning in Brazil.

9.3.3 Extreme Weather Events

Increased air and sea temperatures are expected to translate into more frequent and more severe storms (IPCC 2021). On sandy beaches, such events generate physical disturbances in the sediment, having direct or indirect effects on macrofauna (Diaz and Rosenberg 1995). The expected increase in the frequency and intensity of storms in response to climate change (IPCC 2021) may positively affect benthic macrofauna, resuspending fine sediment particles, which favors the development of organisms that consume suspended material and the benthic-pelagic coupling in the surf zone (Bock and Miller 1995). However, depending on the intensity of the erosive processes that follow the storms, such events can generate sediment defaunation and make it difficult to recolonize the environment (Brauko 2008). Also, storm waves can generate habitat loss, changes in zonation patterns, and a decrease in the abundance and diversity of benthic communities (Scapini et al. 2019). Such impacts can reach more than 20 cm deep into the sedimentary layer, directly influencing the biological communities that are mainly concentrated in the top 50 cm of sediment (Gómez-Pujol et al. 2011).

Studies that aim to understand the effects of climate disturbances such as storms and frontal systems on benthic associations at local and regional scales in coastal regions have been performed worldwide (e.g., Posey et al. 1996; Aagaard et al. 2005; Houser and Greenwood 2007). However, despite the long Brazilian coast, a few studies have evaluated the influence of extreme events, such as flooding and storm waves.

The comparison of the resilience of benthic communities to storm wave events on beaches with distinct human pressures has been the main approach targeting to predict the effect of climate change on Brazilian sandy beach biodiversity (Sola and Paiva 2001; Gallucci and Netto 2004; Cochôa et al. 2006; Brauko 2008; Alves and Pezzuto 2009; Machado et al. 2016, 2019). A study in the state of Santa Catarina (Southern Brazil) showed that upper-shore detritivorous species take advantage of storms due to the increase of natural debris thrown onto the beach by the waves (Alves and Pezzuto 2009). Besides, passive transport by wave action and swashing might be responsible for the increase in species density with limited mobility capability, which is distributed in the lower intertidal and infralittoral fringe, such as the polychaetes *H. californiensis*, *Pisionidens indica* (Aiyar and Alikunhi 1940), and Nemertea (Saloman and Naughton 1977; Hughes et al. 2009). Corte et al. (2017) showed that the community from a tidal flat recovered within a short time (weeks to months) after storms, highlighting that most species typical of sedimentary shorelines are, to some degree, resilient to short-term changes in wave energy. It should be considered, however, that urbanization might impair the communities' resilience to natural stochastic events since recovery after erosion processes seems to be slower on urbanized coasts (Harris et al. 2011; Witmer and Roelke 2014).

Machado et al. (2016) evaluated the effect of the interaction between natural extreme events (storm waves) and urbanization on the macrofauna community on the Rio de Janeiro coast and found that urban beach community has less resilience when compared to nonurban beaches, indicating that high storm wave events and increasing urbanization synergism are a threat to the macrofauna. The authors also showed that the dominant species *E. brasiliensis* was characterized mainly by juveniles after storm events, mainly in a nonurbanized beach, probably as larval development anticipation and recruitment in response to the intensification of hydrodynamics. Similarly, Machado et al. (2019) found that the interaction between storm waves and urbanization impacted the abundance of *O. quadrata* burrows negatively and suggested that this species is more vulnerable to the combined effects of extreme climatic events and urbanization. Table 9.1 shows the main results of several studies evaluating the effects of extreme events on macrofauna on sandy beaches worldwide. The results highlight that the urbanized environment is more susceptible to these events and the importance of natural disturbances in the benthic intertidal dynamics of pristine sandy beaches. More recently, Oliveira and Yokoyama (2021) verified that storm wave events combined with urbanization can reduce the recruitment and increase mortality of *O. quadrata* on the coast of São Paulo. The authors also suggest that storms may increase *O. quadrata* migration to the upper shore, which can constitute an ecological trap, considering that coastal vegetation is usually crossed by roads.

Effects of extreme weather events such as storm waves and cold fronts on sandy beach ecosystems may be even higher in urbanized areas. Storms and urbanization may act synergistically and further affect the environment and biodiversity of sandy beaches. To better understand and predict the impact of urbanization and extreme events on sandy beach ecosystems, gathering data over time, including periods before and after storms, is of utmost importance. Nevertheless, long-term monitoring of sandy beaches is scarce, and these ecosystems are rarely considered in conservation plans and coastal management initiatives (Harris et al. 2014).

9.4 Coastal Erosion, Armoring, and Beach Nourishment

Erosion is a global phenomenon impacting sandy beaches worldwide. It is a consequence of several threats discussed earlier in this chapter such as climate change (sea-level rising and storminess), coastal urbanization, and anthropogenic modifications at local scales (e.g., cleaning and grooming, sand mining, and engineering interventions). From 1984 to 2016, about 24% of the sandy beaches in the world were eroding at a rate greater than 0.5 m $year^{-1}$ (Luijendijk et al. 2018). Increasing erosive processes are registered along the whole Brazilian coast, with higher rates

Table 9.1 List of studies that evaluated the effects of extreme weather events on benthic communities of sandy beaches

Authors	Country	Event	Intensity	Days after each event	Major results
Crocker (1968)	USA	Hurricane	High	2 days (Event I); 16 and 30 days (Event II)	Reduction in amphipods abundance
Saloman and Naughton (1977)	USA	Hurricane	High	1, 2, 3, 6, 9, 14, and 28 days	Increased richness and recruitment of *Emerita talpoida*
Hughes et al. (2009)	USA	Hurricane	High	6 days	Increased richness, diversity, and opportunistic species abundance
Witmer and Roelke (2014)	USA	Hurricane	High	Monthly samplings	Higher resilience of the macrofauna at nonurbanized beach
Jaramillo et al. (2012)	Chile	Tsunami	High	ca. 30 days	Restoration of intertidal habitat followed by rapid colonization of mobile species on armored beaches
Harris et al. (2011)	South Africa	Storm waves	High	47 days (Event I); 15 days (Event II)	Higher resilience of the macrofauna at nonurbanized beach
Cochôa et al. (2006)	Brazil	Cold front	Low/ moderate	1 day	Changes in zonation pattern
Alves and Pezzuto (2009)	Brazil	Cold front	Moderate	2 and 4 days (Event I); 2 days (Event II)	Increasing detritivorous species (*Excirolana braziliensis*) density on reflective beach
Machado et al. (2016)	Brazil	Storm waves	Low/ moderate	15 and 42 days (Event I); 28 and 42 days (Event II)	Event I: Increasing community numerical indicators at nonurbanized site Event II: Increasing detritivorous species (*Excirolana braziliensis*) density at the urbanized site
Rodil et al. (2016)	Chile	Earthquake	High	Monthly samplings	Rapid colonization of crustacean populations after an earthquake in upper and mid-shore armored levels

(continued)

Table 9.1 (continued)

Authors	Country	Event	Intensity	Days after each event	Major results
Corte et al. (2017)	Brazil	Storm waves	High	2 days after events	Decrease in species richness, abundance and biomass of invertebrate assemblages. Recover within a few weeks
Gül and Griffen (2019)	USA	Hurricane	High	Monthly samplings	Population decline of *O. quadrata* after hurricane, mainly adult crabs
Machado et al. (2019)	Brazil	Storm waves	Low/ moderate	4–5 days before event; 15 and 30 days after events	Population decline of *O. quadrata* after events
Mathews and Maccarone (2019)	USA	Hurricane	High	Yearly	Decreased burrows (decline in human activities after the hurricane). Reduction in the size of crabs
Sugumaran et al. (2019)	India	Cyclone	High	Daily, during 10 days	Meiofauna declined after the cyclone with recovery after a week
Carneiro et al. (2020)	Brazil	Heatwaves and cold spells	Moderate/ high	43 days – experimental conditions	Heatwave promoted the population decline of *Anomalocardia flexuosa*
Gunasekaran et al. (2021)	India	Cyclone	High	1 day	Gastropod and bivalve mortality
Oliveira and Yokoyama (2021)	Brazil	Storm waves	Low/ moderate	3–5 days	Population declining of *O. quadrata* after most events
Quintero-Torres and López-Sánchez (2022)	Caribbean	Hurricanes	High	2 months after hurricanes	Decreased diameter burrows of *O. quadrata* after events

Table modified from Machado et al. (2016)

in the northern region near the Amazon River mouth (Luijendijk et al. 2018). An iconic case in Brazil is Atafona Beach located at the mouth of the Paraíba do Sul River, state of Rio de Janeiro, where tens of houses have been destroyed by erosion as well as by foredune migration (Muehe 2006). To protect the coast against erosion, managers usually rely on two engineering initiatives: coastal armoring and beach nourishment. Although both approaches may indeed mitigate coastal erosion or restore the sandy beach dynamics, they are also linked to ecological impacts that threaten sandy beach biodiversity.

Coastal armoring can be defined as the use of physical structures (e.g., breakwaters, groynes, jetties, seawalls, and dikes) to protect shorelines from coastal erosion. It is commonly used in sites where homes, commercial buildings, and infrastructure are located very close to the shore. In the states of Rio Grande do Sul and Espírito Santo, for example, coastal erosion has hampered tourism activities and threatened urban infrastructure, leading to the construction of containment structures (Esteves and Santos 2002). On the coast of the metropolitan region of the Fortaleza municipality (state of Ceará, Northeast Brazil), there are more than 70 coastal engineering interventions for erosion control, approximately half of which are armoring (Paula 2015).

Among the ecological effects of coastal armoring, the most important is the loss of habitats for several upper shore species, such as the ghost crab on the São Paulo coast (Pombo and Turra 2019), and a reduction in the turtle nesting area. Shorebirds can also be harmed by reduced foraging area and prey availability due to habitat loss (Dugan et al. 2003). In addition, it is worth mentioning the visual impacts that armoring structures can generate (Griggs 2005), affecting the tourist potential of a coastal area.

The impacts of armoring depend on the type of material used, the size of the area covered, and the coastal environmental characteristics. Although there is no clear definition, it is considered that the use of vertical concrete walls generates greater impacts than the inclined revetment, especially impacts related to beach access, since seawalls make this process difficult (Griggs 2005). In armored beaches, passive erosion processes can also occur in both seawalls and revetments, culminating in the reduction of the volume of sand on the beach, especially in front of the structures (Griggs 2005). In the long term, this effect becomes increasingly significant, and it is hardly mitigated except by continuous beach nourishment strategies.

Beach nourishment is the artificial widening of the sand stripe by adding sand onto the beach. This process is often used to protect urban infrastructure (e.g., buildings and walkways) from wave impact in sites where the dynamic of sand accretion and erosion is unbalanced. Because hard structures such as breakwaters and seawalls (i.e., coastal armoring see below) may cause more problems than solutions regarding coastal protection (sometimes increasing rather than stopping the erosion process), beach nourishment is considered a "soft" engineering intervention (Martins and Pereira 2014) and preferable for either prompt remediation or long-term protection of shore (Colosio et al. 2007). Sand nourishment is increasingly used as a "nature-based" solution; however, its effectiveness is also limited because the erosive forces of waves, storms, and rising sea levels do not disappear after nourishment takes place, requiring significant natural resources and replenishment over time. Recently, studies have suggested a combination of shoreface nourishment with coastal vegetation restoration to reduce coastal erosion (Chen et al. 2022).

In Brazil, the iconic Copacabana beach (RJ) was artificially nourished by the offshore dumping method between the end of 1960 and the beginning of 1970 (Figs. 9.3 and 9.4), widening the sand stripe from 55 to 140 m for touristic reasons and erosion control (Vera-Cruz 1972). Very recently, large nourishment was done on Camburiú (state of Santa Catarina) and Iracema (state of Ceará) beaches.

Fig. 9.3 View of the walkway of Copacabana beach in 1920s, near the Copacabana Palace Hotel, with the advanced erosive process before the nourishment performed to protect the coast. (Source: https://redesina.com.br/historia-do-bairro-copacabana/)

Fig. 9.4 Copacabana beach before (left) and after (right) the nourishment performed at the end of 1960 and beginning of 1970. (Source: Digital collection Afro-Brasileiro Flickr)

Theoretically, beach invertebrates can tolerate the harsh environment and the impacts caused by nourishment should be minor (e.g., Danovaro et al. 2018). However, impacts related to beach nourishment have been reported worldwide and include the burial of shallow reefs and degradation of other beach habitats, depressing nesting in sandy vertebrates (Chap. 5), and reducing the densities of invertebrate prey for shorebirds, surf fishes, and crab (Peterson and Bishop 2005; Fanini et al. 2009). Unfortunately, studies on the ecological impacts of beach nourishment are scarce along the Brazilian coast. The only study in Brazil (state of Santa Catarina) regarding the environmental impacts of beach nourishment showed acute mass mortality of the beach-dwelling mollusk *T. mactroides* and other 17 species, probably produced by the mechanical impact of the hydrological dredge of sand, burying, increased water turbidity, and chemical contamination (Pezzuto et al. 2006). At Copacabana Beach, people attribute the disappearance of some local species to nourishment or pollution, but the real cause is hard to establish due to the absence of previous studies and monitoring programs. Media reports by ancient beachgoers state that before nourishment activities, the mole crab *Emerita brasiliensis* (Schmitt 1935) was abundant on Copacabana beach, but it is a rare beach dweller nowadays (Instituto Aqualung 2016). Some researchers suggest that nourishment is the cause of mole crab rarity (O Globo 2014), but there are no scientific data to support this hypothesis.

The use of both coastal armoring and beach erosion as alternatives to mitigate coastal erosion remains a very complex issue with many gaps in understanding. Although considered a necessity in many coastal regions, these strategies still divide expert opinions. Considering local stressors and climate change predictions at the continent–ocean interface, greater efforts are necessary to develop management strategies that affect less the beach dynamics and the ecosystem resilience and preclude the coastal squeeze.

9.5 Indicator Species

The use of macroinvertebrates as indicator species for monitoring beaches has been proposed as a low-cost/benefit ratio strategy since their abundance is predictably affected by environmental impacts (ecological-disturbance indicator species) or they accumulate pollutants in their tissue (sentinel species and biomonitors) (Fig. 9.5). Globally, ghost crab species have been used as the main indicator species of urbanization and vehicle traffic impacts. Sandhoppers (Crustacea: Talitridae) are also commonly tested as indicator species of both physical (trampling) and chemical (trace elements) impacts on sandy beaches (Costa and Zalmon 2019b, c; Costa et al. 2021).

The indicator potential of macroinvertebrates has also been tested on the Brazilian coast. Almost all taxa investigated responded negatively to human disturbances, being less abundant in urbanized beaches than in periurban and pristine beaches (Fig. 9.6).

The ghost crab *Ocypode quadrata* is especially sensitive to impacts from vehicle traffic (Neves and Bemvenuti 2006; Magalhães et al. 2009) and beach cleaning (Souza et al. 2017). Trampling is considered the main causative agent of decreasing abundance of the sandhopper *A. brasiliensis* (Veloso et al. 2006, 2008, 2010; Cardoso et al. 2016; Machado et al. 2017) and other sedentary macroinvertebrates, such as the polychaetes *H. californiensis* and *Scolelepis* spp. (Machado et al. 2017), and the crustacean isopod, such as *Excirolana* spp. (Veloso et al. 2011; Vieira et al. 2012). However, the occurrence of all aforementioned species is not necessarily an indication that beach biodiversity is not affected by trampling. The effects of trampling depend on their frequency and intensity in addition to the species' traits; acute

Fig. 9.5 Representatives of the sandy beach fauna, composed primarily of invertebrates, usually considered good bioindicators of chronic and acute impacts. (**a**) Tiger beetle *Cylindera nivea*; (**b**) ghost crab *Ocypode quadrata*; (**c**) polychaete *Hemipodia californiensis* (photos: Leonardo Costa); (**d**) sandhopper *Atlantorchestoidea brasiliensis* (photo: Julyana Madureira)

trampling limited to the high touristic season seems less adverse than chronic ones (Costa et al. 2017; Costa and Zalmon 2019c). It is important to notice, however, that it is difficult to isolate trampling effects from generalized and broader urbanization impacts and those that may act synergically to affect benthic species (Machado et al. 2017; Costa and Zalmon 2019a).

Monitoring of sandy beach biodiversity is especially facilitated by the low-cost sampling of species that construct burrows, such as ghost crabs and ghost shrimps, and surface-active species, such as tiger beetles (Insecta: Cicindelidae) (Costa and Zalmon 2019a). Abundance can be easily estimated by burrow counting (for ghost crabs and ghost shrimps) and direct counting (visual census) during periods of surface activity of large taxa such as ghost crabs and tiger beetles (Costa and Zalmon 2019a, b). Brazilian authors have advanced in proposing novel methods targeting to minimize limitations in counting ghost crab burrows. Measurement of depth and the "burrow resetting method" are the main strategies to deal with spatial-temporal

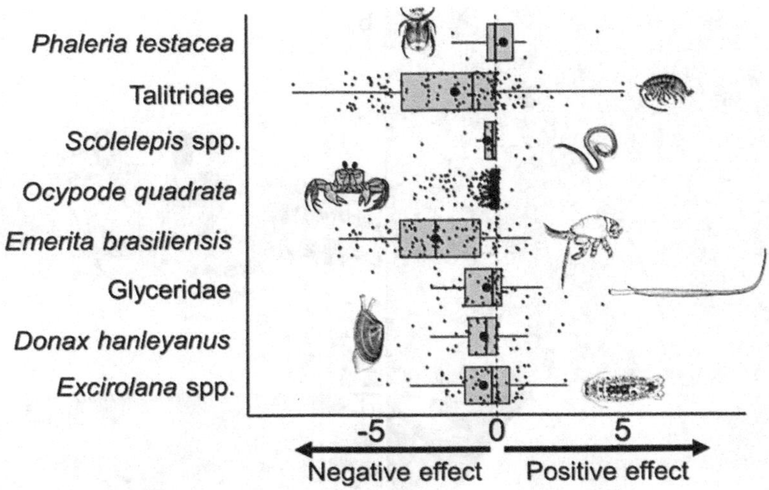

Fig. 9.6 Response of ecological-disturbance indicator taxa to human stressors on sandy beaches. Small dots are response ratios calculated as the log of the quotient of the mean value recorded for impacted sites divided by the mean value at corresponding nonimpacted sites from surveys carried out on Brazilian beaches. Vertical lines inside the boxes are the median values and boxes are the interquartile range. Dots inside boxes are mean values. The vertical dashed line is the null value (RR = 0). Dots on the left and the right of the vertical dashed line represent the negative and positive effects of human disturbances, respectively

variability and unpredictability in burrow occupation rates (Pombo and Turra 2013, 2019). The last method consists of sampling the same areas twice; the first time corresponding to the traditional counting, after then the burrows' surface is gently covered with sand and reassessed after 12–24 h; then only reopened burrows are counted again (Pombo and Turra 2019). Interestingly, by using the burrow resetting method, studies have shown that traditional burrow counting usually overestimates ghost crab population sizes, mainly on low-urbanized beaches (Pombo and Turra 2019; Costa et al. 2021). Other macroinvertebrates still lack alternative tools for easy and nondestructive sampling for effective and rapid monitoring of human impacts on Brazilian sandy beaches, and so far, ghost crabs are the main indicator species in Brazil (Fig. 9.7).

As multiple stressors simultaneously affect beaches, the relative importance of each one has been difficult to reveal using indicator species through correlative studies. Another concern of using indicator species is the scarce information on how human disturbances interact with natural habitat conditions to yield specific faunal responses. Although all species among the main indicator taxa respond negatively to increasing urbanization on Brazilian beaches, the effects can depend on the morphodynamics (Fig. 9.8). In general, for taxa closely related with the intertidal zone and subject to its hydrodynamic harshness, such as the sandhopper *A. brasiliensis*, the cirolanids *Excirolana* spp., and the mole crab *E. brasiliensis*, stronger negative responses have been found on nondissipative beaches than on dissipative ones

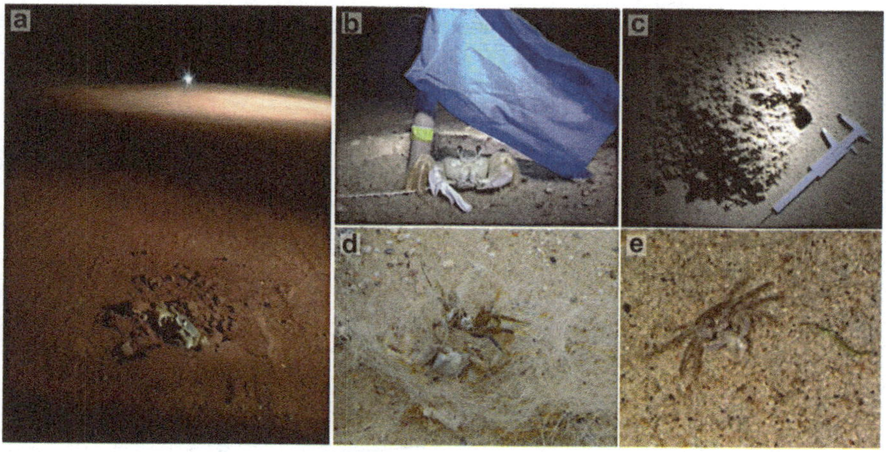

Fig. 9.7 The ghost crab *Ocypode quadrata* is the main indicator of human impacts on Brazilian sandy beaches. (**a**) Ghost crab constructing a burrow in the dirt road (photo: Leonardo Costa); (**b**) ghost crab interacting with marine debris (photo: Julyana Madureira); (**c**) ghost crab burrow at night (photo: Julyana Madueira); (**d**) ghost crabs entangled in fishing gear (photo: Leonardo Costa); (**e**) juvenile ghost crab feeding on the talitrid *Atlantorchestoidea brasiliensis* (photo: Vitor Arueira)

(Fig. 9.8). Contrarily, for the ghost crab, which dramatically depends on the supralittoral to thrive, the negative effect of increasing urbanization is apparently more consistent on dissipative beaches than on nondissipative ones (Fig. 9.8). Costa et al. (2022a, b) found similar patterns at global scale (considering family taxonomic resolution) and proposed the Cumulative Harshness Hypothesis, which predicts higher sensitivity of beach macroinvertebrates to human disturbances when inhabiting a harsher physical environment according to their life histories. The differential effects of human disturbances on macroinvertebrates related to beach morphodynamics suggest that the efficiency of these ecological-disturbance indicators for beach monitoring can be context-dependent.

Macroinvertebrates have been rarely evaluated as biomonitors of pollutants on Brazilian sandy beaches despite their high potential to accumulate metals, microplastics, and other contaminants (Rainbow and Phillipst 1993). Generally, carnivorous species have higher heavy metal concentrations than other trophic guilds, but the accumulation seems to be more related to proximity to urban areas and morphodynamic characteristics along the coast than to trophic features (Cabrini et al. 2017, 2018). A single study showed that the ghost crab *O. quadrata* ingests plastics, but its indicator potential is not evident due to the high trophic plasticity and, consequently, ingestion of MPs from multiple punctual sources (Costa et al. 2019b). Monitoring of pollutants in Brazil using beach species is less applied compared to other parts of the world.

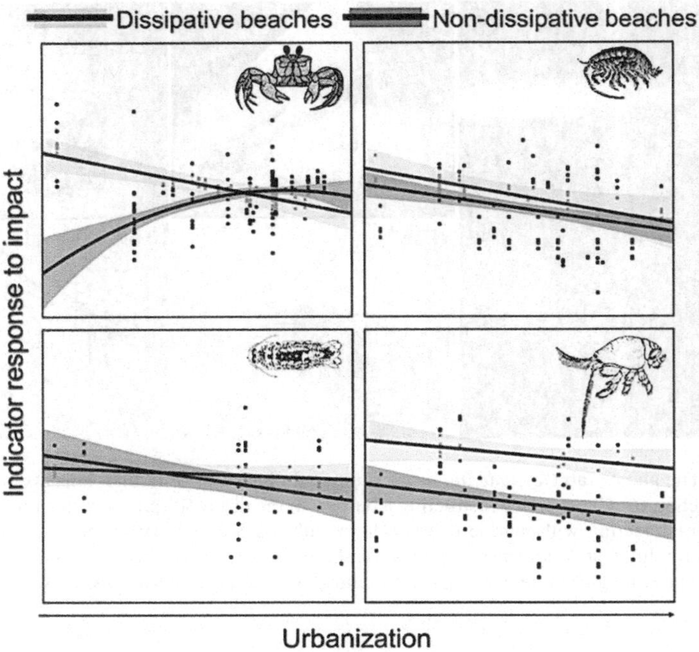

Fig. 9.8 Response ratios of macroinvertebrates (*Ocypode quadrata*, *Atlantorchestoidea brasiliensis*, *Excirolana* spp., and *Emerita brasiliensis*) to human disturbances, discriminated by local beach morphodynamics (dissipative vs. nondissipative). Response ratios were calculated as the log ratio of the mean value of impacted sites and the mean value of the corresponding nonimpacted sites

9.6 Mitigating Actions

Beaches subjected to anthropogenic pressures are more refractory to ecosystem conservation, demanding management interventions to mitigate the negative effects of human actions on their biodiversity and environmental characteristics. Coastal managers are usually not engaged with beach ecologists. Consequently, beach management has focused almost exclusively on interventions to maintain coastal defense (e.g., beach armoring and nourishment) or on the tourism economy (e.g., beach cleaning) worldwide (Schlacher et al. 2008). Unfortunately, Brazil is no exception, and management actions on Brazilian beaches are not based on ecological requirements. Here, we propose mitigation actions that might reduce some impacts listed throughout this chapter, although they have not been tested concerning their efficiency.

(i) Pollution: educational signs and initiatives that raise awareness about the impacts of solid waste and promote the concept of "leave no trace" to beachgoers may reduce the amount of solid waste on sandy beaches. The disposal of more waste containers along the shores could also be beneficial in this matter.

Diffuse sources of organic and chemical pollutants (sewage, metals, and microplastics) are difficult to mitigate and the reduction of these impacts must be based on emission control policies and monitoring (e.g., filters); creating mechanisms aiming to ban the use of plastic items (cups, cutlery, straws) by the beach kiosks and hawkers' commerce leading to achieve a state of "no local plastic littering."

(ii) Recreational activities (tourism): beach zoning to maintain pristine beaches as biodiversity hotspots; areas with restricted access and controlled number of visitors. The control of the number of visitors must be based on studies about the ecological and social carrying capacity of sandy beaches with distinct morphodynamics (Defeo et al. 2009). Although vehicle traffic on beaches is prohibited in Brazil, surveillance of wide coastal strips is challenging. Thus, the presence of surveillance radars to punish vehicle traffic and the use of natural obstructions (e.g., trunks) to reduce vehicle speed and avoid fauna crushing could be alternative mitigation actions.

(iii) Erosion: protection of dunes and backshore areas, restoring programs of beach vegetation; nourishment must be prioritized instead of beach armoring; nourishment project should use similar sediment (grain size) from unpolluted areas to avoid changes in beach dynamics (erosion/accretion equilibrium); and contamination by toxic and pathogenic components from exogenous sand (Peterson et al. 2000; Rippy et al. 2013).

(iv) Beach cleaning: to prioritize manual cleaning by trained professionals and use nonselective tools (e.g., rakes with less space between the tines) to avoid restricting the removal of large litter; not remove natural detritus (wrack, fish, and invertebrates carcasses), only in unusual cases of large accumulation that induce persistence of indicator microbes and some pathogen (Zielinski et al. 2019); when detritus removal are necessary, mechanical grooming must be restricted to small tourist beach stretches (Defeo et al. 2009). In the case of acute pollution by disasters (e.g., oil spill), beach cleaning must be also manual, removing oiled detritus and patches of oil from the sand surface (Borzone and Rosa 2009).

In conclusion, sandy beaches are important ecosystems for maintaining coastal biodiversity and human societies, and many coastal cities depend on their goods and services. Therefore, integrated management plans to reduce the impact of human activities on urbanized beaches and protect pristine areas are necessary to sustain their tourism value and preserve their functioning. To resolve these ecological problems over the short- and medium-term, specific legislation to protect sandy beach ecosystems, combined with fines and enhancement of ecological awareness, is probably the best alternative. A challenging but urgent need is the dissemination of information in a more simplified and accessible speech by beach ecologists, aiming to attract more attention from beach users and funding to protect this threatened ecosystem. Conservation plans based on ecological shortcuts (indicators, keystones, umbrellas, and flagship species) must also be prioritized in future studies facing the scarcity of financial support for long-term research, beach complexity, and the need for fast decisions.

9.7 Final Remarks

The number of Brazilian beach ecology experts has been continuously increasing over the past 20 years, leading to one of the most prolific scientific communities in the world in the number of publications (see Nel et al. 2014). However, the result of the literature review performed to write this chapter warns us of the need for increased efforts to comprehend Brazilian beaches' functioning better. Some threats affecting these ecosystems are still poorly studied, such as those caused by coastal armoring and beach nourishment, whose use is expected to rise because of the increasing erosion rates resulting from direct anthropic activities and climate change. The potential of beach litter to cause diseases and its long-term effect on dune vegetation, the impact of beach cleaning practices, and the role of harvesting and fishing on beach biodiversity are also topics that should be further investigated. Studies on threats and impacts on sandy beach ecosystems, including long-term and multidisciplinary studies, are essential for guiding management and conservation initiatives by stakeholders and necessary to cope with the challenges that global changes are imposing on humankind.

Acknowledgments ASG is supported by CNPq (301475/2017-2). LLC is supported by Fundação de Amparo à Pesquisa do Estado do Rio de Janeiro—FAPERJ (E-26/200.620/2022 and E-26/210.384/2022). IRZ is supported by FAPERJ (259920/2021) and the Brazilian Agency for Research Development—CNPq (301203/2019-9).

References

Aagaard T, Kroon A, Andersen S et al (2005) Intertidal beach change during storm conditions; Egmond, the Netherlands. Mar Geol 218(1):65–80

Aiyar G, Alikunhi KH (1940) On a new pisionid from the sandy beach, Madras. Rec Ind Mus 42:89–107

Albino J, Paiva DS, Machado GM (2001) Geomorfologia, tipologia, vulnerabilidade erosiva e ocupação urbana das praias do litoral do Espírito Santo, Brasil. Aust Geogr 2:63–69

Alves ES, Pezzuto PR (2009) Effect of cold fronts on the benthic macrofauna of exposed sandy beaches with contrasting morphodynamics. Braz J Oceanogr 57(2):73–96

Amaral ACZ, Corte GN, Denadai MR et al (2016) Brazilian sandy beaches: characteristics, ecosystem services, impacts, knowledge and priorities. Braz J Oceanogr 64(SPE2):5–16

Andrades R, Pegado T, Godoy BS et al (2020) Anthropogenic litter on Brazilian beaches: baseline, trends and recommendations for future approaches. Mar Pollut Bull 151:110842

Aragonés L, López I, Palazón A et al (2016) Evaluation of the quality of coastal bathing waters in Spain through fecal bacteria *Escherichia coli* and *Enterococcus*. Sci Total Environ 566:288–297

Baeyens W, Mirlean N, Bundschuh J et al (2019) Arsenic enrichment in sediments and beaches of Brazilian coastal waters: a review. Sci Total Environ 681:143–154

Barboza CAM, Mattos G, Soares-Gomes A et al (2021) Low densities of the ghost crab Ocypode quadrata related to large scale human modification of sandy shores. Front Mar Sci 8:1–11

Barile PJ (2004) Evidence of anthropogenic nitrogen enrichment of the littoral waters of east central Florida. J Coast Res 20(4):1237–1245

Barnes DK (2002) Invasions by marine life on plastic debris. Nature 416(6883):808–809

Barrella W, Ramires M, Rotundo MM et al (2016) Biological and socio-economic aspects of recreational fisheries and their implications for the management of coastal urban areas of south-eastern Brazil. Fish Manag Ecol 23(3–4):303–314
Bender MG, Machado GR, Silva PJA et al (2014) Local ecological knowledge and scientific data reveal overexploitation by multigear artisanal fisheries in the Southwestern Atlantic. PLoS One 9:e110332
Benjarano AC, Michel J (2016) Oil spills and their impacts on sand beach invertebrate communities: a literature review. Environ Pollut 218:709–722
Bird BL, Branch LC, Miller DL (2004) Effects of coastal lighting on foraging behavior of beach mice. Conserv Biol 18:1435–1439
Bock MJ, Miller DC (1995) Storm effects on particulate food resources on an intertidal sandflat. J Exp Mar Biol Ecol 187:81–101
Bom FC, Colling LA (2020) Impact of vehicles on benthic macrofauna on a subtropical sand beach. Mar Ecol 41:e12595
Borzone CA, Rosa LC (2009) Impact of oil spill and posterior clean-up activities on wrack-living talitrid amphipods on estuarine beaches. Braz J Oceanogr 57:315–323
Brauko KM (2008) Efeitos da passagem de sistemas frontais sobre a macrofauna bêntica de praias arenosas do Paraná. Dissertation, University of Paraná
Cabrini TM, Barboza CA, Skinner VB et al (2017) Heavy metal contamination in sandy beach macrofauna communities from the Rio de Janeiro coast, Southeastern Brazil. Environ Pollut 221:116–129
Cabrini TM, Barboza CA, Skinner VB et al (2018) Investigating heavy metal bioaccumulation by macrofauna species from different feeding guilds from sandy beaches in Rio de Janeiro, Brazil. Ecotoxicol Environ Saf 162:655–662
Cardonha AM, Vieira RH, Rodrigues DP et al (2004) Fecal pollution in water from storm sewers and adjacent seashores in Natal, Rio Grande do Norte, Brazil. Int Microbiol 7(3):213–218
Cardoso RS, Barboza CA, Skinner VB et al (2016) Crustaceans as ecological indicators of metropolitan sandy beaches health. Ecol Indic 62:154–162
Carneiro AP, Soares CHL, Manso PRJ (2020) Impact of marine heat waves and cold spell events on the bivalve *Anomalocardia flexuosa*: a seasonal comparison. Mar Environ Res 156:104898
Carvalho DG, Neto JAB (2016) Microplastic pollution of the beaches of Guanabara Bay, Southeast Brazil. Ocean Coast Manag 128:10–17
Checon HH, Silva M, Corte GN et al (2021) Night underwater rides: the activity of a sandy beach gastropod is affected by interactive effects of light availability and water level. Mar Biol Res 17:523–528
Chen WL, Muller P, Grabowski RC et al (2022) Green nourishment: an innovative nature-based solution for coastal erosion. Front Mar Sci 8:814589
Chua EM, Shimeta J, Nugegoda D et al (2014) Assimilation of polybrominated diphenyl ethers from microplastics by the marine amphipod, *Allorchestes compressa*. Environ Sci Technol 48:8127–8134
Cochôa AR, Lorenzi L, Borzone CA (2006) A influência da passagem de uma frente meteorológica na distribuição da macrofauna bentônica mesolitoral de uma praia arenosa exposta. Trop Oceanogr 34(2):59–71
Cole M (2016) A novel method for preparing microplastic fibers. Sci Rep 6:34519
Cole M, Lindeque P, Halsband C et al (2011) Microplastics as contaminants in the marine environment: a review. Mar Pollut Bull 62(12):2588–2597
Colosio F, Abiatti M, Airoldi L (2007) Effects of beach nourishment on sediments and benthic assemblages. Mar Pollut Bull 54:1197–1206
Corcoran PL, Biesinger MC, Grifi M (2009) Plastics and beaches: a degrading relationship. Mar Pollut Bull 58:80–84
Corte GN, Schlacher TA, Checon HH et al (2017) Storm effects on intertidal invertebrates: increased beta diversity of few individuals and species. PeerJ 5:e3360

Corte GN, Checon HH, Shah Esmaeili Y et al (2022) Evaluation of the effects of urbanization and environmental features on sandy beach macrobenthos highlights the importance of submerged zones. Mar Pollut Bull 182:113962

Costa LL, Zalmon IR (2019a) Sensitivity of macroinvertebrates to human impacts on sandy beaches: a case study with tiger beetles (Insecta, Cicindelidae). Estuar Coast Shelf Sci 220:142–151

Costa LL, Zalmon IR (2019b) Multiple metrics of the ghost crab *Ocypode quadrata* (Fabricius, 1787) for impact assessments on sandy beaches. Estuar Coast Shelf Sci 218:237–245

Costa LL, Zalmon IR (2019c) Issues to consider when sampling the sandhopper *Atlantorchestoidea brasiliensis* (Dana, 1853) for impact assessments on sandy beaches. Mar Biol Res 15(8–9):451–457

Costa MF, Do Sul JAI, Silva-Cavalcanti JS et al (2010) On the importance of size of plastic fragments and pellets on the strandline: a snapshot of a Brazilian beach. Environ Monit Assess 168(1–4):299–304

Costa LL, Tavares DC, Suciu MC (2017) Human-induced changes in the trophic functioning of sandy beaches. Ecol Indic 82:304–315

Costa LL, Madureira JF, Zalmon IR (2018a) Changes in the behaviour of *Ocypode quadrata* (Fabricius, 1787) after experimental trampling. J Mar Biol Assoc U K 99(5):1135–1140

Costa LL, Rangel DF, Zalmon IR (2018b) Evidence of marine debris usage by the ghost crab *Ocypode quadrata* (Fabricius, 1787). Mar Pollut Bull 128:438–445

Costa LL, Arueira VF, Da Costa MF et al (2019a) Can the Atlantic ghost crab be a potential biomonitor of microplastic pollution of sandy beaches sediment? Mar Pollut Bull 145:5–13

Costa LL, Madureira JF, Di Beneditto APM et al (2019b) Interaction of the Atlantic ghost crab with marine debris: evidence from an in situ experimental approach. Mar Pollut Bull 140:603–609

Costa LL, Secco H, Arueira VF (2020) Mortality of the Atlantic ghost crab *Ocypode quadrata* (Fabricius, 1787) due to vehicle traffic on sandy beaches: a road ecology approach. J Environ Manag 260:110168

Costa LL, Soares-Gomes A, Zalmon IR (2021) Burrow occupation rates and spatial distribution within habitat of the ghost crab *Ocypode quadrata* (Fabricius, 1787): implications for impact assessments. Reg Stud Mar Sci 44:101699

Costa LL, Constantino WD, Ferreira KA et al (2022a) Sandy beach ecology in the Anthropocene: ghost shrimp (Crustacea: Callianassidea) as a study model. Estuar Coast Shelf Sci 275:107999

Costa LL, Machado PM, Barboza CAM, Soares-Gomes A, Zalmon IR (2022b) Recovery of ghost crabs metapopulations on urban beaches during the Covid-19 "anthropause". Mar Environ Res 180:105733

Crocker RA (1968) Distribution and abundance of some intertidal sand beach amphipods accompanying the passage of two hurricanes. Chesap Sci 9(3):157–162

da Rosa LC (2022) Sandy beach macroinfauna response to the worst oil spill in Brazilian coast: no evidence of an acute impact. Mar Pollut Bull 180:113753

Danovaro R, Nepote E, Lomartire M (2018) Limited impact of beach nourishment on macrofaunal recruitment/settlement in a site of community interest in coastal area of the Adriatic Sea (Mediterranean Sea). Mar Pollut Bull 128:259–266

Dantas NCFM, Duarte OS, Ferreira WC (2020) Plastic intake does not depend on fish eating habits: identification of microplastics in the stomach contents of fish on an urban beach in Brazil. Mar Pollut Bull 153:110959

Davino AMC, Melo MBD, Caffaro Filho RA (2015) Assessing the sources of high fecal coliform levels at an urban tropical beach. Braz J Microbiol 46(4):1019–1026

Defeo O (2003) Marine invertebrate fisheries in sandy beaches: an overview. J Coast Res 35:56–65

Defeo O, Mclachlan A, Schoeman DS et al (2009) Threats to sandy beach ecosystems: a review. Estuar Coast Shelf Sci 81:1–12

Defeo O, McLachlan A, Armitage D et al (2021) Sandy beach social–ecological systems at risk: regime shifts, collapses, and governance challenges. Front Ecol Environ 10:564–573

Denadai MR, Jacobucci DFC, Fontana I et al (2015a) Assessment of contamination of the beach clam *Tivela mactroides*: implications for food safety of a recreational and subsistence marine resource in Caraguatatuba Bay, Brazil. Int J Food Contam 2:6
Denadai MR, Pombo M, Bernadochi LC et al (2015b) Harvesting the beach clam *Tivela mactroides*: shortand long-term dynamics. Mar Coast Fish 7(7):103–115
Derraik JG (2002) The pollution of the marine environment by plastic debris: a review. Mar Pollut Bull 44(9):842–852
Diaz RJ, Rosenberg R (1995) Marine benthic hypoxia: a review of its ecological effects and the behavioral responses of benthic macrofauna. Oceanogr Mar Biol 33:245–303
Dominguez JML (2009) The coastal zone of Brazil. In: Geology and geomorphology of Holocene coastal barriers of Brazil, Lecture notes in earth sciences, vol 107. Springer, Berlin, Heidelberg
Duarte C, López J, Benítez S et al (2016) Ocean acidification induces changes in algal palatability and herbivore feeding behavior and performance. Oecologia 180:453–462
Duarte C, Quintanilla-Ahumada D, Anguita C et al (2019) Artificial light pollution at night (ALAN) disrupts the distribution and circadian rhythm of a sandy beach isopod. Environ Pollut 248:565–573
Dugan JE, Hubbard DM, Mccrary MD et al (2003) The response of macrofauna communities and shorebirds to macrophyte wrack subsidies on exposed sandy beaches of southern California. Estuar Coast Shelf Sci 58:25–40
Espinel-Velasco N, Hoffmann L, Agüera A et al (2018) Effects of ocean acidification on the settlement and metamorphosis of marine invertebrates and fish larvae. Mar Ecol Prog Ser 606:237–257
Esteves LS, Santos IR (2002) Impacto Econômico da erosão na praia do Hermenegildo (RS), Brasil. Pesqui Geocienc 28(2):393–403
Fanini L, Marchetti GM, Scapini F et al (2009) Effects of beach nourishment and groynes building on population and community descriptors of mobile arthropodofauna. Ecol Indic 9(1):167–178
Fanini L, Zampicinini G, Tsigenopoulos CS et al (2017) Life-history, substrate choice and Cytochrome Oxidase I variations in sandy beach peracaridans along the Rio de la Plata estuary. Estuar Coast Shelf Sci 187:152–159
Fanini L, Defeo O, Elliott M, Paragkamian S, Pinna M, Salvo VS (2021) Coupling beach ecology and macroplastics litter studies: current trends and the way ahead. Mar Pollut Bull 173:112951
Fisner M, Taniguchi S, Moreira F et al (2013) Polycyclic aromatic hydrocarbons (PAHs) in plastic pellets: variability in the concentration and composition at different sediment depths in a sandy beach. Mar Pollut Bull 70(1–2):219–226
Gallucci F, Netto SA (2004) Effects of the passage of cold fronts over a coastal site: an ecosystem approach. Mar Ecol Prog Ser 281:79–92
Gaston KJ, Davies TW, Bennie J et al (2012) Reducing the ecological consequences of night-time light pollution: options and developments. J Appl Ecol 49(6):1256–1266
GESAMP, Kershaw PJ, Rochman CM (2016) Sources, fate and effects of microplastics in the marine environment: part two of a global assessment, vol 93. IMO/FAO/UNESCO-IOC/UNIDO/WMO/IAEA/UN/UNEP/UNDP Joint Group of Experts on the Scientific Aspects of Marine Environmental Protection, London, p 220
Gómez-Pujol L, Jackson DWT, Cooper JAG et al (2011) Spatial and temporal patterns of sediment activation depth on a high energy microtidal beach. J Coast Res 64:85–89
Griggs G (2005) The impacts of coastal armoring. Shore Beach 73(1):13–22
Gül MR, Griffen BD (2019) Combined impacts of natural and anthropogenic disturbances on the bioindicator *Ocypode quadrata* (Fabricius, 1787). J Exp Mar Biol Ecol 519:151185
Gunasekaran K, Karthikeyan P, Yosuva M et al (2021) Niver ciclonic impacts on mollusk habitat destruction in Parangipettai, southeast coast of Tamil Nadu, India: a case study. Mar Pollut Bull 173:113022
Gusmão F, Di Domenico M, Amaral ACZ et al (2016) In situ ingestion of microfibres by meiofauna from sandy beaches. Environ Pollut 216:584–590

Harris L, Nel R, Smale M et al (2011) Swashed away? Storm impacts on sandy beach macrofaunal communities. Estuar Coast Shelf Sci 94(3):210–221

Harris L, Nel R, Holness S et al (2014) Setting conservation targets for sandy beach ecosystems. Estuar Coast Shelf Sci 150:45–57

Haynes D, Quinn GP (1995) Temporal and spatial variability in community structure of a sandy intertidal beach, Cape Paterson, Victoria, Australia. Mar Freshw Res 46(6):931–942

Hernáez P, Hereman MJ, Pimenta CER (2019) La efectividad de una ley de protección al servicio de la conservación de un recurso marino: El ejemplo del camarón fantasma *Callichirus major* (Decapoda, Callianassidae) de la costa de Brasil. Inheringia Ser Zool 109:e2019001

Houser C, Greenwood B (2007) Onshore migration of a swash bar during a storm. J Coast Res 23(1):1–14

Hubbard DM, Dugan J, Schooler NK (2014) Local extirpations and regional declines of endemic upper beach invertebrates in Southern California. Estuar Coast Shelf Sci 150:67–75

Hughes C, Richardson CA, Luckenbach M et al (2009) Difficulties in separating hurricane induced effects from natural benthic succession: Hurricane Isabel, a case study from Eastern Virginia, USA. Estuar Coast Shelf Sci 85(3):377–386

Instituto Aqualung (2016) O misterioso desaparecimento dos tatuís. http://www.institutoaqualung.com.br/Site/Conteudo/Artigo.aspx?C=7bo8Rz88SAI%3D. Accessed in 14 May 2019

Intergovernmental Panel on Climate Change (IPCC) (2013) Contribution of Working Group I to the Fifth assessment report of the Intergovernmental Panel on Climate Change. Climate change 2013: The physical science basis. Cambridge University Press, Cambridge

Intergovernmental Panel on Climate Change (IPCC) (2021) Summary for policymakers. In: Zhai P, Pirani A, Connors SL et al (eds) Climate change 2021: the physical science basis. Contribution of Working Group I to the Sixth assessment report of the Intergovernmental Panel on Climate Change. Cambridge University Press, Cambridge. (In press)

Islam MS, Tanaka M (2004) Impacts of pollution on coastal and marine ecosystems including coastal and marine fisheries and approach for management: a review and synthesis. Mar Pollut Bull 48(7–8):624–649

Ivar Do Sul JÁ, Costa MF (2007) Marine debris review for Latin America and the Wider Caribbean Region: from 1970s until now, and where do we go from here? Mar Pollut Bull 54:1087–1104

Jambeck JR, Geyer R, Wilcox C et al (2015) Plastic waste inputs from land into the ocean. Science 347(6223):768–771

Jaramillo E, Dugan JE, Hubbard DM et al (2012) Ecological implications of extreme events: footprints of the 2010 earthquake along the Chilean coast. PLoS One 7(5):1–8

Kamrowski RL, Limpus C, Moloney J et al (2012) Coastal light pollution and marine turtles: assessing the magnitude of the problem. Endanger Species Res 19(1):85–98

Klein AHF, Short AD (2016) Brazilian beach systems: introduction. In: Short AD, Klein AHF (eds) Brazilian beach systems. Springer, Cham

Krelling AP, Williams AT, Turra A (2017) Differences in perception and reaction of tourist groups to beach marine debris that can influence a loss of tourism revenue in coastal areas. Mar Policy 85:87–99

Laurino IRA, Checon HH, Corte GN et al (2020) Flooding affects vertical displacement of intertidal macrofauna: a proxy for the potential impacts of environmental changes on sandy beaches. Estuar Coast Shelf Sci 245:106882

Leduc AOHC, Nunes JACC, Araújo CB et al (2021) Land-based noise pollution impairs reef fish behavior: a case study with a Brazilian carnival. Biol Conserv 253:108910

Lopez GG, Saliés EDC, Lara PH et al (2015) Coastal development at sea turtles nesting ground: efforts to establish a tool for supporting conservation and coastal management in northeastern Brazil. Ocean Coast Manag 116:270–276

Luarte T, Bonta CC, Silva-Rodriguez EA et al (2016) Light pollution reduces activity, food consumption and growth rates in a sandy beach invertebrate. Environ Pollut 218:1147–1153

Lucey NM, Lombardi C, DeMarchi L et al (2015) To brood or not to brood: are marine invertebrates that protect their offspring more resilient to ocean acidification? Sci Rep 5:12009

Luijendijk A, Hagenaars G, Ranasinghe R et al (2018) The state of the world's beaches. Sci Rep 8:6641
Lynn KD, Quintanilla-Ahumada D, Anguita C et al (2021) Artificial light at night alters the activity and feeding behaviour of sandy beach amphipods and pose a threat to their ecological role in Atlantic Canada. Sci Total Environ 780:146568
Machado PM, Costa LL, Suciu MC et al (2016) Extreme storm wave influence on sandy beach macrofauna with distinct human pressures. Mar Pollut Bull 107:125–135
Machado PM, Suciu MC, Costa LL et al (2017) Tourism impacts on benthic communities of sandy beaches. Mar Ecol 38(4):e12440
Machado PM, Tavares DC, Zalmon IR (2019) Synergistic effect of extreme climatic events and urbanization on population density of the ghost crab *Ocypode quadrata* (Fabricius, 1787). Mar Ecol 40:e12525
Magalhães WF, Lima JB, Barros F (2009) Is *Ocypode quadrata* (Fabricius, 1787) a useful tool for exposed sandy beaches management in Bahia State (Northeast Brazil)? Braz J Oceanogr 57(2):153–155
Marques WC, Stringari CE, Kirinus EP et al (2017) Numerical modelling of the Tramandaí beach oil spill, Brazil – case study for January 2012 event. Appl Ocean Res 65:178–191
Martinelli-Filho JE, Monteiro RCP (2019) Widespread microplastics distribution at an Amazon macrotidal sandy beach. Mar Pollut Bull 145:219–223
Martins KA, Pereira OS (2014) Coastal erosion at Pau Amarelo Beach, Northeast of Brazil. J Coast Res 71(SI):17–23
Mascarenhas R, Santos RD, Santos AD et al (2004) Nesting of hawksbill turtles in Paraiba-Brazil: avoiding light pollution effects. Mar Turt Newsl 104:1–3
Mathews PL, Maccarone AD (2019) Responses by Atlantic ghost crab (*Ocypode quadrata*) populations to hurricane impacts on a Texas beach: a 10-year study. Trans Kans Acad Sci 122(1–2):59–68
Matsuoka T, Nakashima T, Nagasawa N (2005) A review of ghost fishing: scientific approaches to evaluation and solutions. Fish Sci 71(4):691
McLachlan A, Defeo O (2018) The ecology of sandy shores. Elsevier, Academic Press, London
Merwe VJP, West EJ, Ibrahim K (2012) Effects of off-road vehicle tire ruts on the beach dispersal of green sea turtle *Chelonia mydas* hatchlings. Endanger Species Res 18(1):27–34
Moss D, Mcphee DP (2006) The impacts of recreational four-wheel driving on the abundance of the ghost crab (*Ocypode cordimanus*) on a subtropical sandy beach in SE Queensland. Coast Manag 34(1):133–140
Muehe D (2003) Beach morphodynamics research in Brazil: evolution and applicability. J Coast Res SI35:32–42
Muehe D (2006) Erosion in the Brazilian coastal zone: an overview. J Coast Res 39:43–48
Naylor E (2010) Chronobiology of marine organisms. Cambridge University Press, Cambridge
Nel R, Campbell EE, Harris L et al (2014) The status of sandy beach science: past trends, progress, and possible futures. Estuar Coast Shelf Sci 150:1–10
Neves FM, Bemvenuti CE (2006) The ghost crab *Ocypode quadrata* (Fabricius, 1787) as a potential indicator of anthropic impact along the Rio Grande do Sul coast, Brazil. Biol Conserv 133(4):431–435
Neves RA, Seixas JT, Rodrigues N et al (2022) Impacts of the COVID-19 pandemic restrictions on solid waste pollution in the worldwide iconic Copacabana Beach (Rio de Janeiro, Brazil). Mar Pollut Bull 181:113865
O Globo (2014) Tatuís estão sumidos das praias mais movimentadas do Rio. http://g1.globo.com/rio-de-janeiro/noticia/2014/09/tatuis-estao-sumidos-das-praias-mais-movimentadas-do-rio.html. Accessed in 14 May 2019
Oliveira FRF, Yokoyama LQ (2021) Response of *Ocypode quadrata* to storm waves on an urbanized sandy beach. Ocean Coast Res 69:21005
Oliveira AJFC, França PTR, Pinto AB (2010) Antimicrobial resistance of heterotrophic marine bacteria isolated from seawater and sands of recreational beaches with different organic

pollution levels in southeastern Brazil: evidences of resistance dissemination. Environ Monit Assess 169:375–384

Owczarzak IG (2015) Ocorrência e uso do *Hemipodia californiensis* (Hartman 1938) (Polychaeta, Glyceridae) em praias arenosas de três localidades da Ilha de Santa Catarina, Brasil. Undergradute thesis, Federal University of Santa Catarina

Paula DP (2015) Erosão costeira e estruturas de proteção no litoral da região metropolitana de Fortaleza (Ceará, Brasil): um contributo para artificialização do litoral. REDE 9(1):73–86

Peters A, Verhoeven KJF (1994) Impact of artificial lighting on the seaward orientation of hatchling loggerhead turtles. J Herpetol 28:112–114

Peterson CH, Bishop MJ (2005) Assessing the environmental impacts of beach nourishment. Bioscience 55(10):887–896

Peterson CH, Hickerson DHM, Johnson GG (2000) Short-term consequences of nourishment and bulldozing on the dominant large invertebrates of a sandy beach. J Coast Res 16:368–378

Pezzuto PR, Resgalla C Jr, Abreu JGN et al (2006) Environmental impacts of the nourishment of Balneário Camboriú beach, SC, Brazil. J Coast Res 39(SI):863–868

Pinheiro LM, Monteiro RC, do Sul JAI, Costa MF (2019) Do beachrocks affect microplastic deposition on the strandline of sandy beaches? Mar Pol Bul 141:569–572

Pinto KC, Hachich EM, Sato MIZ et al (2012) Microbiological quality assessment of sand and water from three selected beaches of South Coast, São Paulo State, Brazil. Water Sci Technol 66(11):2475–2482

Pombo M, Turra A (2013) Issues to be considered in counting burrows as a measure of Atlantic ghost crab populations, an important bioindicator of sandy beaches. PLoS One 8(12):e83792

Pombo M, Turra A (2019) The burrow resetting method, an easy and effective approach to improve indirect ghost-crab population assessments. Ecol Indic 104:422–428

Pombo A, Baptista T, Granada L et al (2018) Insight into aquaculture's potential of marine annelid worms and ecological concerns: a review. Rev Aquac 12:107–121

Posey M, Lindberg W, Alphin T et al (1996) Influence of storm disturbance on an offshore benthic community. Bull Mar Sci 59(3):523–529

Quintero-Torres E, López-Sánchez B (2022) Short-term effects of hurricane events on the bioindicator species *Ocypode quadrata* inhabiting an island in essence free of anthropogenic disturbances. J Exp Mar Biol Ecol 548:151684

Ragagnin MN, Turra A (2022) Imposex incidence in the sandy beach snail *Hastula cinerea* reveals continued and widespread tributyltin contamination after its international ban. Reg Stud Mar Sci 49:102118

Rainbow PS, Phillipst DJH (1993) Cosmopolitan biomonitors of trace metals. Mar Pollut Bull 26:593–601

Rech TF, Soto GAT, Turra A (2021) Species with insufficient data and red lists: the dilemma of the beach trigonal clam *Tivela mactroides*. J Nat Conserv 62:126024

Rêgo JCL, Soares-Gomes A, Silva FS (2018) Loss of vegetation cover in a tropical island of the Amazon coastal zone (Maranhão Island, Brazil). Land Use Policy 71:593–601

Richardson AJ, Schoeman DS (2004) Climate impact on plankton ecosystems in the Northeast Atlantic. Science 305(5690):1609–1612

Rippy MA, Franks PJS, Feddersen F et al (2013) Beach nourishment impacts on bacteriological water quality and phytoplankton bloom dynamics. Environ Sci Technol 47:6146–6154

Rochman CM, Tahir A, Williams SL et al (2015) Anthropogenic debris in seafood: plastic debris and fibers from textiles in fish and bivalves sold for human consumption. Sci Rep 5:14340

Rodil IF, Jaramillo E, Acuña E et al (2016) Long-term responses of sandy beach crustaceans to the effects of coastal armouring after the 2010 Maule earthquake in South Central Chile. J Sea Res 108:10–18

Rosa LC, Guilheme PDB (2014) Exploited resources in the live bait trade on the coastal shore of Paraná, southern Brazil. Sci Plena 10:101001-1

Roth F, Lessa GC, Wild C (2016) Impacts of a high-discharge submarine sewage outfall on water quality in the coastal zone of Salvador (Bahia, Brazil). Mar Pollut Bull 106:46–48

Rutz C, Loretto MC, Bates AE et al (2020) COVID-19 lockdown allows researchers to quantify the effects of human activity on wildlife. Nat Ecol Evol 4:1156–1159

Rydell J (1992) Exploitation of insects around streetlamps by bats in Sweden. Funct Ecol 6:744–750

Saloman CH, Naughton SP (1977) Effect of hurricane Eloise on the benthic fauna of Panama City Beach, Florida, USA. Mar Biol 42(4):357–363

Santos TMT, Petracco M, Venekey V (2022) Effects of vehicle traffic and trampling on the macrobenthic community of Amazonian macrotidal sandy beaches. J Mar Biol Assoc U K 102:285. https://doi.org/10.1017/S0025315422000480

Scapini F, Degli EI, Defeo O (2019) Behavioral adaptations of sandy beach macrofauna in face of climate change impacts: a conceptual framework. Estuar Coast Shelf Sci 225:106236

Schlacher TA, Thompson LM (2008) Physical impacts caused by off-road vehicles to sandy beaches: spatial quantification of car tracks on an Australian barrier island. J Coast Res 24(sp2):234–242

Schlacher TA, Thompson L, Price S (2007) Vehicles versus conservation of invertebrates on sandy beaches: mortalities inflicted by offroad vehicles on ghost crabs. Mar Ecol 28(3):354–367

Schlacher TA, Schoeman DS, Dugan J et al (2008) Sandy beach ecosystems: key features, sampling issues, management challenges and climate change impacts. Mar Ecol 29:70–90

Schlacher TA, Lucrezi S, Connolly RM et al (2016) Human threats to sandy beaches: a meta-analysis of ghost crabs illustrates global anthropogenic impacts. Estuar Coast Shelf Sci 169:56–73

Schmitt WL (1935) Mud shrimps of the Atlantic coast of North America. Smithsonian Misc Collect 93(2):1–21

Schoeman D, Schlacher TA, Defeo O (2014) Climate-change impacts on sandy-beach biota: crossing a line in the sand. Glob Chang Biol 20:2383–2392

Serafini TZ, Carneiro K, Lima MF et al (2010) Identifying and mitigating hatchling disorientation on nesting beaches. Mar Turt Newsl 129:14–16

Seuront L, Cribb N (2017) Fractal analysis provides new insights into the complexity of marine mammal behavior: a review, two methods, their application to diving and surfacing patterns, and their relevance to marine mammal welfare assessment. Mar Mammal Sci 33(3):847–887.

Shimizu RM, Rosso S (2000) Influence of an oil spill on the abundance of *Callichirus major* (Say, 1818) on a sandy beach in southeastern Brazil (Crustacea: Decapoda: Thalassinidea). Nauplius 1:63–72

Silva E, Marco A, Da Graça J et al (2017) Light pollution affects nesting behavior of loggerhead turtles and predation risk of nests and hatchlings. J Photochem Photobiol B 173:240–249

Soares OM, Teixeira CEP, Bezerra LEA et al (2020) Oil spill in South Atlantic (Brazil): environmental and governmental disaster. Mar Policy 115:103879

Soares MCR, Barros EL, Guerra RGP (2022) Not just sand: the folly of dismantling the environmental protection of dunes in Brazil. Land Use Policy 112:105803

Sola MCR, Paiva PC (2001) Variação temporal da macrofauna bentônica sublitoral da praia da Urca (RJ) após a ocorrência de ressacas. Rev Bras Oceanogr 49(1–2):137–142

Soto EH, Botero CM, Milanés CB et al (2021) How does the beach ecosystem change without tourists during COVID-19 lockdown? Biol Conserv 255:108972

Southward AJ, Hawkins SJ, Burrows MT (1995) Seventy years' observations of changes in distribution and abundance of zooplankton and intertidal organisms in the western English Channel in relation to rising sea temperature. J Therm Biol 20:127–155

Souza JRB, Borzone CA (2003) A extração de corrupto, *Callichirus major* (Say) (Crustacea, Thalassinidea), para uso como isca em praia do litoral do Paraná: as populações exploradas. Rev Bras Zool 20:625–630

Souza GN, Oliveira CA, Tardem AS (2017) Counting and measuring ghost crab burrows as a way to assess the environmental quality of beaches. Ocean Coast Manag 140:1–10

Suciu MC, Tavares DC, Costa LL et al (2017) Evaluation of environmental quality of sandy beaches in southeastern Brazil. Mar Pollut Bull 119(2):133–142

Sugumaran J, Padmasai R, Altaff K (2019) The effects of tropical cyclone Gaja on sandy beach meiofauna of Palk Bay, India. Reg Stud Mar Sci 31:100747
Suzuki Y, Shimizu H, Kuroda T (2021) Plant debris are hotbeds for pathogenic bacteria on recreational sandy beaches. Sci Rep 11:11496
Teixeira TP, Neves LM, Araújo FG (2012) Thermal impact of a nuclear power plant in a coastal area in Southeastern Brazil: effects of heating and physical structure on benthic cover and fish communities. Hydrobiologia 684:161–175
Tsukada E, Fernandes E, Vidal C, Salla RF (2021) Beach morphodynamics and its relationship with the deposition of plastic particles: a preliminary study in southeastern Brazil. Mar Pollut Bull 172:112809
Valiela I, Foreman K, Lamontagne M et al (1992) Couplings of watersheds and coastal waters: sources and consequences of nutrient enrichment in Waquoit Bay, Massachusetts. Estuaries 15(4):443–457
Vedolin MC, Teophilo CYS, Turra A et al (2018) Spatial variability in the concentrations of metals in beached microplastics. Mar Pollut Bull 129(2):487–493
Veloso VG, Silva ES, Caetano CH et al (2006) Comparison between the macroinfauna of urbanized and protected beaches in Rio de Janeiro State, Brazil. Biol Conserv 127(4):510–515
Veloso VG, Neves G, Lozano M et al (2008) Responses of talitrid amphipods to a gradient of recreational pressure caused by beach urbanization. Mar Ecol 29:126–133
Veloso VG, Sallorenzo IA, Ferreira BCA et al (2010) *Atlantorchestoidea brasiliensis* (Crustacea: Amphipoda) as an indicator of disturbance caused by urbanization of a beach ecosystem. Braz J Oceanogr 58(1):13–21
Veloso VG, Neves G, De Almeida Capper L (2011) Sensitivity of a cirolanid isopod to human pressure. Ecol Indic 11(3):782–788
Vera-Cruz D (1972) Artificial nourishment of Copacabana beach. Coast Eng 1:76
Vieira JV, Borzone CA, Lorenzi L (2012) Human impact on the benthic macrofauna of two beach environments with different morphodynamic characteristics in southern Brazil. Braz J Oceanogr 60(2):135–148
Vieira JV, Ruiz-Delgado MC, Reyes-Martínez MJ et al (2016) Assessment the short-term effects of wrack removal on supralittoral arthropods using the M-BACI design on Atlantic sandy beaches of Brazil and Spain. Mar Environ Res 119:222–237
Vousdoukas MI, Ranasinghe R, Mentaschi L, Plomaritis TA, Athanasiou P, Luijendijk A, Feyen L (2020) Sandy coastlines under threat of erosion. Nat Clim Change 10:260–263
Watson GJ, Murray JM, Schaefer M et al (2018) Bait worms: a valuable and important fishery with implications for fisheries and conservation management. Fish Fish 18(2):374–388
Witmer AD, Roelke DL (2014) Human interference prevents recovery of infaunal beach communities from hurricane disturbance. Ocean Coast Manag 87:52–60
World Wildlife Fund (WWF) (2019) Solving plastic pollution through accountability
Zielinski S, Botero CM, Yanes A (2019) To clean or not to clean? A critical review of beach cleaning methods and impacts. Mar Pollut Bull 139:390–340

Chapter 10
Beach Management and Conservation in Brazil: Challenges and Opportunities

Luciana Yokoyama Xavier, Leandra Regina Gonçalves, Mayara Oliveira, Marina Ribeiro Corrêa, Nicole Malinconico, Marcus Polette, and Alexander Turra

10.1 The Importance of Sandy Beaches

Before you go through this chapter, try a quick exercise: Picture yourself looking at the sea. Take a moment to listen to the sound of the waves. Notice the shades of blue/green and the seeming infinity of the ocean. Where are you standing? Most people would answer: "The beach." However, even if you are one of the few that would instead look from a different point of view, like an elevated cliff or mountain, or through a higher window or balcony, it would not be risky to assume that a beach is also part of the scenario. Be it pristine and remote, or lively and full of people; be its waves low and soothing, or high and energetic; be it a place for spiritual practices, sports training, daily work, leisure, or purely inspiration, beaches populate our imagination and are the coastal ecosystem more densely used by humankind (Schlacher et al. 2007).

L. Y. Xavier (✉) · N. Malinconico · A. Turra
Instituto Oceanográfico da Universidade de São Paulo, Praça do Oceanográfico, São Paulo, SP, Brazil
e-mail: lyxavier@usp.br

L. R. Gonçalves
Instituto do Mar da Universidade Federal de São Paulo, R. Carvalho de Mendonça, Santos, SP, Brazil

M. Oliveira
University of Queensland, Brisbane, QLD, Australia

M. R. Corrêa
Instituto de Energia e Ambiente da Universidade de São Paulo, São Paulo, SP, Brazil

M. Polette
Universidade do Vale do Itajaí, Campus Itajaí, Itajaí, SC, Brazil

A. C. Z. Amaral et al. (eds.), *Brazilian Sandy Beaches*, Brazilian Marine Biodiversity, https://doi.org/10.1007/978-3-031-30746-1_10

From a geomorphological point of view, sandy beaches can be defined as sedimentary deposits, predominantly sandy fractions, formed by the action of waves and/or tides (Mclachlan and Defeo 2017). However, beaches are so much more than that. They support ecosystem services related to coastal protection and maintenance of biodiversity and conservation, provide for human leisure and recreation, and have an undisputed role in human well-being (Williams and Micallef 2009; Sardá and Azcárate 2018; Chap. 8). As the gateways to the sea, beaches have a prominent role in fostering ocean literacy and making humans understand the importance of conserving the ocean (Santoro et al. 2017).

Human activities on the coastline have shaped beaches throughout history (Willians and Micallef 2009). These activities and how society understands and perceives beaches affect their management (Williams and Micallef 2009; Cervantes et al. 2018). According to Brazilian National Coastal Management Plan, beaches are "*the area periodically covered and exposed by water, including the subsequent strip of deposited material such as sand, gravel, pebbles, and boulders, up to the limit where natural vegetation starts or, in its absence, another ecosystem begins*" and are an asset for common use and free access to the people (Brasil 1988a). This definition, focusing on the emerged part of the beach, is the base for Brazilians' relationship with beaches, which makes them one of the most democratic spaces in the country.

Brazil has many beaches with a diverse combination of morphodynamic features, conservation status, uses, and cultural occupation patterns (Chap. 1). Despite their numerous benefits and economic relevance, Brazilian beaches are under different chronic pressures and high risks (see Chap. 9). Coastal cities are growing as people are attracted to coastal areas. Major enterprises modify shoreline and coastal dynamics. Climate change increases the number of storm surges ravaging our coasts. In 2019, the largest oil spill to reach the Brazilian coast uncovered the fragile situation of Brazilian beaches' management: not only were management agencies slow in taking action, but they were also unprepared to deal with the crude oil, increasing impacts on biodiversity and marginalized and vulnerable coastal communities (Magris and Giarizzo 2020; Soares et al. 2020). Such a situation raises the questions that inspired this chapter: How are Brazilian beaches being taken care of? Can current practices guarantee that future generations experience them as we do? Will we have healthy beaches providing benefits to people over this century?

This chapter presents an overview of the environmental status and the scientific, legal, and institutional basis for beach management and conservation in Brazil. By reviewing management approaches and four case studies, we identify the main challenges compromising beach conservation strategies and the main opportunities for novel approaches. These general principles are the foundation of ecosystem-based management (EBM), which can foster a more comprehensive and robust beach management system in Brazil.

10.2 Conservation Status and Threats to Brazilian Sandy Beaches

Brazil has one of the most diverse beach collections in the world. The Brazilian coastline extends for over 9000 km, and beaches occupied 82,778 ha of coastal ecosystems (0.5% of the total coastal ecosystems area) according to a 2012 data survey (Prates et al. 2012).[1] Diversity strikes as their dominant characteristic, extending over tropical and subtropical zones and subjected to a range of waves, wind, and tide regimes (Chap. 1). Brazilian sandy beaches vary from tide-dominated to wave-dominated, with multiple intermediate classes (Prates et al. 2012; Amaral et al. 2016; Chap. 1); have different conservation statuses (Prates et al. 2012); support diverse economic, cultural, and recreational uses (MMA 2015; Chap. 8); and face different threats (Strohaecker 2008; Nicolodi and Petermann 2010; Chap. 9) (Fig. 10.1).

By 2012, protected areas regulated by the National Protected Areas System (from Portuguese, Sistema Nacional de Unidades de Conservação, SNUC – Brasil 2000) covered 24% of the Brazilian beaches area (Table 10.1). This percentage has increased since (ICMBio 2020), but by then, it had already surpassed the 10% national target for coastal ecosystem protection,[2] the Aichi Targets,[3] and the Sustainable Development Goals (SDG 14.5).[4]

Despite the degree of protection, the conservation of Brazilian beaches still faces many obstacles. Implementation and management of marine protected areas are two of the most significant challenges to marine conservation in Brazil (Silva 2019), especially within sustainable-use protected areas (equivalent to IUCN Category VI[5]), which represent the majority of protected areas on beaches. As multiecosystem areas, beach management is often pushed into the background of the

[1] Giving coastal dynamics, we understand that actual values diverge from this estimate.

[2] Conabio Normative Resolution 03/2013. Available at: https://www.mma.gov.br/images/arquivo/80049/Conabio/Documentos/Resolucao_06_03set2013.pdf (Accessed: 02.mar.2020)

[3] Aichi Target 11: By 2020, at least 17 percent of terrestrial and inland water, and 10 percent of coastal and marine areas, especially areas of particular importance for biodiversity and ecosystem services, are conserved through effectively and equitably managed, ecologically representative and well-connected systems of protected areas and other effective area-based conservation measures, and integrated into the wider landscapes and seascapes. Available at: https://www.cbd.int/aichi-targets/target/11 (Accessed: 02.mar.2020)

[4] Sustainable Development Goal 14: Conserve and sustainably use the ocean, seas and marine resources for sustainable development. Target 14.5: By 2020, conserve at least 10 percent of coastal and marine areas, consistent with national and international law and based on the best available scientific information. Available at: https://sustainabledevelopment.un.org/sdg14 (Accessed: 02.mar.2020)

[5] IUCN Category VI are Protected areas that conserve ecosystems and habitats, together with associated cultural values and traditional natural resource management systems. They are generally large, encompassing many ecosystems, with most of the area in a natural condition, where a proportion is under sustainable natural resource management and where low-level nonindustrial use of natural resources compatible with nature conservation is seen as one of the main aims. Available

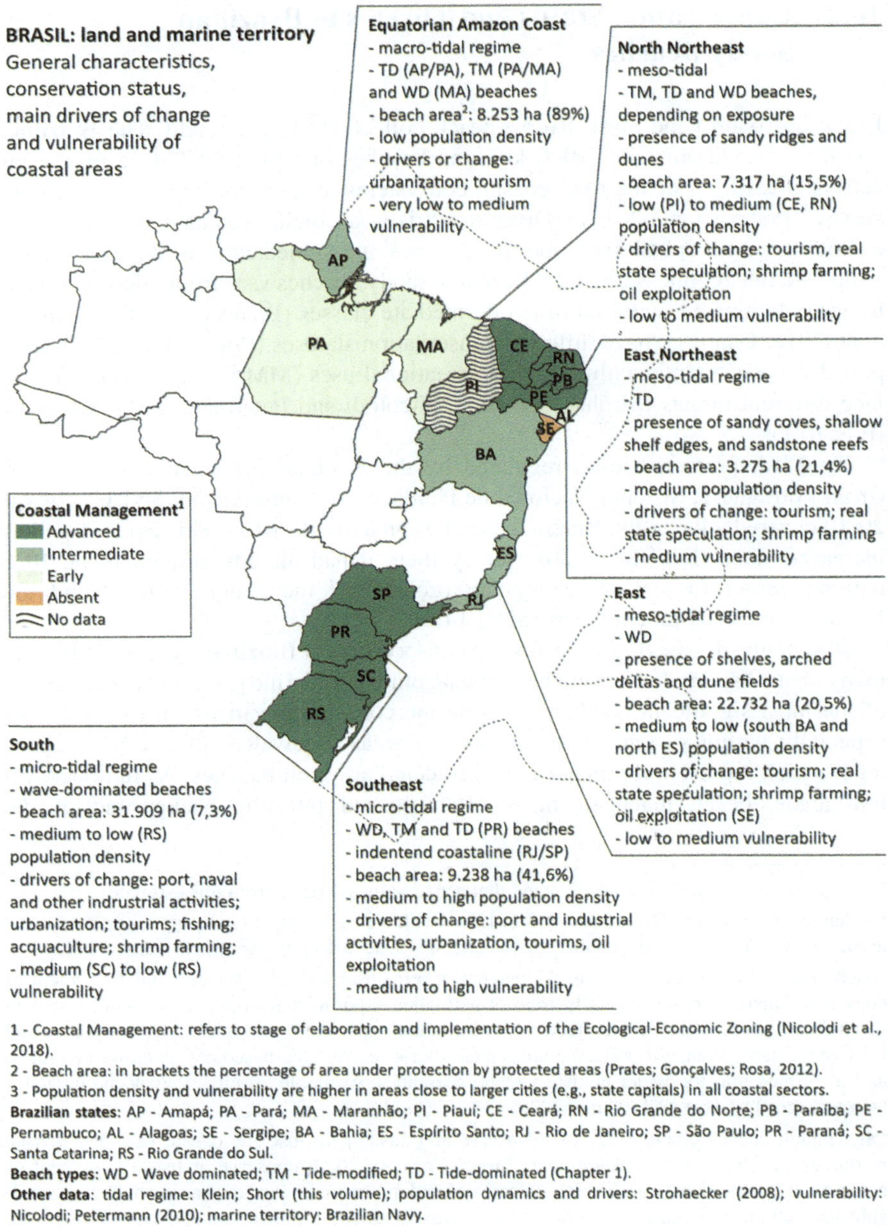

Fig. 10.1 Brazilian six coastal sectors and marine territory (dotted line representing the Exclusive Economic Zone and the extension of Brazilian Continental Shelf) indicating: beach ecosystems' general characteristics, conservation status, main drivers of change, vulnerability classifications, and degree of implementation of the Ecological-Economic Zoning in coastal states. (*Source:* Made by the authors based on references listed in the figure)

Table 10.1 Total beach area (Beach Area, ha) and beach area included in protected areas considering no-take (No-take), sustainable use (Sust. Use), and the total amount (total area and as a percentage of beach area) of protected beach areas in each Brazilian Coastal State (BCS)

BCS	Beach area (ha)	Beach area as protected area (ha)			
		No-Take	Sust. Use	IUCN VI	Total (%)
AP	183	143	–	–	143 (78.2)
PA	1560	–	831	–	831 (53.3)
MA	6510	102	1051	5222	6375 (97.9)
PI	786	–	–	455	455 (57.8)
CE	4009	131	48	86	265 (6.6)
RN	2576	–	161	256	417 (16.2)
PB	754	14	11	95	120 (15.9)
PE	993	–	11	275	287 (28.9)
AL	1528	–	56	236	292 (1.1)
SE	987	257	–	286	543 (55.1)
BA	6374	52	897	2585	3533 (55.4)
ES	15,371	–	–	579	579 (3.8)
RJ	4251	509	125	778	1412 (33.2)
SP	4126	309	302	1520	2131 (51.7)
PR	861	226	–	71	297 (34.4)
SC	5986	185	–	1875	2061 (34.4)
RS	25,923	271	–	–	27 (1.0)
Total	82,778	2199	3493	14,319	20,011 (24.17)

Emphasis is given to the area covered by extensive multiple-use and multiple ecosystems protected areas (IUCN VI), which include beaches but do not focus on their conservation

Translated from Prates et al. (2012)

Brazilian Coastal States: Amapá (AP), Pará (PA), Maranhão (MA), Piauí (PI), Ceará (CE), Rio Grande do Norte (RN), Paraíba (PB), Pernambuco (PE), Alagoas (AL), Sergipe (SE), Bahia (BA), Espírito Santo (ES), Rio de Janeiro (RJ), São Paulo (SP), Paraná (PA), Santa Catarina (SC), e Rio Grande do Sul (RS)

management of such extensive sustainable use areas (Vianna and Xavier 2014). Therefore, not only is it necessary to increase the effectiveness and implementation of protected areas but also promote integrated management since Brazilian beaches face many threats related to terrestrial and marine activities (Amaral et al. 2016; Chap. 9) outside SNUC ruling.

Coastal development (the land-use changes related to urbanization and coastal infrastructures) is one of the main threats to sandy beaches (MMA 2015; Amaral et al. 2016). Coastal areas concentrate most of Brazil's largest human aggregations (Oliveira and Nicolodi 2012), and the growth rate of coastal cities is higher when compared to other areas (De Andrés et al. 2018; Gonçalves et al. 2020a). Coastal infrastructure related to real estate activities, marine ports, and oil production has altered the shoreline, affected coastal dynamics, and modified landscapes and

at: https://www.iucn.org/theme/protected-areas/about/protected-area-categories (Accessed: 03. jun.2020) ecosystems (Strohaecker 2008; Amaral et al. 2016; Chap. 9). Tourism development, a significant income for coastal cities (MTur 2019), intensifies the demand for infrastructures such as restaurants, hotels, roads, and second housing, fostering urbanization and its impacts (Strohaecker 2008; Willians and Micallef 2009; Chap. 9).

Coastal development is also a significant source of land-based marine pollution related to urban and industrial waste (including marine litter) and mining and agricultural activities in the river basins that drain into the sea (Harris et al. 2015). In Brazil, 55% of the population has access to proper sewage treatment, 18% have their sewage collected but not treated, and 27% have no sewage treatment (ANA 2017). Tourism season increases up to ten times the number of people at the coast (Cabral 2015) and overloads sewage and waste treatment plants (MMA 2015). Income asymmetry and real estate speculation intensify this problem as they drive the most vulnerable people (including the traditional population) away from areas closer to the sea to vulnerable or risk areas (Cristiano et al. 2018). Unplanned coastal occupation threatens the coastal environment and the population that lives in areas subjected to flooding, landslides, and poor sanitation conditions (Nicolodi and Petermann 2010; Polette and Lins-de-Barros 2012).

Marine-based activities – such as oil exploration, maritime transport, and the recreational use of motor vehicles and boats – are additional pollution sources (Defeo et al. 2009). Marine pollution is diffuse and demands coordinated local, national, and international management. International litter driven by ocean currents to Brazilian beaches (Possatto et al. 2015) and the 2019 oil spill (Magris and Giarizzo 2020; Soares et al. 2020) demonstrate the influence of transboundary activities and processes on sandy beaches. Beach pollution impacts tourism activities (Krelling et al. 2017) and other relevant provisioning (as the ones related to fishing and harvesting) and cultural services (as the maintenance of traditional practices) (Amaral et al. 2016; Magris and Giarizzo 2020).

Climate change impacts inflict additional challenges to beach management. Sea-level rise; regime shifts for wind, wave, and precipitation; storms and temperature extremes; and decreased sedimentary contribution of river basins threaten beach uses and services (PBMC 2016). Climate change can also increase beach erosion (McLachlan and Defeo 2017). Many beaches risk disappearing, trapped between a rising sea and urban structures (Schlacher et al. 2007; Pontee 2013). Impacts related to climate change affect beaches directly and are anticipated to reinforce the existing urbanization and pollution impacts. As ocean water advances toward urbanized areas, coastal communities have their territory flooded, water and sewage pipelines are damaged, buildings structures are compromised, and water supplies are prone to salinization (Nicolodi and Petermann 2010; Polette and Lins-de-Barros 2012).

Threats to beaches are diverse, complex, and have multiple sources and overlapping and synergic impacts (Chap. 9). Such a scenario demands an approach to beach management that is equally complex, diverse, and integrated to provide long-term sustainability for beach ecosystems.

10.3 Beach Management in Brazil

Brazil has 17 coastal states and 443 coastal municipalities,[6] of which 295 have maritime beaches.[7] The beach management system comprises a collection of policies and initiatives that will be as efficient as their capacity to integrate efforts toward beach sustainability. Brazilian law defines beaches as national heritage/patrimony (Federal Constitution/1988, Article 20 – Brasil 1988b) and assets for common use and of free access to the people (Brasil 1988a) that cannot be sold or transferred (Brasil 2002). As part of the coastal zone, beach management is under the umbrella of the integrated coastal management system coordinated by the Brazilian Ministry for Environment. It is also subject to the General Office of National Heritage/Patrimony of the Ministry of Planning, Budget, and Management as a national patrimony. As sites for other uses and activities, sectoral policies and regulations affect beach management (e.g., water resources, mining activities, or waste management). At the local level, the municipal ruling affects beaches, and social initiatives toward coastal and beach conservation are gaining momentum. This section presents some of the central Brazilian policies and initiatives related to beach management and case studies that illustrate their challenges (Fig. 10.2).

10.3.1 Brazilian Public Policies Directly Related to Beach Management

10.3.1.1 The National Coastal Management Plan

The National Coastal Management Plan (Plano Nacional de Gerenciamento Costeiro – PNGC, Federal Law 7.661/1988) (Brasil 1988a) is the core institution[8] for coastal zone[9] management in Brazil. It sets the foundation for an integrated management and establishes its objectives, principles, instruments, goals, and responsibilities. The Coastal Management Integration Group (Grupo de Integração do Gerenciamento Costeiro – GI-GERCO) coordinates the plan implementation at the federal level. It comprises representatives from federal ministries, research and

---

[6] Instituto Brasileiro de Geografia e Estatística, list of coastal municipalities: https://www.ibge.gov.br/geociencias/organizacao-do-territorio/estrutura-territorial/34330-municipios-costeiros.html?edicao=34336&t=o-que-e. (Accessed: 20.jul.2021)

[7] List by the General Office of National Heritage/Patrimony of the Ministry of Planning, Budget. Available at: https://www.gov.br/economia/pt-br/assuntos/patrimonio-da-uniao/destinacao-de-imoveis/gestao-de-praias/tagp-e-normativos/tagp-e-normativos. (Accessed: 20.jul.2021)

[8] Institutions are defined here as formal or informal rules, policies, norms, conventions, and codes used by society and government to organize and/or regulate their actions (Epstein et al. 2015).

[9] The Brazilian Coastal Zone is the area in the air-sea-land interface, including renewable and non-renewable resources. It comprises the geographical area of coastal municipalities (terrestrial part) and the Territorial Sea (12 nautical miles from the water baseline) (Brasil 2004).

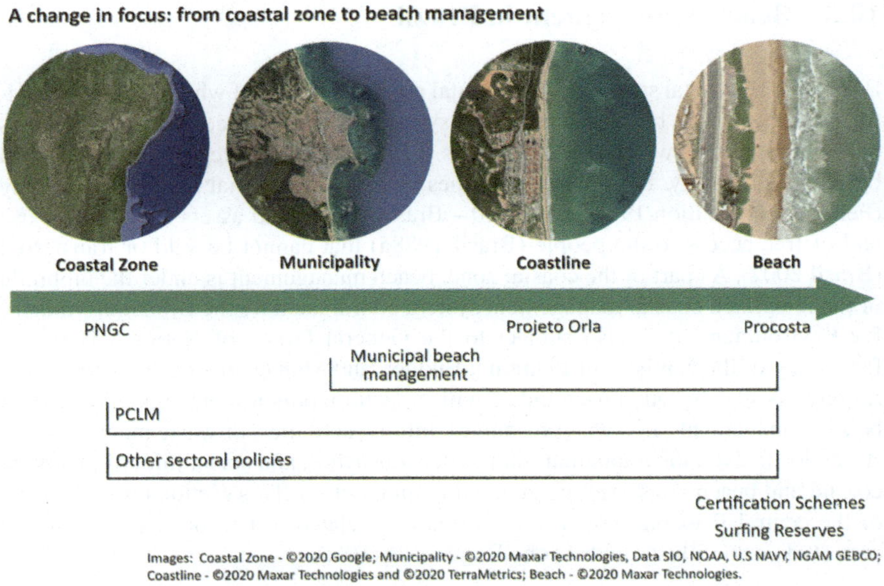

Fig. 10.2 Coastal and beach management approaches described in Sect. 10.3. Public policies: PNGC – National Coastal Management Plan (Plano Nacional de Gerenciamento Costeiro, Federal Law 7.661/1988); Projeto Orla – Project for the Integrated Management of the Coastline (Projeto de Gestão Integrada da Orla Marítima, Federal Decree 5.300/2004); Beach Municipalization Process (Federal Law 13.240/2015 and SPU Ordinance 113/2017); Procosta – National Program for Coastline Conservation (Programa Nacional de Conservação da Costa (Ordinance of the Ministry of Environment 76/2018); and PCLM – Plan to Combat Marine Litter (Plano de Combate ao Lixo no Mar – Brasil 2019b), and other sectoral policies that apply to the whole country; and social initiatives: Certification schemes and Surf Reserves. (*Source:* MP personal files)

education bodies, the private sector, environmental nongovernmental organizations, and municipal and state governments (Gonçalves et al. 2021). Its diverse and inclusive composition targets sectoral integration, decentralization, and democratization. GI-GERCO plans and implements coastal management through the Federal Action Plans for the Coastal Zone (Planos de Ação Federal para a Zona Costeira – PAF-ZC), a strategic instrument to define actions and priorities, foster shared responsibility, strengthen institutions, and promote capacity building and adaptive coastal management. PAF-ZC is periodically accessed and revised, contributing to an adaptive coastal management system. The most recent edition, the IV PAF-ZC,[10] includes actions related to beach management policies described in the following sections.

[10] For the versions of the PAF-ZC visit: https://www.mma.gov.br/biomas/amaz%C3%B4nia/item/8962-plano-de-a%C3%A7%C3%A3o-federal-para-a-zona-costeira-paf_zc.html (Accessed: 17.mar.2020).

Over 30 years, PNGC has increased the institutionalization and democratization of coastal management at the lower administrative levels (MMA 2015). However, some obstacles remain: PNGC instruments are unevenly implemented or have never been implemented, as in the Environmental Quality Report for the coastal zone.[11] While some states have advanced in implementing the Ecological-Economic Coastal Zoning, others are in very early phases (Nicolodi et al. 2018, 2021). Even when implemented, low social engagement and financial and power constraints challenge the effectiveness of management instruments and plans (Klumb-Oliveira and Souto 2015; MMA 2015; Santos et al. 2019).

Within GI-GERCO and PAF-ZC, divergent sectoral perceptions and interests challenge institutional articulation and collaboration leading to sectorized and contentious actions (MMA 2015; Santos et al. 2019). GI-GERCO's works with Sectoral integration was compromised by the Presidential Decree n° 9.759/2019 (Brazil 2019a), which extinguished the vast majority of decision-making or public participation forums. Therefore, GI-GERCO has been, since 2019, nonoperant (Gonçalves et al. 2021).

10.3.1.2 The Project for the Integrated Management of the Coastline

The Project for the Integrated Management of the Coastline (Projeto de Gestão Integrada da Orla Marítima – Projeto Orla – Federal Decree 5.300/2004 – Brasil 2004) is a planning instrument for integrating environmental, urban, and heritage policies; articulating national, state, and municipal governments with civil society; and promoting new mechanisms for social participation and control for coastal sustainability (MMA and MPO 2006). It also represents an effort toward beach management decentralization. Working at the coastline,[12] Projeto Orla directly affects beach management and connects a beach's submerged and emerged areas. The Ministry for Environment and the General Office of National Heritage/Patrimony jointly coordinate the Projeto Orla as one of the priorities for coastal management (see Goals 9 and 13 of the IV PAF-ZC).

Projeto Orla's implementation starts with the voluntary application of the coastal municipality and follows the phases of elaboration, approval, implementation, evaluation, and revision of the Coastline Integrated Management Plan, including the celebration of the necessary cooperation agreements between the federal and local government (MMA and MPO 2005). The plan builds on an assessment of the coastline to develop possible future scenarios; once a desirable scenario is chosen for

[11] Scattered information about the coastal zone is presented on the overall Brazilian Environmental Quality Report (Relatório de Qualidade do Meio Ambiente - RQMA) periodically published by the Brazilian Environment Ministry. RQMA available at: https://www.gov.br/ibama/pt-br/assuntos/noticias/2020/ibama-convida-especialistas-para-participar-da-elaboracao-do-rqma-brasil-2020/20201229EscopodoRQMA2020.pdf

[12] Coastline in Brazil ranges from 10 m deep (seaward) to 50 m (urbanized areas) or 200 m (non-urbanized areas) inland, measured from the mean high tide or coastal ecosystem edge (Brasil 2004).

each coastline stretch, the municipality outlines the strategies for implementing, monitoring, evaluating, and performing necessary changes to the identified solutions to local conflicts (MMA and MPO 2006).

Projeto Orla implementation is a multilevel collaboration. The Coastline National Coordination in the Ministry of Environment and the Coastline State Committees assist municipal governments by providing technical support, monitoring the process, and analyzing the Coastline Integrated Management Plans. At the municipal level, governmental actors work with local stakeholders to elaborate the plan and establish the local Coastline Management Committee (the local participatory group responsible for implementing and coordinating the plan), which must be validated by local society in a public hearing and later formalized by a municipal regulatory act.

Although the Projeto Orla has been the leading national policy for beach management for over 20 years (Cristiano et al. 2018), its implementation is relatively poor: few State Committees are formalized (MMA and MPO 2008), few municipalities have implemented the Coastline Integrated Management Plans (MMA and MPO 2008; Oliveira and Nicolodi 2012), and greater social involvement is required (Oliveira and Nicolodi 2012; Silva and Soriano-Serra 2016). The lack of financial resources, trained personnel, and incentives for extensive development of the Projeto Orla through the Brazilian coast represent significant implementation challenges (Oliveira and Nicolodi 2012; Scherer 2013; Klumb-Oliveira and Souto 2015). Additionally, a more robust institutional basis is needed to force and foster local governments to go through with the process and overcome the lack of institutionalization at the local level that tends to limit processes to the elaboration phase (see Case Study 10.1; Scherer 2013).

Case Study 10.1 – Projeto Orla in Itapema, Santa Catarina

The city of Itapema, in the state of Santa Catarina, started a process to implement Projeto Orla in 2003 and has since then revised, expanded, and updated it.

The first Coastline Integrated Management Plan (Plano de Gestão Integrada da Orla, PGI), proposed for the "Canto da Praia" region in 2003, was overruled for failing to involve social actors (Pinto 2014). In 2008, after two years of participatory workshops and meetings, a new PGI was elaborated, including the central area of the municipality, the Parque Linear Calçadão (Silva and Soriano-Serra 2016), and the Orla Management Committee was created (but not institutionalized). Unfortunately, this PGI was also not implemented.

From 2009 to 2010 and in 2012, meetings were held to implement the Orla Management Committee and revise the PGI. A new version was elaborated for the whole coast of Itapema. This new process' implementation was hampered by social demobilization related to the noninstitutionalization of the Municipal Committee and by changes in municipal government, which paralyzed the process (Silva and Soriano-Serra 2016).

(continued)

Case Study 10.1 (continued)
In 2019, the PGI underwent a new revision within the municipalization of beach management. The municipal government has approved the new PGI. However, it still needs to be approved by the State and National Orla Committees (Itapema 2019). Up to July 2022, the Orla Management Committee was not created (personal information).

Through the years, the Projeto Orla in Itapema has faced several challenges that hampered its implementation, from the lack of social participation in the first PGI (in 2003) to the lack of institutional measures for implementing the municipal Orla Management Committee and the PGI (2007–2012). Affected by limited resources and governmental changes, Project Orla has not yet completed an entire implementation cycle, after which other challenges related to monitoring and adapting the proposed actions are expected.

10.3.1.3 The Second Wave of the Beach Management Municipalization Process

The municipalization process is based on the transfer of the right and responsibility for oceanic (not continental freshwater) beaches management from the federal to the municipal government (Federal Law 13.240/2015 – Brasil 2015; SPU Ordinance 113/2017 – Brasil 2017). Given the limited success of the Projeto Orla toward municipalization, an additional effort was promoted by the Federal Ordinance 113/2017 (Brasil 2017), which formalized and regulated the transfer of beach management to municipalities. The process is coordinated by the General Office of National Heritage/Patrimony in collaboration with the Ministry for Environment, academia, municipalities, and civil society as part of the IV PAF-ZC (Goal 13).

The municipalization promotes an unprecedented focus on beach management. It offers financial incentives by granting municipalities the rights to revenues from sanctions and the onerous uses of beaches and federal properties under municipal administration. As part of the process, municipal governments must implement/review Projeto Orla, which is expected to trigger a new and more stable cycle for this project, consequently fostering social participation and control in beach management.

Most challenges faced by the municipalization processes are the same as for Projeto Orla, but the process is gradually being structured as it gains momentum. Since 2017, 139 municipal governments have requested and 75 have obtained management rights over their oceanic beaches.[13] A series of training workshops have

[13] Official data from MPOG/SPU, last updated on 27/06/2022. Available at: https://www.gov.br/economia/pt-br/assuntos/patrimonio-da-uniao/destinacao-de-imoveis/gestao-de-praias/56-planilha-municipios-tagp-27-06-2022.pdf. (Accessed: 20.jul.2022)

been organized (through 2018–2019) to clarify the rights and responsibilities entailed by the municipalization and the role of each partner in the process, bringing different government levels close to each other. The official yearly report template was disclosed in January 2019, and – although many municipalities are behind scheduled to deliver it[14] – delivered reports are expected to monitor the process and identify bottlenecks. Some municipalities have already started implementation and/or revision of the Projeto Orla (Itapema 2019).

10.3.1.4 The National Program for Coastline Conservation

The National Program for Coastline Conservation (Programa Nacional para Conservação da Linha de Costa – Procosta, Ordinance of the Ministry of Environment 76/2018) (Brasil 2018a) is a permanent program for territorial planning and management that aims to foster adaptation measures based on quality information to minimize climate change impacts and preserve coastal ecosystem (MMA 2017). It is an instrument of the Brazilian National Plan for Adaptation to Climate Change (Plano Nacional de Adaptação à Mudança do Clima – PNA) (MMA 2016) that operationalizes the plan at different levels (Xavier et al. 2022). The program is a priority action (Goal 5 from the IV PAF-ZC) to promote climate change adaptation in Brazil, stated as a voluntary commitment[15] in the 2017 United Nations World Ocean Conference to comply with SDG 14 of the 2030 Agenda.

The Ministry for Environment planned to coordinate Procosta through a working group within the Coastal Management Integration Group. The Procosta working group integrates different federal agencies (including the ministries, secretariats, and the navy), states and municipal governments, civil society (including the private sector, nongovernmental organizations, and sectoral representatives), and research and data management organizations. Municipalities and research organizations were considered a fundamental part of the program, acting as partners for developing instruments and procedures and testing them in situ.

Procosta is structured by four strategic subprojects with specific goals, as follows:

- *Alt-Bat*: define a reference level for the measurement and representation of terrestrial (altimetry – Alt) and submarine (bathymetry – Bat) landscapes. From this, the project was planned to establish the exact limit of the country's coastline, the first step for characterizing maritime land[16] and monitoring the coastline and sea-level rise.

---

[14] Official data from MPOG/SPU, last updated on 27/06/2022. Available at: https://www.gov.br/economia/pt-br/assuntos/patrimonio-da-uniao/destinacao-de-imoveis/gestao-de-praias/extratos-dos-termos-de-adesao. (Accessed 20.jul.2022)

[15] OCEAN ACTION 19679. UN Website. Available at: https://oceanconference.un.org/commitments/?id=19679. (Access 15 Mar. 2020).

[16] According to PNGC, maritime lands are federal lands, located between the imaginary line of mean tides and 33 meters inland.

- *Projection of future coastlines and identification of hazards:* apply computer modeling to predict changes to the Brazilian coastline under different climate change scenarios.
- *Coastal risks and adaptation strategies*: assess potential socio-environmental and economic risks for the coastal zone and propose strategies for preventing and adapting to these risks.
- *Monitoring and management for the conservation of the coastline*: support long-term monitoring and management of the coastline and foster a procedural culture that prioritizes safeguard measures based on scientific data and uncertainties.

Procosta is being implemented along the Brazilian coast at a slow pace. Researchers have been developing methods to establish the reference levels for altimetry and bathymetry (Quetzalcóatl et al. 2019) and testing modeling software to predict future coastline impacts (Ohz et al. 2020). Pilot studies mapped coastal attributes as geodiversity, morphodynamic, and coastal development risks for specific sites in the states of Rio de Janeiro, Pará, Espírito Santo, Pernambuco, and Maranhão (Maia 2019). Two guiding publications were launched: the "Panorama of Coastal Erosion in Brazil" (Brasil 2018b), which updates the national data on erosive processes and applied solutions to control and prevent beach erosion; the "Guidelines for Prevention and Protection of Coastal Erosion" (Brasil 2018c), which provides guidelines and criteria for coastal interventions based on prevalent coastal process, indicating where intervention is more likely to bring positive results (e.g., beach nourishment, coastal protection interventions).

Procosta, launched in 2018, was affected by changes in federal government orientation and priority. Research projects are responsible for most of the reported advances, carried out as local and regional study cases in specific sites. The historical and ongoing budget cuts to research (Angelo 2019) might jeopardize Procosta's implementation and the country's capability to adapt to climate change. A strategy to foster and broaden its implementation to all Brazilian states and municipalities is necessary and must consider ensuring qualified personnel, technical and financial resources, incentives and support from the federal government, and technology and information transfer among states and municipalities.

10.3.1.5 The Plan to Combat Marine Litter

The Brazilian Plan to Combat Marine Litter (Plano de Combate ao Lixo no Mar – PCLM) (Brasil 2019b), coordinated by the Ministry for Environment, aims to reduce the amount and impact of marine litter[17] from multiple sources; to propel scientific research on new methods and technology to combat marine litter, and to inform and engage decision-makers and citizens in the process. It is a voluntary

[17] Marine litter is defined as solid waste from anthropic activities that reach the coastal and marine environment (Brasil 2019b).

commitment[18] stated in the 2017 United Nations World Ocean Conference to comply with the 2030 Agenda and a goal of the IV PAF-ZC (Goal 1).

The plan assesses marine litter in Brazil and states objectives, guidelines, and short-, middle-, and long-term actions to be implemented at multiple administrative levels and by different governmental and social sectors. The elaboration of the plan was initiated as a collaborative and participatory process led by academia. However, the participatory group leading the discussion was dissolved in early 2019, and the plan was finished and launched by the Ministry for Environment with little engagement from other sectors.

Reports of its implementation convey that the PCLM has been reduced to timely and dispersed cleanups, the only result advertised on the Plan's official webpage.[19] Although such campaigns are relevant to raise awareness of the impacts of marine litter in coastal and marine ecosystems (Wyles et al. 2017), they are not efficient for combating marine litter (Loizidou et al. 2018). The structural actions that should support the whole plan – such as defining and applying the methods and strategies to the national marine litter assessment and creating an online portal to register and advertise best practices to combat marine litter – are still on hold. Thus, PLMC implementation takes a dubious path and a slow pace.

10.3.2 Sectoral Public Policies That Impact Beach Management

Beaches interact with terrestrial, coastal, and marine ecosystems by fluxes of people, energy, species, sand, water, nutrients, pollutants, and other substances (Schlacher and Connolly 2009). Any activity impacting these fluxes affects beaches' characteristics, processes, and ecosystem services (De Andrés et al. 2018). Although the sectoral policies regulating these activities do not target beach conservation (see Table 10.2 for some examples), they affect it (Case study 10.2). This institutional interplay amplifies the number of policies that should be considered for beach management under a holistic and integrated approach.

[18] OCEAN ACTION 19679. UN Website. Available at: https://oceanconference.un.org/commitments/?id=19679. (Access 15 Mar. 2020).

[19] Due to the COVID Pandemic, activities related to the PCLM were interrupted and the results reported are related to cleanups. Data available at: https://app.powerbi.com/view?r=eyJrIjoiMjYyMzY0NTMtMTI3Zi00NmZkLWI2YzgtMmQ0NmJhNDI3YTI1IiwidCI6IjM5NTdhMzY3LTZkMzgtNGMxZi1hNGJhLTMzZThmM2M1NTBlNyJ9&pageName=ReportSection3fe695e367970de07103 (Access 20.jul.2022)

Case Study 10.2 – Licensing and Beach Sustainability

A poorly executed licensing process, including post-licensing monitoring and action, ensues drastic consequences, as exemplified by the tragedy of the collapse of the Samarco iron mining waste dam in the state of Minas Gerais, southeastern Brazil, in November 2015. Circa 60 million m^3 of tailings polluted more than 600 km of the Rio Doce, the second-longest river in the Southwestern Atlantic, and reached the coast spreading over more than 770 km^2 (Fig. 10.3).Case Study 10.2 (continued)

Around 170 km of beaches were contaminated (Do Carmo et al. 2017), including protected areas (Fioravanti 2016). Pollutants affected water quality and led to the closure of beaches for bath, surfing, and fishing. Restrictions to beach access resulted in economic loss for the tourism sector, local trade, and traditional fishing communities (ANA 2016; Creado and Helmeric 2018).

The environmental impact assessment of the licensing process is a tool to promote the conservation of natural ecosystems that anticipates risks and impacts and the planning of mitigation and compensation measures. The environmental impact assessment for the Samarco dam presented many technical problems that precluded preventing the collapse and exacerbated environmental and social impacts (Wanderley et al. 2016). Despite having options where the risks and impacts would be lower (i.e., not upstream from the urban area), locational choice focused on minimizing company costs. No alternative technological option was presented, even though it existed. The post-licensing monitoring did not trigger actions to prevent the episode (Fonseca et al. 2017).

This tragedy shows that challenges reside in the enforcement of current legislation. The company used inadequate methods for monitoring impacts, presenting documents in which several relevant physical and chemical parameters were not reported (Do Carmo et al. 2017). The total impact of this tragedy is yet to be accounted for, particularly for beach biodiversity and ecosystem services. However, records of defaunated beaches in areas close to the mouth of Rio Doce evidence the acute effects of the contaminated tailing on the beach biodiversity (Ekos Brasil in press).

Table 10.2 Examples of sectoral policies that can affect beaches and their management

Sectoral legislation	Relation to the beach
Mining Code – Decree-law 227/1967	Creates the regulatory framework for mining activities and establishes the competencies for their management, including deep-sea mining and oil extraction.
National Environmental Policy – Law 6.938/1981	Establishes the environmental policy system and its main instruments (e.g., quality reports, zoning, activities licensing, and impact assessment). Also creates the participatory National Environmental Council that nominates the civil society representative at the Coastal Management Integration Group.

(continued)

Table 10.2 (continued)

Sectoral legislation	Relation to the beach
National Agricultural Policy – Law 8.171/1991	Establishes the framework for agricultural development, including norms for chemical discharges.
National Maritime Policy – Law 1.265/1994	Guides the development of maritime activities, including sectoral plans, ocean literacy programs, international maritime affairs, scientific and technological research, and regulation of economic activities. A ramification of the policy, the National Maritime Resources Plan is currently in its tenth version (Federal Decree 10.544/2020).
National Water Resources Policy – Law 9.433/1997	Establishes the main instruments for water resources management and creates the National Water Resources Management System and the associated governance structure.
National Energy Policy – Law 9.478/1997	Regulates the development of the energy sector, including norms for dams, offshore drilling (oil), and the expansion of renewable energies in the coastal zone.
Protected Areas National System – Law 9.985/2000	Creates the national protected area system and defines its categories and the procedures to create and implement protected areas, including in the coastal and marine zones.
City Statute – Law 10.406/2002	Establishes the urban development policy and creates the local master plan, which is the main regulatory instrument for land-use regulation in urban areas.
National Policy for Sea Resources – Decree 5.377/2005	Regulates, guides, coordinates, and controls marine scientific research, the preservation and conservation of the marine environment, and the exploration and sustainable use of living, mineral, and energy resources in the areas under national jurisdiction and the continental shelf.
National Basic Sanitation Policy – Law 1.445/2007	Establishes the set of services, infrastructure, and water supply facilities, sanitary sewage and urban cleaning, solid waste management, and rainwater drainage.
National Tourism Policy – Law 11.771/2008	Establishes the guidelines and goals for tourism development, including decentralized management aiming to expand tourism to different regions, such as the beaches.
National Policy for the Sustainable Development of Aquaculture and Fisheries – Law 11.959/2009	Regulates the fisheries and aquaculture policies and establishes competencies for regulating fishery activities.
National Policy on Solid Waste – Law 12.305/2010	Defines shared cross-scale responsibility of waste generators: manufacturers, importers, distributors, traders, and citizens. Since most of the litter found at sea comes from land-based activities, coordination among them and the Plan to Combat Marine Litter is a prerequisite to successfully implementing the latter.
National Transport Policy – Law 12.587/2012	Regulates transport system expansion and establishes the competencies for regulating its modes, such as roads, trails, and maritime shipping.
National Policy of Protection and Civil Defense – Law 12.608/2012	Establishes norms for prevention, mitigation, preparedness, response, and recovery of natural disasters, including coastal areas.
Port Law – Law 12.815/2013	Creates the regulatory framework for port expansion, establishes the competencies for port management, and the guidelines to start a new enterprise.

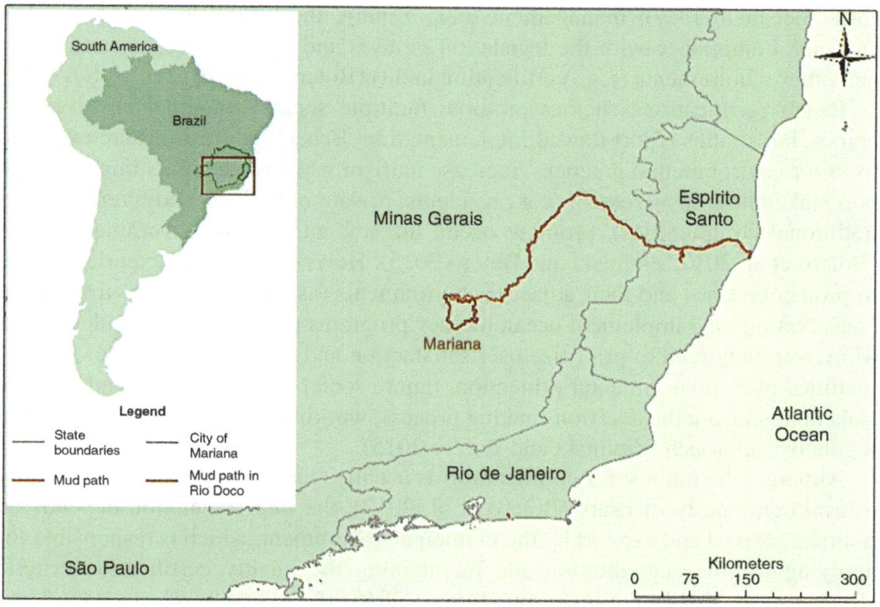

Fig. 10.3 The destruction route: mud path from the rupture of Samarco's dam in Mariana municipality, state of Minas Gerais, down to Rio Doce and the Atlantic Ocean. (*Source:* Elaborated by the authors)

10.3.3 Other Management Initiatives

In addition to the broader federal and sectoral institutions for beach management, local municipal regulations also concern beach uses and activities. For example, in Recife (Pernambuco), Law N° 9837/1967 prohibits soccer practices on beaches during weekends (Recife 1967). In Caraguatatuba (São Paulo), Municipal Law 1.298/2006 (Caraguatatuba 2006) prohibits the presence of domestic animals on the beach. In São Sebastião (São Paulo), Municipal Decree 8.087/2021 (São Sebastião 2021) rules over urban noise from sound equipment on roads, squares, beaches, and other public places. In Balneário Camboriú (Santa Catarina), municipal Law 4.609/2022 (Balneário Camboriú 2022) makes it mandatory for the municipal authority to promote access to beaches for people with physical disabilities and/or limited movement. These are some examples of the diversity of municipal institutions that affect beach management.

Still, at the local level, beach certification schemes are labels granted to beaches, usually touristic ones, for achieving a management status that is supposed to guarantee environmental, social, and economic sustainability. Beach certifications assess and monitor beaches considering a set of criteria: environmental (e.g., water quality, landscape); services and security (e.g., installation of public use equipment, free access, cleaning, lifeguards); education/information (e.g., code of conduct/practice, foster ocean literacy); management (e.g., zoning, the establishment of carrying capacity, compliance with the legislation on uses and activities in the sand strip); and other requirements (e.g., certification audit) (Botero et al. 2012).

Beach certification schemes promote multiple social-ecological positive outcomes, but studies report flawed implementation. Beach certifications are expected to foster environmental practices, increase tourism while reducing its impacts, support stakeholder involvement (e.g., residents, private initiatives, entrepreneurs, and traditional communities), promote ocean literacy, and ensure government action (Botero et al. 2012; Zielinski and Botero 2015). However, reported experiences fail to protect beaches and their adjacent environments (as expected for environmental conservation) and implement ocean literacy programs (Willian and Micallef 2009). Moreover, they tend to prioritize user satisfaction and the improvement of tourism facilities over environmental protection, ignore social-cultural issues, and exclude stakeholders from the decision-making process, working, in practice, as a top-down regulatory approach (Zielinski and Botero 2015).

Although the interest for certifications is usually stirred at the local level, i.e., by a local community of users (Botero et al. 2012), the implementation depends on political interest and support by the municipal government, which is responsible for applying for the accreditation and maintaining the quality certification criteria (Botero et al. 2012; Zielinski and Botero 2015). Certification maintenance and implementation are time and resource-consuming since they demand environmental studies, installing necessary infrastructure, and professional expertise for monitoring, thus requiring political-institutional support, investments, and external assistance (Bernardi and Pires 2015; Zielinski and Botero 2015). A broader understanding of how to apply the different beach awards to specific governance contexts is also required (Botero et al. 2012; Bombana and Polette 2013).

Up to 2020, the international Blue Flag program was the only certification scheme implemented on Brazilian beaches, although implementing other schemes is being prospected (see Case Study 10.3). The Blue Flag program aims to promote sustainability by ensuring long-term high standards and should strengthen the competitive capacity of a beach by boosting its image as a tourism destination and increasing tourists' and residents' quality of life (FEE 2006). The International Foundation for Environmental Education coordinates the program, which has representatives from around the world.

In Brazil, the Blue Flag label was granted to 22 beaches by 2022 (Blue Flag 2022), and its implementation and maintenance are challenged by inadequate basic sanitation; illegal occupation of the coastline; inefficient organizational structure; seasonal variation of beach users; jurisdictional conflicts for beach management; little involvement of beach users; faulty ocean literacy programs; and inadequate inspection (Neto et al. 2011; Bombana and Polette 2013; Scherer 2013). Furthermore, aside from not consistently delivering the expected tourist boost (Mir-Gual et al. 2015), researchers criticize the Blue Flag program for having a narrow and static view of beach ecosystems (Bombana and Polette 2013).

Besides certification schemes, other society-led initiatives emerge as approaches to beach conservation. For example, Surfing Reserves (see Case Study 10.4) represent a worldwide strategy to complement ongoing conservation initiatives, preserve waves and adjacent areas, connect people, and promote environmental management. They are comparable to protected areas as coastal management instruments, reinforcing the existing legal and formal provisions but not necessarily demanding a legal or formal institutional process.

Case Study 10.3 – ISO 13.009:2015

The International Organization for Standardization (ISO) is the leading global standards network. ISO is recognized worldwide for developing standards to endorse innovation and solve global challenges. The standard ISO 13.009:2015 was the only ISO created to promote sustainable beach management as a tool for disseminating and boosting the local economy and tourism.

ISO 13.009:2015 establishes requirements and recommendations to promote environmental protection, sustainable tourism infrastructure and services, security, information, and communication and aims to help beach managers to foster sustainable tourism activities. To be granted the certification, the degradation process of beaches with tourist potential needs to be pointed out, corrected, and continuously monitored to preserve beach quality.

Guarapari, in the Brazilian state of Espírito Santo, is widely known for its beaches' tourism activities, especially medicinal sand (monazite sand). The most famous beach in this municipality is the Areia Preta beach, which receives significant tourists annually. However, increased tourism on this beach led to the impacts caused by uncontrolled growth, which affected the beach's environmental quality, which is relevant to support tourist activities.

ISO 13.009:2015 requires an efficient, inclusive, and collaborative governance system, as well as long-term planning, that guarantees the maintenance of the tourism infrastructure and beach quality status through consulting and informing stakeholders. A preliminary study to implement ISO 13.009:2015 in Areia Preta beach (Felippe 2017) pointed out that establishing such a governance system might be an obstacle. The government sector tends to centralize decision-making and limits the stakeholders' participation. Problems related to overlapped responsibility and miscommunication among federal, state, and municipal governments make the decision-making process inefficient. Governmental discontinuity forfeits long-term beach management activities, resulting in limited time and budget for creating and developing beach management plans.

Case Study 10.4 – Surfing Reserves – A New Approach to Coastal Management

Surfing Reserves are expected to contribute to coastal protection by discussing and implementing rules for beach use, maintaining social, cultural, and economic attributes (e.g., by protecting a river), and preserving and supporting essential ecosystem services (Silva et al. 2016; Figueiredo and Almeida 2019). In addition, Surfing Reserves have the potential to mobilize and build capacity for local communities to engage in the implementation of other formal and informal institutional arrangements, such as protected areas and coastal management plans.

The Surfing Reserve title is granted by the NGO Save the Waves Coalition. It involves a process of engaging nongovernmental and governmental actors in the environmental preservation of a territory (Silva et al. 2016). Implementing International Surfing Reserves has been carried out as a bottom-up process without a formal institutional arrangement.

Up to 2019, there were 11 Surfing Reserves worldwide (Figueiredo and Almeida 2019). Guarda do Embaú beach (state of Santa Catarina) was granted the "International Surfing Reserve" title in 2019. The process occurred amid numerous discussions with the local community, entrepreneurs, and surfers. This led to strong community engagement, later relying on the support of the local government. The environmental quality of the beach has been monitored in partnership with local authorities and private organizations.

Meanwhile, the nongovernmental organizations Aprender Ecologia Institute and Ecosurf Institute, the leaders of this movement in Brazil, have been discussing the topic in meetings and forums, working to create the Brazilian National Program of Surfing Reserves.

The challenges for implementing Surfing Reserves in Brazil reside in the lack of community buy-in, local authorities' political will, and the need to coordinate action to reinforce existing arrangements without overlapping efforts. It is also necessary to define measurable criteria to guide the implementation and evaluation process (Figueiredo and Almeida 2019).

10.4 Challenges for Beach Management

Understanding the flaws and challenges to the application of beach management in Brazil is the first step to improving the management system. Based on the analysis of the management approaches and case studies reported in this chapter, we highlight six main challenges:

10.4.1 Promote Multilevel Governance

Multilevel governance (i.e., decision- and policy-making that involve multiple actors and take place across multiple jurisdictions and sectors) represents an excellent approach to dealing with the multidimensionality of environmental issues (Termeer et al. 2010), but it is often challenging in complex political-institutional environments (as reported for the National Coastal Management Plan, the Projeto Orla, ISO 13.009: 2015). In Brazil, the 1988 National Constitution (Brazil 1988b) favors administrative decentralization and establishes a multilevel governance system, but practice shows that there is still a long way ahead toward such principles.

Many strategies exist in the coastal and beach management system to foster multi- and cross-level governance. The federal Coastal Management Integration Group includes multiple organizations at all administrative levels (e.g., representative of the subgroup that integrates the 17 coastal states, G17). The National Coastal Management Plan sets specific instruments for coastal management at each administrative level that should be integrated across levels (i.e., state and local plans should be implemented at each coastal state/municipality following federal directives). The Projeto Orla establishes committees at the state and federal levels to support municipal action.

Multilevel integration and support for local government are essential to implementing top-down policies. The poor implementation of the National Coastal Management Plan instruments at the state and municipal level evidence an implementation gap that must be overcome. Beach management is mainly exercised at the local (municipal) and sublocal (community/specific beaches) administrative levels (Botero and Diaz 2009; Azuz-Adeath et al. 2018), where capacity building is necessary.

Multilevel integration is also necessary for bottom-up approaches. Beach certification schemes require the local government to apply for the certification. Although Surfing Reserves can be implemented without local government involvement, the integration between municipal administration and the local community is highly recommended (Figueiredo and Almeida 2019). Multilevel governance is crucial to reach the level of coordinated action necessary for beach management, either for implementing federal policies at the local level or for supporting local actions toward beach conservation.

10.4.2 Foster Policy Integration

In Brazil, administrative decentralization was accompanied by a highly sectoral political and administrative framework, leading to conflicts among instruments from different public policies (Abrucio and Sydow 2018). Policy integration is a goal of the National Coastal Management Plan and related instruments, as exemplified by the Federal Action Plan for the Coastal Zone (PAF-ZC), which focuses on implementing and articulating coastal management instruments among each other and with other policies. For instance, Goals 9 (Integrate and articulate the Projeto Orla with the Municipal Master Plans) and 13 (Evaluate the effective actions of the Orla Management Committees and Coastline Integrated Coastal Management Plan) of the IV PAF-ZC (2017–2019) are related to the process of municipalization and evidence the federal focus on beach management for the period.

To foster sectoral integration, policies need common and integrated goals and guidelines. Unfortunately, sectoral policies tend to target economic development and profit maximization over environmental protection (Lafferty and Hovden 2003), and their (mis)application to the coastal zone can impair coastal management. The Samarco tragedy (Case study 10.2) is one of many examples. Furthermore, different sectoral agencies tend to work together only when formal procedures demand, as for Environmental Impact Assessment approvals. Even then, collaboration is procedural and fails to promote and maintain integration (De Oliveira 2002). In this regard, a common goal toward environmental protection can be the starting point for connecting different instruments, sectoral policies, and agencies (De Oliveira 2002). This is even more relevant for beaches where sectoral activities, such as recreational and tourism, depend on a healthy ecosystem.

10.4.3 Implementation of Existing Policies and Instruments

Although Brazil has a consistent legal and institutional framework for coastal management, long-standing legislative instruments are poorly implemented (e.g., National Coastal Management Plan, Projeto Orla), and recent ones show a low implementation rate, if any (e.g., Municipalization, Procosta, Plan to Combat Marine Litter). Coastal management in Brazil is an inherent testing task due to the continental dimension of the Brazilian coast; the poor integration among governmental agencies and policies increases the challenges. The coastal management discrepancies among states (Nicolodi et al. 2018, 2021) evidence the existing discrepancies throughout the country and represent an opportunity for sharing experiences and best practices to guide implementation at the national level through collaboration among state/municipal governments, society, and the private sector.

10.4.4 Promote Social Participation

The active involvement of social actors is based on the premise that those affected by decision-making should have the opportunity to influence it. In Brazil, public participation is considered a democratic right (Brazil 1988b) and is both a requisite and a challenge for the beach management approaches and case studies presented here. Since beach management has a solid local component (Botero and Diaz 2009; Azuz-Adeath et al. 2018), engaging local stakeholders may be crucial to its implementation. The multilevel governance demonstrates the growing fluidity of political power (Abrucio and Sydow 2018) and the possibility of involving different actors, mainly at the local level, but it is still challenging to reach a high level of engagement for local authorities and society.

Public participation in environmental management is expected to promote better informed, context-adapted measures (Seixas et al. 2019), but its outcomes depend on how actors are involved and processes are conducted (Xavier et al. 2019). There are many obstacles to social participation in management processes, such as coping with the lack of representativeness, overcoming language barriers, sharing knowledge and power among participants, and empowering and engaging individuals, especially the poorest and marginalized ones (Seixas et al. 2019). Furthermore, in contexts of solid structural inequalities, additional attention to process development must be given to prevent participatory processes from reproducing inequalities and fostering marginalization processes (de Oliveira 2002).

10.4.5 Guarantee the Necessary Resources

Limited resources often result in policy failures and environmental degradation (Klumb-Oliveira and Souto 2015; MMA 2015; Santos et al. 2019). Insufficient financial resources hamper implementing and maintaining policies, programs, and projects. When funding is not an obstacle, a lack of personnel, expertise, guidelines, and general operational infrastructure (i.e., poor political-institutional capacity) challenge management (e.g., Projeto Orla, Municipalization, Procosta, Blue Flag, Surfing Reserves).

Coastal and beach management depends on governmental priorities, and few coastal states and municipalities put aside the necessary budget for their implementation. Brazilian law indicates possible sources of public funding for coastal management but does not regulate the inclusion of implementation costs in government budgets at any level (Klumb-Oliveira and Souto 2015), which also affects beach governance. While some actions are supported by mid-term budget planning (Santos et al. 2020) that should be less affected by changes in governments, it is urgent to develop alternatives for raising funds, especially at the municipal level, which tends to be the one with fewer resources and political-institutional capacity to implement beach management actions (Oliveira and Nicolodi 2012).

One alternative for raising funds is fostering intersectoral articulation (de Oliveira 2002; Santos et al. 2019) with more substantial institutional support and governance structure. Previous attempts to align Projeto Orla with other sectoral agendas (e.g., Oil and gas, Tourism, and City Planning) have failed (Oliveira and Nicolodi 2012), indicating that besides alignments, binding regulations should be implemented. Since 2017, revenues related to municipalization can fund the Projeto Orla. However, although revenues from sanctions are to be applied in the areas affected by the infraction (the coastline), the application of revenues related to public space/properties to foster beach management is not legally binding. Therefore, the decision shall be discussed and articulated in each municipality. Considering the expected increase in the implementation of Projeto Orla in the next few years, investing in capacity building to coordinate the process and elaborate the Integrated Coastline Management Plan and its implementation, monitoring, and evaluation guidelines is an urgent demand.

10.4.6 Overcome Processes Discontinuity

When management initiatives are designed to be implemented during single mandates, public policy discontinuity is frequently the rule rather than the exception. This is the case with many of the approaches to beach management in Brazil, from Projeto Orla to the implementation of Beach Certification Schemes. Some proposals, as discussed here, barely overcame the planning phase before being gradually discontinued. The discontinuity of processes derives from personnel changes (i.e., new elections), lack of formalization and institutionalization (e.g., Projeto Orla, Blue Flag), and low political-institutional support and buy-in (e.g., Blue Flag, Surfing Reserve).

Whenever a process is stopped before its implementation, a loss of financial resources, time, and social energy is observed (Pereira and Asmus 2013; Silva and Soriano-Serra 2016). Additionally, personnel changes affect political-institutional memory (Pereira and Asmus 2013; Seixas et al. 2019) and impact future projects. To overcome such discontinuity, democratization, and social engagement are strategies (Vianna and Xavier 2014). An articulated, engaged, informed, and empowered civil society can withhold the mechanisms for beach management regardless of political-institutional changes and contradictions (Azuz-Adeath et al. 2018).

10.5 Moving Forward: A New Beach Management Approach

The challenges explored in the previous section show that, although Brazil has regulations that can foster beach sustainability, a stronger governance system is necessary to advance beach management. Such a governance system should be more inclusive of different stakeholders and based on long-term planning for its implementation, monitoring, and evaluation to adapt to a changing environment. Additionally, research shows the need for better integration of ecological data to move the focus of beach management from managing human activities to a more comprehensive approach that can promote a healthy and productive beach ecosystem. As expressed by Sardá et al. (2015), we need a new integrated beach management system. In this section, we discuss some paths for building such a system.

Beach management can be described as the management of human activities that affect this environment by changing fluxes among beaches and their surrounding ecosystems (Williams and Micallef 2009). Through the years, beach management approaches have been shifting their focus from cultural and regulating services (i.e., tourism and recreation, erosion and disturbance protection) to a broader understanding of beaches as multidimensional, interconnected, and complex systems (McLachlan et al. 2013; Sardá et al. 2015). In a nutshell, beaches are social-ecological systems, i.e., coupled, coevolving, and dynamic human–nature systems, with reciprocal and interdependent feedback (McLeod and Leslie 2009). To be more effective in maintaining the beach's multiple functions, management must consider biophysical characteristics (structure and functioning) and the influence that social and political dimensions exert on the natural environment.

In this regard, ecosystem-based management (EBM) is a promising approach to beach management (Sardá et al. 2015; Corrêa et al. 2021). Beach management demands institutional and governance approaches that are more holistic, integrative, and participatory (McLachlan et al. 2013; Sardá et al. 2015), capable of delivering innovative solutions that involve articulation and cooperation among social actors and considering different spatial, temporal, and jurisdictional scales. EBM entails a holistic view that encompasses beach complexity by maintaining both environmental integrity and socioeconomic prosperity, offering new opportunities for beach sustainability (Long et al. 2015; Sardá et al. 2015). This approach seeks to overcome sectoral fragmentation and acknowledges the need for social participation and long-term adaptive planning based on the use of multiple knowledge systems (Long et al. 2015).

EBM results in positive outcomes for environmental management, but its implementation in real-world contexts is still embryonic (Leslie et al. 2015; Sardá et al. 2015). For Brazil, although some instruments, such as PNA and Pro-Costa, are based on EBM principles, operational guidelines are still required to translate such principles into sustainable management actions (Xavier et al. 2022). EBM can be the guideline for instituting a comprehensive beach management system that complements existing practices and overcomes recurrent obstacles (see above) through application, evaluation, and adaptation (Leslie and McLeod 2007; Asmus et al. 2018). EBM can foster ongoing processes, adapting and qualifying them. For this, attention should be given to some strategies that may facilitate this transition, which are presented and discussed below.

10.5.1 Develop a New Approach to Science and Policy

As a holistic approach, EBM relies on sound multidisciplinary and integrated information and faces the obstacles of lack of quality information to support actions (Leslie and McLeod 2007; Asmus et al. 2018). This challenges the scientific community to innovate in producing integrated and contextualized information. Managing beaches as social-ecological systems demand studies from many disciplines (e.g., ecology, geophysics, oceanography, and sociology) and information delivered in an integrated and context-oriented approach. Although many biological/ecological or geophysical studies exist, interdisciplinary and social-political studies are scarce (Amaral et al. 2016; Nel et al. 2014; Xavier et al. 2020) and must be boosted. Moreover, fewer studies analyze the needs of management processes (Nel et al. 2014; Xavier et al. 2020), which hamper the application of scientific information in decision-making (Arkema et al. 2006).

Besides new scientific approaches, a new science–policy relationship is needed. For this, involving scientists in the discussion of complex, risky, and uncertain management processes fosters social participation and leads to positive outcomes for sustainability, equity, knowledge sharing and production (Leslie et al. 2015; Jacobi et al. 2019). Focusing on problem-solving and involving scientists and nonscientists in knowledge production and discussion, approaches such as Post-Normal Science (Jacobi et al. 2019), transdisciplinary science (Luks and Siebenhüner 2007), or citizen science (Jarvis et al. 2015) are gaining momentum and represent an opportunity for strengthening local-level beach management.

Integrating new stakeholders in knowledge production and management also increases the knowledge system considered for management. Knowledge can be decomposed into different and complementary components, such as scientific, traditional, and environmental information (which can be learned) and perception (how the information is understood) (Xavier et al. 2018). Traditional knowledge systems (or local traditional knowledge) incorporate information, perceptions, practices, norms, and beliefs derived from empirical experiences, accumulated and transmitted through generations (Stori et al. 2019). Given the stronger local and sublocal components of beach management, locally based knowledge systems provide key information for beach management.

10.5.2 Find a Shared Working Language

Integrating different knowledge systems in the co-production of knowledge and decision-making demands a common language. The ecosystem services concept can be a useful tool for communication in the implementation of EBM (Sardá et al. 2015; Sardá and Azcárate 2018). Efficient and sustainable management, which guarantees the resilience of the ecosystem and consequent human well-being, depends on making the integration of human and nature dimensions evident for all involved partners (McLeod and Leslie 2009). Because ecosystem services result in ecosystem processes that permeate and sustain the social dimension, they represent a key connection between the two dimensions (McLeod and Leslie 2009).

Ecosystem services comprise the relationship between environmental attributes and social concepts such as the sense of belonging, perception, power, interest and legitimacy (Hicks et al. 2015). It can support the improvement of integrated planning with social participation and a governance system that ensures sustainability

(Galler et al. 2016). Applying the ecosystem service concept to decision-making emphasizes the trade-offs of management alternatives and their consequences on the social-ecological system (Granek et al. 2009), encourages actors to pragmatically assess the links between ecosystems and human well-being (Potschin and Haines-Young 2011), and allows the identification and inclusion of missing variables in the decision-making process (Danley and Widmark 2016).

10.5.3 Apply a Holistic Approach to Identify Threats

Identifying potential human threats and their impacts on the ecosystems is a required step in beach management (Harris et al. 2015; Sardá et al. 2015). The Driver-Pressure-State-Impact-Response (DPSIR) framework is a problem-structuring framework that assists to identify and understand the social-ecological processes and accounts for the cumulative effect of multiple human activities on ecosystem services (Kelble et al. 2013). In this framework, driving forces (social demands) result in actions that stress the system (pressures) and change the status of the natural environment (state changes). These changes result in social, economic, and environmental quality loss (impacts). Finally, responses (such as public policies and management actions are discussed, proposed, and applied to prevent adverse changes) (Atkins et al. 2011).

The framework has been adapted and adjusted in different contexts and for various reasons. One of its adaptations, the DAPSI(W)R(M)D, considers Activities (A) from different economic sectors such as tourism and real estate as a consequence of the drivers. It also highlights the social dimension of Welfare (W) and quality of life (Atkins et al. 2011). In this application, Measures (M) would also be required that relate to the feasibility of the technology, the viability of the economic situation based on marginal social cost and benefit comparisons, and so on to successfully manage the marine environment (Elliot et al. 2017).

Another adaptation of the diagnostic framework was to include Ecosystem Services (E) and combine it with EBM (Kelble et al. 2013). The EBM-DPSER promotes a more holistic and integrated management strategy (Elliot et al. 2017) and assists in setting the limits of management by evidencing endogenous and exogenous pressures and their potential and real impacts on ecosystem services (Atkins et al. 2011; Sardá and Azcárate 2018). It fosters participatory decision-making by providing a structured action that stresses the link between environmental problems and human threats (Cooper 2013) and discusses the trade-offs among ecosystem services (Kelble et al. 2013). A more recent adaptation, the Cross-scale Ecosystem-Based Assessment (DIET) (Gonçalves et al. 2020b) included the relevance of levels and scales to broaden the existing assessment frameworks to strengthen the ecosystem approach and promote an integrated cross-scale perspective.

10.5.4 Apply Spatial Management Tools

Of the different area-based management tools, marine spatial planning can be generally described as the zoning applied to a marine/coastal space to order, regulate and address conflicts among the different economic, social, and conservation interests in such space (Ehler and Douvere 2009). It shares many EBM principles – such as social-ecological sustainability, adaptability, integration, transparency, and social participation – and can foster and add value to the latter (Gilliland and Laffoley 2008). Applying marine spatial planning fosters social participation; promotes the use of tangible, intangible, complex, and variable data; enables wide characterization of fragmented spaces; and facilitates the definition of typologies for sustainable use (Mclachlan et al. 2013; Munro et al. 2019).

Designing a coherent structure for MSP-based coastal management is a work in progress. Studies focusing on the application of spatial planning to beaches address the management of marine invertebrate harvesting, protection of bird nesting sites, and tourism development (Harris et al. 2015; Munro et al. 2019). In Brazil, discussions for the implementation of marine spatial planning date back to 2011, but implementation is still incipient (Gerhardinger et al. 2019). Beaches are usually considered as homogeneous features within a single zoning class, and there is no proposal of specific areas within a beach for the development of different activities, as illustrated in Ecological-Economical Zoning (São Paulo 2017) and the zoning of Marine Protected Areas (São Paulo 2019).

10.5.5 Promote Interplay Management

Interplay management – the political efforts to purposefully shape and improve institutional interaction (Stokke 2001) – addresses and discusses the policy integration needed for beach management. The absence of interplay management leads to a lack of coordination in decision-making and drives conflicts related to multiple issues and interests (Therville et al. 2019).

Interplay management varies from exchanging information between decision-making bodies to coordination and implementation of policies (Stokke 2001). Even in a fragmented environment, interplay management enables synergies and convergence between and across policies and sectoral institutions. When the possibility to change or review institutions is absent, practitioners can exchange experiences and try to improve and adapt through practices in functionally linked policies. Such a process may foster social learning and qualify decision-making.

10.5.6 *Promote Learning and Adaptation*

EBM should enhance the collective capacity of agencies and social actors for learning and devising innovative and improved solutions to environmental problems. This collective and collaborative learning process, named social learning (Garmendia and Stagl 2010), is well aligned with EBM: it considers complex social-ecological systems; it is participatory and integrative; it is adaptive; and it is oriented toward long-term social-ecological sustainability (Garmendia and Stagl 2010; Xavier et al. 2019).

Social learning is expected to increase social capital, promote collaboration and integration, and promote adaptation toward sustainability (Luks and Siebenhüner 2007; Garmendia and Stagl 2010). Studies report that social learning influences the continuity of participatory processes and can be fostered by process design (Garmendia and Stagl 2010; Xavier et al. 2019). Fostering social learning and increasing social capital at a lower level may overcome challenges related to structuring a compatible governance system and promoting the continuity of beach management processes.

10.6 Final Remarks

The overview of beach management in Brazil shows that the country seems to be imprisoned in a vicious cycle that compromises beach sustainability. Although many legal instruments for beach management exist and other sectoral instruments also relate to beach management, assertive action is still needed to integrate instruments and practices. EBM is a promising approach to overcoming challenges for integrated beach management; however, the approach faces obstacles of its own in moving from theory to practice, many of which are recurrent in Brazil's beach management system.

Setting an adequate governance system to enable and support EBM requires transitional practices that can, at the same time, provide a solid structure for EBM implementation and enough flexibility for adaptation through the implementation process. It is a learning-by-doing practice to concomitantly implement EBM and structure the necessary governance system that it requires. It should tackle four critical domains: build governance capacity, avoid panaceas, promote inter and transdisciplinarity, and establish partnerships and networks (Gonçalves and Xavier 2021).

Acknowledgments The authors would like to acknowledge the support of the São Paulo Research Foundation (FAPESP: LYX: 2017/21797-5 and 2019/13851-5, LRG: 2018/00462-8 and 2019/04481-0; MRC 2018/13238-9 and 2019/13898-1), the National Council for Scientific and Technological Development (AT 309697/2015-8 and 310553/2019-9), and the Coordenação de Aperfeiçoamento de Pessoal de Nível Superior – Brasil (CAPES) – Finance Code 001 (MO, NM). The work is part of the activities of the ongoing thematic project, "Environmental Governance in the São Paulo Macrometropolis, due to climate variability" (FAPESP: 2015/03804-9), linked to the FAPESP Global Climate Change Research Program; of the regular project "Fostering the ecosystem-based approach in beach spatial planning and conservation" (FAPESP: 2018/19776-2) and of the research project "Será que vai dar praia? Inovações para a sustentabilidade dos oceanos" (Grupo Fundação Boticário: 1133_20182).

References

Abrucio FL, Sydow CT (2018) Federalismo e governança multinível em regiões metropolitanas: o caso brasileiro. In: Carneiro JMB, Frey K, (Orgs.) (eds) Governança multinível e desenvolvimento regional sustentável. Experiências do Brasil e da Alemanha, 1st edn. Oficina Municipal, São Paulo, pp 47–68

Amaral ACZ, Corte GN, Rosa-Filho JS et al (2016) Brazilian sandy beaches: characteristics, ecosystem services, impacts, knowledge and priorities. Braz J Oceanogr 64(SI2):5–16

ANA - Agência Nacional de Águas (Brasil) (2016) Encarte Especial sobre a bacia do Rio Doce - Rompimento da Barragem em Mariana (MG). In: Conjuntura dos Recursos Hídricos no Brasil - Informe 2015, 50. Superintendência de Planejamento de Recursos Hídricos, Brasília

ANA - Agência Nacional de Águas (Brasil) (2017) Atlas esgotos: despoluição de bacias hidrográficas. Brasília - DF

Angelo C (2019) Brazil freezes science spending. Nature 568:155–156

Arkema KK, Abramson SC, Dewsbury BM (2006) Marine ecosystem-based management: from characterization to implementation. Front Ecol Environ 4(10):525–532

Asmus ML, Nicolodi J, Scherer MEG et al (2018) Simples para ser útil: base ecossistêmica para o gerenciamento costeiro. Desenvolv Meio Ambient 44:4–19

Atkins JP, Burdon D, Elliott M et al (2011) Management of the marine environment: integrating ecosystem services and societal benefits with the DPSIR framework in a systems approach. Mar Pollut Bull 62(2):215–226

Azuz-Adeath I, Muñoz-Sevilla NP, Rivera-Arriaga E et al (2018) Microscale governance and temporal regulations in beach management. In: Botero CM, Cervantes O, Finkl CW (eds) Beach management tools - concepts, methodologies and case studies. Springer, Cham, pp 659–678

Balneário Camboriú (2022) Câmara municipal. Lei Ordinária n°4.609 de 28 de janeiro de 2022. Ensures the full exercise of the individual and social rights of people with disabilities, and their effective social integration on the beaches of Balneário Camboriú and other measures. Diário Municipal. Publicação N° 3586228

Blue Flag (2022) Brazilian beaches with the Blue Flag certification. Blueflag site. Available at https://www.blueflag.global/all-bf-sites. Accessed 22 Sept 2022

Bernardi LP, Pires PS (2015) O programa bandeira azul de certificação para praias na percepção dos gestores nacionais dos países participantes. R Turismo - Visão e Ação (online) 17(3):542–568

Bombana B, Polette M (2013) Governança de praias urbanas: Adaptação do sistema de gestão ambiental de praias da UNE-EN 150.104: 2008 (ISO 14001 PRAIAS) para a praia Central de Balneário Camboriú, SC–Brasil. R. Costas 2(1):1–18

Botero C, Diaz LH (2009) La playa como espacio costero particular en la Gestión Integrada Costera, revisión desde la bibliografía especializada. R Medio Ambiente, Turismo y Sustent 2:99–107

Botero CMS, Zielinski S, Castro LAN (2012) Esquemas de certificación de playas en América Latina: Diagnóstico de una herramienta de Manejo Integrado Costero. R Costas 1:49–63

Brasil (1988a, 18 May) Lei n° 7.661, de 16 de Maio de 1988. Institutes the National Coastal Management Plan. Diário Oficial da União, Brasília - DF, p 8,633

Brasil (1988b) Constituição (1988). Constituição da República Federativa do Brasil. Brasília - DF: Senado Federal

Brasil (2000, 19 July) Lei n° 9.985, de 18 de Julho de 2000. Institutes the National Protected Areas System and makes other provisions. Diário Oficial da União, Brasília - DF, p 1

Brasil (2002, 11 Nov) Lei n° 10.406, de 10 de Janeiro de 2002. Institute the Civil Code. Diário Oficial da União, Brasília - DF, p 1

Brasil (2004, 8 Dec) Decreto n° 5.300, de 07 de Dezembro de 2004. Regulates Law n° 7.661/88, which institutes the National Coastal Management Plan, establishes rules for land use on the coastal zone, criteria for shoreline management, and makes other provisions. Diário Oficial da União, Brasília - DF, p 2

Brasil (2015, 31 Dec) Lei n° 13.240, de 30 de Dezembro de 2015. Regulates the administration, sales and transfer of federal property and their use for fundraising, and makes other provisions. Diário Oficial da União, Brasília - DF, p 1, extra ed

Brasil (2017) Ministério do Planejamento, Desenvolvimento e Gestão. Secretaria do Patrimônio da União. Portaria n° 113, de 12 de Julho de 2017. Formalize beach management transfer and institutes the terms for beach management municipalization via the Contract of Adhesion. Diário Oficial da União, Brasília - DF, 13 Jul. 2017, n° 113, Section 1, p 153

Brasil (2018a, 28 March) Ministério do Meio Ambiente. Gabinete do Ministro. Portaria n° 76, de 26 de Março de 2018. Institutes the National Program for Coastline Conservation. Diário Oficial da União, Brasília - DF, n° 60, section 1, p 161

Brasil (2018b) Panorama da erosão costeira no Brasil. [on line] Ministério do Meio Ambiente, Secretaria de Recursos Hídricos e Qualidade Ambiental, Departamento de Gestão Ambiental Territorial; Muehe, D (Org). Brasília - DF: MMA. 761p. Available at: https://www.mma.gov.br/images/arquivo/80342/Panorama_erosao_costeira_Brasil.pdf. Accessed 15 March 2020

Brasil (2018c) Grupo de Integração do Gerenciamento Costeiro (GI-GERCO/CIRM). Guia de Diretrizes de Prevenção e Proteção à Erosão Costeira. Brasília - DF: GI-GERCO/CIRM. 114 pp. Available at: https://www.mma.gov.br/images/arquivo/80342/Final_Guia_de_Diretrizes_09112018.pdf. Accessed 15 March 2020

Brasil (2019a, 11 April) Decreto n° 9.759 de 11 de Abril de 2019. Extinguishes and establishes guidelines, rules and limitations for collegiate bodies of the federal public administration. Diário Oficial da União, Brasília - DF, p 1

Brasil (2019b) Ministério do Meio Ambiente. Agenda Nacional de Qualidade Ambiental Urbana: Plano de Combate ao Lixo no Mar [recurso eletrônico]. Brasília - DF. Available at: https://www.marinha.mil.br/dellaguna/node/104. Accessed 25 March 2020

Cabral ER (2015) Gestão ambiental em espaços de lazer e turismo: As praias urbanas da Amazônia brasileira. Rosa dos Ventos 7(2):269–287

Caraguatatuba (2006) Câmara Municipal. Lei ordinária n°1.298, de 13 de setembro de 2006. Provides for the promotion of public health and animal health, as well as the preservation of the environment, through the control of animal populations in the Municipality of Caraguatatuba. Diário Municipal - Executivo. Caraguatatuba - SP

Cervantes O, Botero C, Finkl C (2018) State-of-the-art users' perception on beaches from the tree of science platform. In: Botero CM, Cervantes O, Finkl CW (eds) Beach management tools - concepts, methodologies and case studies. Springer, Cham, pp 861–873

Cooper P (2013) Socio-ecological accounting: DPSWR, a modified DPSIR framework, and its application to marine ecosystems. Ecol Econ 94:106–115

Correa MR, Xavier LY, Gonçalves LR (2021) Desafios para promoção da abordagem ecossistêmica à gestão de praias na América Latina e Caribe. Estud Avançados 35:219–236

Creado ESJ, Helmreich S (2018) A wave of mud: the travel of toxic water, from Bento Rodrigues to the Brazilian Atlantic. Rev Inst Estud Bras 69:33–51

Cristiano SC, Portz L, Nasser P et al (2018) Strategies for the Management of the Marine Shoreline in the Orla Ararangua Project (Santa Catarina, Brazil). In: Botero CM, Cervantes O, Finkl CW (eds) Beach management tools - concepts, methodologies and case studies. Springer, Cham, pp 735–754

Danley B, Widmark C (2016) Evaluating conceptual definitions of ecosystem services and their implications. Ecol Econ 126:132–138

De Andrés M, Barragán JM, Scherer M (2018) Urban centers and coastal zone definition: which area should we manage? Land Use Policy 71:121–128
Defeo O, McLachlan A, Schoeman DS et al (2009) Threats to sandy beach ecosystems: a review. Estuar Coast Shelf Sci 81(1):1–12
De Oliveira JAP (2002) Implementing environmental policies in developing countries through decentralization: the case of protected areas in Bahia, Brazil. World Dev 30(10):1713–1736
Do Carmo FF, Kamino LHY, Tobias Junior R et al (2017) Fundão tailings dam failures: the environment tragedy of the largest technological disaster of Brazilian mining in global context. Perspect Ecol Conserv 15(3):145–151
Ehler C, Douvere F (2009) Marine spatial planning: a step-by-step approach toward ecosystem-based management. Intergovernmental Oceanographic Commission and man and the biosphere Programme. IOC manual and guides No. 53, ICAM Dossier No. 6. UNESCO, Paris. (English)
Elliott M, Burdon D, Atkins JP et al (2017) "And DPSIR begat DAPSI (W) R (M)!"-a unifying framework for marine environmental management. Mar Pollut Bull 118(1–2):27–40
Ekos Brasil (In press) Avaliação dos Impactos do Rompimento da Barragem de Fundão das Unidades de Conservação: Reserva Biológica de Comboios. São Paulo: Instituto Ekos Brasil
Epstein G, Pittman J, Alexander SM et al (2015) Institutional fit and the sustainability of social-ecological systems. Curr Opin Environ Sustain 14:34–40
FEE - Foundation for Environmental Education (2006) Awards for Improving the Coastal Environment: The example of the Blue Flag. 34p
Felippe E (2017) Avaliação de Aplicação da ISO 13.009:2015 na Praia da Areia Preta: uma contribuição sustentável para valorização turística de Guarapari. Dissertation, Universidade Nova de Lisboa, Lisboa
Figueiredo M, Almeida F (2019) O estado da arte sobre as reservas de surf: uma visão escalar, do global à proposta de um programa nacional. Brasil, [S.e.]
Fioravanti C (2016) Impactos visíveis no mar. Pesquisa FAPESP 242:42–47
Fonseca A, Sánchez LH, Ribeiro JCJ (2017) Reforming EIA systems: a critical review of proposals in Brazil. Environ Impact Assess 62:90–97
Galler C, Albert C, Von Haaren C (2016) From regional environmental planning to implementation: paths and challenges of integrating ecosystem services. Ecosyst Serv 18:118–129
Garmendia E, Stagl S (2010) Public participation for sustainability and social learning: concepts and lessons from three case studies in Europe. Ecol Econ 69:1712–1722
Gerhardinger LC, Quesada-Silva M, Gonçalves LR et al (2019) Unveiling the genesis of a marine spatial planning arena in Brazil. Ocean Coast Manag 179:104825
Gilliland PM, Laffoley D (2008) Key elements and steps in the process of developing ecosystem-based marine spatial planning. Mar Policy 32:787–796
Gonçalves LR, Xavier LY (2021) Promoting coastal and ocean governance through ecosystem-based management. In: Leal Filho W (ed) Life below water. Encyclopedia of the UN sustainable development goals. Springer International Publishing, Cham
Gonçalves LR, Xavier LY, Turra A et al (2020a) O litoral da Macrometrópole: tão longe de Deus e tão perto do Diabo. Desenvolv Meio Ambiente 54:40–65
Gonçalves LR, Oliveira M, Turra A (2020b) Assessing the complexity of social-ecological systems: taking stock of the cross-scale dependence. Sustainability 12(15):6236
Gonçalves LR, Gerhardinger LC, Pollete M et al (2021) An endless endeavor: the evolution and challenges of multi-level coastal governance in the global south. Sustainability 13:10413
Granek EF, Polasky S, Kappel CV et al (2009) Ecosystem services as a common language for coastal ecosystem-based management. Conserv Biol 24(1):207–216
Harris L, Nel R, Holness S et al (2015) Quantifying cumulative threats to sandy beach ecosystems: a tool to guide ecosystem-based management beyond coastal reserves. Ocean Coast Manag 110:12–24
Hicks CC, Cinner JE, Stoeckl N et al (2015) Linking ecosystem services and human-values theory. Conserv Biol 29:1471–1480

ICMBIO - Instituto Chico Mendes De Conservação Da Biodiversidade (2020) Painel Dinâmico de Informação sobre Unidades de Conservação. Website. Available at: https://www.cbd.int/doc/legal/cbd-en.pdf Accessed 17 March 2020

Itapema (Prefeitura) (2019) Plano de Gestão Integrada da Orla (PGI): Adequação e Atualização

ISO - International Organization for Standardization (2015) ISO 13.009:2015, Tourism and related services - Requirements and recommendations for beach operation. Genève

Jacobi PR, Toledo RF, De Giatti LL (2019) Ciência Pós-normal: ampliando o diálogo com a sociedade diante das crises ambientais contemporâneas, 1st edn. Faculdade de Saúde Pública da USP, São Paulo

Jarvis RM, Breen BB, Krägeloh CU et al (2015) Citizen science and the power of public participation in marine spatial planning. Mar Policy 57:21–26

Kelble CR, Loomis DK, Lovelace S et al (2013) The EBM-DPSER conceptual model: integrating ecosystem services into the DPSIR framework. Plos One 8(8):e70766

Klumb-Oliveira LA, Souto RD (2015) Integrated coastal management in Brazil: analysis of the National Coastal Management Plan and selected tools based on international standards. Rev Gestão Costeira Integrada 15(3):311–323

Krelling AP, Williams AT, Turra A (2017) Differences in perception and reaction of tourist groups to stranded marine debris influence the loss of tourism revenue in coastal areas. Mar Policy 85:87

Lafferty W, Hovden E (2003) Environmental policy integration: towards an analytical framework. Environ Polit 12(3):1–22

Leslie HM, Mcleod KL (2007) Confronting the challenges of implementing marine ecosystem-based management. Front Ecol Environ 5(10):540–548

Leslie HM, Sievanen L, Crawford TG et al (2015) Learning from ecosystem-based management in practice. Ocean Coast Manag 43(5):471–497

Loizidou XI, Loizides MI, Orthodoxou DL (2018) Persistent marine litter: small plastics and cigarette butts remain on beaches after organized beach cleanups. Environ Monit Assess 190(7):414

Long RD, Charles A, Stephenson RL (2015) Key principles of marine ecosystem-based management. Mar Policy 57:53–60

Luks F, Siebenhüner B (2007) Transdisciplinarity for social learning? The contribution of the German socio-ecological research initiative to sustainability governance. Ecol Econ 63(2–3):418–426

Magris RA, Giarrizo T (2020) Mysterious oil spill in the Atlantic Ocean threatens marine biodiversity and local people in Brazil. Mar Pollut Bull 153:110961

Maia MAM (2019) Ações do Serviço Geológico do Brasil na prevenção de desastres naturais, geologia ambiental e ordenamento territorial. (Programa de conservação da Linha de Costa - Procosta). Oral Presentation In: Petrobras meeting on The Rio Grande Rise. Rio de Janeiro, 02 de outubro de 2019. Available at: https://www.cprm.gov.br/publique/media/cprm_divulga/palestra_adelaide.pdf. Accessed 25 March 2020

Mclachlan A, Defeo O (2017) The ecology of Sandy Shores, 3rd edn. Academic Press, Amsterdam

Mclachlan A, Defeo O, Jaramillo E et al (2013) Sandy beach conservation and recreation: guidelines for optimizing management strategies for multi-purpose use. Ocean Coast Manag 71:256–268

Mcleod KL, Leslie M (2009) Ecosystem-based management for the oceans. Island Press, Washington DC

Mir-Gual M, Pons GX, Martín-Prieto JA et al (2015) A critical view of the Blue Flag beaches in Spain using environmental variables. Ocean Coast Manag 105:106–115

MMA - Ministério Do Meio Ambiente (2005) Secretaria de Qualidade Ambiental; MPO - Ministério Do Planejamento, Orçamento E Gestão. Secretaria do Patrimônio da União. Projeto Orla: Guia de implementação. MMA, Brasília

MMA - Ministério Do Meio Ambiente (2006) Projeto Orla: Fundamentos para gestão integrada. MMA, Brasília

MMA - Ministério Do Meio Ambiente (2008) Avaliação do estado atual de implementação do Projeto Orla

MMA - Ministério Do Meio Ambiente (2015) In: Pereira FC, Oliveira MRL, (Orgs.) (eds) Plano Nacional de Gerenciamento Costeiro - 25 anos do gerenciamento costeiro no Brasil. Brasília - DF
MMA - Ministério do Meio Ambiente (2016) Plano Nacional de Adaptação à Mudança do Clima - Volume I: Estratégia Geral. Ministério do Meio Ambiente (MMA), Brasília-DF, Brazil
MMA - Ministério do Meio Ambiente (2017) Plano Nacional de Adaptação à Mudança do Clima: 1o Relatório de Monitoramento e Avaliação 2016–2017. Brasília - DF
MTur - Ministério Do Turismo (2019) Anuário Estatístico de Turismo 2019 - Ano Base 2018, 2nd edn. Brasília - DF
Munro J, Korbyn H, Palmer D et al (2019) Charting the coast: spatial planning for tourism using public participation GIS. Curr Issues Tour 22(4):486–504
Neto LM, Pereira J, Felippe S (2011) Turismo sustentável: um estudo de caso referente à manutenção do Programa Bandeira Azul, em Jurerê Internacional. Rev Acadêmica Observatório de Inovação do Turismo 6(3):1–5
Nel R, Campbell E, Harris L et al (2014) The status of sandy beach science: past trends, progress, and possible futures. Estuar Coast Shelf Sci 150:1–10
Nicolodi JL, Petermann RM (2010) Mudanças Climáticas e a Vulnerabilidade da Zona Costeira do Brasil: Aspectos ambientais, sociais e tecnológicos. Rev Gestão Costeira Integrada 10(2):151–177
Nicolodi JL, Asmus M, Turra A et al (2018) Avaliação dos Zoneamentos Ecológico- Econômicos Costeiros (ZEEC) do Brasil: proposta metodológica. Desenvolv Meio Ambient 44:378–404
Nicolodi JL, Asmus M, Polette M et al (2021) Critical gaps in the implementation of Coastal Ecological and Economic Zoning persist after 30 years of the Brazilian coastal management policy. Mar Policy 128:104470
Oliveira MRL, Nicolodi JL (2012) A Gestão Costeira no Brasil e os dez anos do Projeto Orla. Uma análise sob a ótica do poder público. Rev Gestão Costeira Integr 12:91–100
Ohz A, Klein AHF, Franco DA (2020) Multiple linear regression-based approach for storm surge prediction along South Brazil. In: Leal Filho W (ed) Climate change, hazards and adaptation options. Springer, Cham, pp 27–50
PBMC (2016) Painel Brasileiro De Mudanças Climáticas. In: Marengo JA, Scarano FR (eds) Impacto, vulnerabilidade e adaptação das cidades costeiras brasileiras às mudanças climáticas: Relatório Especial do Painel Brasileiro de Mudanças Climáticas. PBMC, COPPE - UFRJ, Rio de Janeiro
Pereira BCPC, Asmus ML (2013) Perspectivas atuais da implementação da política ambiental brasileira em âmbito local: O plano ambiental do município de Rio Grande, RS Brasil, em foco. Rev Costas 2(2):28–38
Pinto AC (2014) O Projeto Orla no litoral do Estado de Santa Catarina. 2014. Dissertation Universidade do Estado de Santa Catarina, Florianópolis
Polette M, Lins-De-Barros F (2012) Os desafios urbanos na zona costeira brasileira frente às mudanças climáticas. Rev Costas 1(1):165–180
Pontee N (2013) Defining coastal squeeze: A discussion. Ocean Coast Manag 84:204–207
Possatto FE, Spach HL, Cattani AP et al (2015) Marine debris in a World Heritage Listed Brazilian estuary. Mar Pollut Bull 91(2):548–553
Potschin MB, Haines-Young RH (2011) Ecosystem services: exploring a geographical perspective. Prog Phys Geogr 35(5):575–594
Prates APL, Gonçalves MA, Rosa MR (2012) Panorama da conservação dos ecossistemas costeiros e marinhos no Brasil, 2nd edn. Ministério do Meio Ambiente, Brasília
Quetzalcóatl O, González M, Cánovasa V et al (2019) SMCε, a coastal modeling system for assessing beach processes and coastal interventions: application to the Brazilian coast. Environ Model Softw 116:131–152
Recife (1967) Câmara Municipal. Lei ordinária n°9.837, de 11 de outubro de 1967. Provides about football practice on the beaches. Diário Municipal - Executivo. Recife - PE, 11 out 1967

Santoro F, Santin S, Scowcroft G et al (eds) (2017) Ocean literacy for all – a toolkit. IOC/UNESCO & UNESCO Venice Office, Paris (IOC Manuals and Guides, 80)
Santos CR, Polette M, Vieira RS (2019) Gestão e Governança Costeira no Brasil: O Papel Do grupo de Integração do Gerenciamento Costeiro (Gi-Gerco) e Sua Relação com O plano de Ação Federal (PAF) de Gestão da Zona Costeira. Rev Costas 1(2):135–162
Sardá R, Azcárate JPL (2018) A DEcision MAking (DEMA) tool to be used in Ecosystem-Based Management System (EBMS) applications. In: Botero C, Cervantes O, Finkl C (eds) Beach management tools - concepts, methodologies and case studies. Springer, Cham, pp 21–40
Sardá R, Valls JF, Pintó J et al (2015) Towards a new Integrated Beach management system: the ecosystem-based management system for beaches. Ocean Coast Manag 118:167–177
São Paulo (State) (2017, 9 Nov) Assembleia legislativa. Decreto n° 69.913, de 8 de Novembro de 2017. Diário Oficial - Executivo, São Paulo, p 1
São Paulo (State) (2019) Fundação para a Conservação e a Produção Florestal do Estado de São Paulo. Minuta do Plano de Manejo da Área de Proteção Ambiental Marinha do Litoral Norte de São Paulo. São Paulo. Available at https://www.sigam.ambiente.sp.gov.br/sigam3/Default.aspx?idPagina=15387. Accessed 17 March 2020
São Sebastião (2021, 29 Jan) Câmara Municipal. Decreto municipal n° 8087, de 29 de janeiro de 2021. Amends decree 7277/2018, which provides on urban noise and protection of the Good Living and Public Peace within the scope of São Sebastião city. Diário Municipal - Executivo. São Sebastião - SP
Schlacher TA, Connolly RM (2009) Land-ocean coupling of carbon and nitrogen fluxes on sandy beaches. Ecosystems 12(2):311–321
Schlacher TA, Dugan J, Schoeman DS et al (2007) Sandy beaches at the brink. Divers Distrib 13(5):556–560
Scherer M (2013) Gestão de Praias no Brasil: Subsídios para uma Reflexão. Rev Gestão Costeira Integr 13(1):3–13
Seixas CS, Davidson-Hunt I, Kalikoski DC et al (2019) Collaborative coastal Management in Brazil: advancements, challenges and opportunities. In: Salas S, Barrágan-Palladines MJ, Chuenpagdee R (eds) Viability and sustainability of small-scale fisheries in Latin America and the Caribbean. MARE Publication series, 1st edn. Springer International Publishing, pp 425–451
Silva AP (2019) Brazilian large-scale marine protected areas: other "paper parks"? Ocean Coast Manag 169:104–112
Silva MEM, Soriano-Serra EJ (2016) O processo de implementação do Projeto Orla em Itapema, Santa Catarina-Brasil. Desenvolv Meio Ambiente 36:315–330
Silva ST, Santos MD, Dutra C (2016) Reservas de surf e a proteção da sociobiodiversidade. Rev Nomos 36(2):345–367
Soares MDO, Teixeira CEP, Bezerra LEA et al (2020) Oil spill in South Atlantic (Brazil): environmental and governmental disaster. Mar Policy 115:103879
Stokke OS (2001) The interplay of international regimes: putting effectiveness theory to work. The Fridtj of Nansen Institute 14:1–35
Stori FT, Mani-Peres C, Turra A et al (2019) Traditional ecological knowledge supports ecosystem-based management in disturbed coastal marine social-ecological systems. Front Mar Sci 6:571
Strohaecker TM (2008) Dinâmica Populacional. In: MMA - Ministério do Meio Ambiente. Macrodiagnóstico da Zona Costeira e Marinha. MMA, Brasília, pp 58–75
Termeer CJAM, Dewulf A, Lieshout MV (2010) Disentangling scale approaches in governance research: comparing monocentric, multilevel, and adaptive governance. Ecol Soc 5:29
Therville C, Brady U, Barreteau O et al (2019) Challenges for local adaptation when governance scales overlap. Evidence from Languedoc, France. Reg Environ Chang 19(7):1865–1877
Vianna LP, Xavier LY (2014) A APA Marinha do Litoral Norte (APAMLN) de São Paulo: do conflito à gestão participativa. In: Bensusan N, Prates AP (eds) A diversidade cabe na unidade? Áreas protegidas no Brasil, 1st edn. Mil Folhas, Brasília, pp 608–621

Wanderley LJ, Mansur MS, Milanez B et al (2016) Desastre da Samarco/Vale/BHP no Vale do Rio Doce: aspectos econômicos, políticos e sócio ambientais. Cienc Cult 68(3):30–35

Wyles KJ, Pahl S, Holland M et al (2017) Can beach cleans do more than clean-up litter? Comparing beach cleans to other coastal activities. Environ Behav 49(5):509–535

Williams AT, Micallef A (2009) Beach management: principles and practice, 1st edn. Earthscan, London

Xavier LY, Jacobi PR, Turra A (2018) On the advantages of working together: social learning and knowledge integration in the management of marine areas. Mar Policy 88:139–150

Xavier LY, Jacobi PR, Turra A (2019) Local agenda 21: planning for the future, changing today. Environ Sci Pol 101:7–15

Xavier LY, Gonçalves LR, Checon HH et al (2020) Are we missing the bigger picture? An analysis of how science can contribute to an ecosystem-based approach to beach management on the São Paulo macrometropolis. Ambient Soc 23(SI):e01411

Xavier LY, Guilhon M, Gonçalves LR et al (2022) Waves of change: towards ecosystem-based management to climate change adaptation. Sustainability 14:1317

Zielinsk S, Botero C (2015) Are eco-labels sustainable? Beach certification schemes in Latin America and the Caribbean. J Sustain Tour 23(10):1550–1572

Index

A. C. Z. Amaral et al. (eds.), *Brazilian Sandy Beaches*, Brazilian Marine Biodiversity, https://doi.org/10.1007/978-3-031-30746-1

SPRINGER NATURE

GPSR Compliance

The European Union's (EU) General Product Safety Regulation (GPSR) is a set of rules that requires consumer products to be safe and our obligations to ensure this.

If you have any concerns about our products, you can contact us on ProductSafety@springernature.com

In case Publisher is established outside the EU, the EU authorized representative is:

Springer Nature Customer Service Center GmbH
Europaplatz 3
69115 Heidelberg, Germany

The manufacturer's authorised representative in the EU is Springer Nature Customer Service Centre GmbH, Europaplatz 3, 69115 Heidelberg, Germany. If you have any concerns regarding our products, please contact ProductSafety@springernature.com

Printed and bound by CPI Group (UK) Ltd, Croydon, CR0 4YY
30/09/2026
02225509-0011